INTRODUCTION TO WEAPONS OF MASS DESTRUCTION

INTRODUCTION TO WEAPONS OF MASS DESTRUCTION

Radiological, Chemical, and Biological

R. EVERETT LANGFORD

WILEY-INTERSCIENCE
A JOHN WILEY & SONS, INC., PUBLICATION

Published by John Wiley & Sons, Inc., Hoboken, New Jersey.
Published simultaneously in Canada.

For general information on our other products and services please contact our Customer Care Department within the U.S. at 877-762-2974, outside the U.S. at 317-572-3993 or fax 317-572-4002.

Wiley also publishes its books in a variety of electronic formats. Some content that appears in print, however, may not be available in electronic format.

Library of Congress Cataloging-in-Publication Data is available.

ISBN 0-471-46560-7

Printed in the United States of America.

10 9 8 7 6 5 4 3 2 1

To my wife, Cecilia Son-Hee Langford.

ACKNOWLEDGMENTS

Appreciation is expressed to Lewis Publishers/CRC Press for granting permission to use copyrighted material (Figures 1.6 through 1.13) that was originally published in *Introduction to Hazardous Materials Incidents* by Reginald L. Campbell and Roland E. Langford, ISBN 0-87371-362-1 (1991).

Appreciation is also expressed to Victoria Roberts of Lone Wolf Enterprises, Ltd. for all her help with advice, editing, proofreading, and general help during the production this book.

Appreciation is also expressed to the editorial staff at John Wiley and Sons and the compositor at Lone Wolf Enterprises, Ltd. for making the transition from manuscript to book as smooth as it was.

I also thank my family for their understanding on those evenings and weekends when I was at the computer rather than being with them.

Finally, thanks to all those who were generous with their time in sharing their experiences to make this a better book. There are too many to recognize individually (I would certainly leave someone out by error), but I sincerely appreciate all their contributions.

ABOUT THE AUTHOR

Dr. Langford received his Ph.D. in physical geochemistry and mass spectrometry from the University of Georgia in 1974 under Professor Charles E. Melton and a second Ph.D. in environmental health physics from the University of North Carolina—Chapel Hill in 1996 under Professor James Watson, Jr. He had earlier received a Master of Science degree in thermodynamics and physical chemistry from the University of Georgia in 1971, and Bachelor of Science in physics and chemistry from Georgia Southern University in 1967. He is a Certified Industrial Hygienist, a Certified Safety Professional, a Registered Hazardous Substances Professional, a Fellow of the American Institute of Chemists, a Diplomate of the American Academy of Sanitarians, a Registered Sanitarian, a Certified Professional Environmental Auditor, and an Engineer-in-Training. He is currently the Manager of Industrial Hygiene and Product Stewardship at Huntsman LLC, the world's largest privately-held chemical company and fourth largest chemical company in the United States. Prior to joining Huntsman in 1998, he spent twenty years on active duty with the U.S. Army as an Environmental Science Officer, retiring in the rank of Lieutenant Colonel after having commanded the U.S. Army Toxicology Research Unit of Walter Reed Army Institute of Research at Wright-Patterson Air Force Base for six years. Other military assignments included command of the Environmental Sanitation Detachment in Taegu, Korea; Preventive Medicine Officer for NATO during Operation Joint Endeavor in Bosnia-Herzegovina; Environmental Scientist at Fort Huachuca, Arizona; Sanitary Engineer at the U.S. Army Environmental Hygiene Agency at Aberdeen Proving Ground; and Branch Chief for Clinical Chemistry at the U.S. Army Academy of Health Sciences, Fort Sam Houston, Texas. He is a member of the American Chemical Society, Health Physics Society, American Institute of Chemical Engineers, American Conference of Governmental Industrial Hygienists, American Industrial Hygiene Association, American Society of Safety Engineers, Association of Military Surgeons of the United States, and National Environmental Health Association. Dr. Langford is listed in a number of references, including *Who's Who in the World*, *Who's Who in America*, *Who's Who in Business and Finance,* and *Who's Who in Science and Engineering*.

CONTENTS

PREFACE

The purpose of this book is to provide an introduction to radiological, chemical, and biological (NBC) weaponry, a brief review of their histories, summaries of their effects, suggested protection from their use, comments on safe storage and handling, simple decontamination, and possible medical treatments. It is designed to be both a textbook for those new to the subject as well as a summary reference to the more experienced practitioner. The goal is to provide clear, technically accurate, concise information to the public, industrial hygiene and other safety professionals, first responders, and writers in the news media.

Following 9/11, the threat of NBC weapons use has increased after the period of relative calm following the collapse of the Soviet Union. It is known that Iraq (which listed in detail the quantities of all the chemical and biological agents it possessed in 1991 as a condition of the cease-fire—evidence of their destruction could not be found by U.N. inspectors prior to the 2003 attack on that country), Iran, North Korea, Libya, and a number of other states have been developing at least some weapons using nuclear, biological, or chemical materials. The possibility of their use is real, but there is much information disseminated which is either false or excessively alarming. The public, safety professionals, emergency responders, and those in the media all need facts upon which to base both policy and possible response. This book is intended to perform these functions.

Introduction to Radiological, Biological, and Chemical Warfare Agents is divided into three basic sections, one for each of the weapon types. Each of these sections is further subdivided into parts addressing more focused topics within the umbrella of the sectional topic.

The text does not present any classified or otherwise restricted information nor give sufficient detail to allow a criminal to access facilities or produce a weapon. The book is not intended to replace medical treatment protocols, standard operating procedures, or more detailed and specialized manuals or other documents.

It has been reported that Ben Jonson, shortly after publishing his first Dictionary of the English language, was informed he had omitted a word; his excuse was "Sheer ignorance, Ma'am". Likewise, for any omissions or errors in this book, the author is solely to blame due to sheer ignorance.

1

WEAPONS OF MASS DESTRUCTION (WMD)

INTRODUCTION TO WMD

The term "weapons of mass destruction" or "WMD" has been thrust upon the collective minds of people around the world by recent horrible events like the Japanese subway attack using Sarin nerve agent, the abortive Russian hostage release at the Moscow theater using fentanyl, and threats from North Korea and the Hussein regime of Iraq for use of a variety of biological and chemical weapons. While most human beings would consider the term as a collective for nuclear, biological, and chemical (NBC) weapons, this may not be technically correct. While a nuclear bomb or warhead would indeed bring about mass destruction, some applications of chemical weapons, in particular, might be very limited in scope, even limited to one individual. All weapons of mass destruction may not cause extensive destruction. A better definition for these weapons might be: those things which kill people in more horrible ways than bullets or trauma, or which cause effects other than simply damaging or destroying buildings and objects, with an element of fear or panic included.

This book is intended as an introduction to nuclear, biological, and chemical weapons in both warfare and terrorism. The intent is not to make the reader an expert in these frightening weapons, but to present clear facts of their destructive ability as well as their limitations. The more a person knows about these weapons, the better informed they are as a voter, as a citizen, possibly as a reporter, and as a human being. The information contained herein is intended for first responders (fire fighters, police, and emergency medical personnel) so that they may be more knowledgeable should such weapons ever be used. It is also hoped that members of the various news media might find information here so that their reports, whether in theory or upon use, are more scientifically correct and well-rounded. Members of the public, too, should find the information valuable in planning for their response in the unlikely event that such weapons should be used.

NUCLEAR WEAPONS

Most people in today's world are all too aware of what nuclear weapons are, but few understand the complexities of design and construction that have allowed only a few nations to become members of the "nuclear club." Nuclear weapons utilize the physical process of fission by which a radioactive atom can be caused to split into two or more fragments while releasing atomic energy. Nuclear weapons produce great heat and light, a strong pressure wave, and radioactivity. The chapters on nuclear weapons will describe the historical background, some simple designs, and the effects of both warfare and terrorist weapons. No classified or nuclear sensitive material is contained in these pages; only information that has been made public is presented.

BIOLOGICAL WEAPONS

When the term "biological weapon" is used, there are many different natural and man-made agents, such as anthrax, smallpox, and plague, included. These weapons produce either disease or affect the body's normal functioning. Most are bacteria, viruses, fungi, or toxins; the vast majority occurring naturally, although some research has been conducted into making natural pathogens and toxins more potent or less resistant to vaccines and medicines. A brief history of the use of pathogens and toxins will be presented as well as information on known or suspected enhancements to natural disease-causing agents. Limitations on the use of biological weapons will be stressed. The chapters on biological weapons will describe those biological agents considered most likely for use as warfare or terrorist weapons.

CHEMICAL WEAPONS

While "chemical agents" have been used for thousands of years in wars around the world, most people would define "chemical weapon" as a relatively new concept in warfare and terrorism. Chemical weapons are generally those substances that adversely affect the body's natural functioning. The most common types include those chemicals which prevent the blood from carrying oxygen, those that irritate or destroy the respiratory system, those which cause the nervous system to malfunction, those which cause uncontrolled bleeding, and those which cause skin and other lesions. The chapters on chemical weapons will include a history of the employment of chemicals to harm or kill an opponent as well as modern discoveries with greatly increased toxicity. A summary of the most likely chemical agents will be presented, along with their advantages and disadvantages as weapons of war and terrorism.

SIMILARITIES OF NUCLEAR, BIOLOGICAL, AND CHEMICAL WEAPONS

Most people would likely not consider any similarities among these weapon systems, but they possess a few features in common. First, all three, unlike more common weapons, have a psychological component beyond their physical, chemical, and biological make-up and use: fear and anxiety. All three have a fear factor beyond what are considered as "normal" weapons of war. While a high explosive or rifle bullet can kill just as readily as a nuclear bomb, for example, the fear of the unknown with respect to radiation and radioactivity causes a greater response in the majority of people. Also, there are certain laws of physics, including meteorology, which govern the movement of materials, whether radioactive fallout, chemicals, toxins, or pathogens. The basic concepts of dispersion of materials will be presented in somewhat detail in this chapter with brief reminders in each weapon section.

DIFFERENCES AMONG NUCLEAR, BIOLOGICAL, AND CHEMICAL WEAPONS

Each of the broad categories of warfare and terrorist weapons has unique characteristics, making them individually frightening. The three general classes of WMD, chemical, biological, and nuclear, are intended to produce fear, but some are intended to destroy people and facilities while others are intended simply to kill people or even just make them so ill that they cannot perform required functions. The area of destruction or harm can be quite different for each of these weapon systems. These differences will be discussed in detail in later chapters.

Nuclear weapons, like high explosive weapons, produce heat, light, and pressure. While the magnitude may be quite different, the effects are similar, although reduced in conventional explosives. The uniqueness of nuclear weapons, whether a nuclear bomb or warhead or "dirty bomb," is radiation. The possible long-term effects of radiation compound the fear of these weapons. And, chief among the effects of radiation, is the possibility of cancer and birth defects, even after many years have passed following the exposure. It is well known that the fetus and children are more sensitive to the effects of radiation than adults; thus, fear of the effects of nuclear weapons on one's children and grandchildren is something not present when a conventional high explosive bomb is detonated. The employment of a true nuclear weapon would cause mass destruction due to the required size of the weapon by the laws of physics.

Nuclear bombs and warheads truly meet the definition of weapons of mass destruction because, by their very nature, these weapons produce large amounts of heat, light, blast, and radiation. In fact, it is not possible to make a very small nuclear bomb, as will be seen in the chapter on bomb design. A so-called "dirty bomb," that is, the use of explosives to spread radioactivity, does not suffer from such design limitations, but would not be a true "weapon of mass destruction" because of limitations on the extent of contamination. Of the three categories of nuclear/biological/chemical weapons, true nuclear weapons are those of mass destruction. Interestingly, of the three, only nuclear weapons have not been banned by international treaties.

Biological weapons are largely agents of normal disease, although they might be enhanced technologically. The uniqueness of biological weapons, both pathogens and toxins, lies in the fact that modern science and medicine have largely convinced us that diseases can often be cured. Even horrible diseases like various cancers have been successfully treated so that many people survive and even go on the lead normal lives. Agents, on the other hand, are selected for the reason that they are impossible or difficult to treat using modern medicine. Our faith in our science can be shaken by biological weapons. The psychological dread of seeing a loved one sick without being able to help is a major factor in the use of biological weapons. Our respected medical facilities may be overwhelmed by sheer numbers of victims. Fortunately, the widespread dispersion of most biological (and many chemical) agents is difficult in practice. International treaties and agreements have banned all use of biological weapons.

Chemical weapons today are human-designed substances designed to kill or harm an enemy. However, there are difficulties in exactly defining "chemical weapon." For example, a narrow reading of international conventions banning chemical warfare weapons leads to the prohibition against the use of riot control agents (like tear gas) in war. The illogical result is: a nation can use tear gas to contain demonstrators and rioters while being prevented from employing exactly the same chemical in war against an enemy. Also, some materials designed to be less than lethal against an enemy might be banned—because they are chemicals—while shooting that enemy or using a flamethrower is acceptable.

The uniqueness of chemical weapons lies in the fact that these substances are usually designed for employment against a large number of unprotected people. Most militaries have trained their soldiers to be able to function in an area contaminated by chemical agents. It is against unsuspecting and innocent victims that most such weapons would be employed. This generates another paradox: chemical weapons of war would most likely be used against civilians, not soldiers.

AEROSOLS

Throughout this book, there will be references to aerosols; these are minute droplets of solids or liquids suspended in the air. Because of their very small size, the particles do not immediately fall to the ground, but can stay in the air for possibly long times, held aloft by air currents. Imagine a sneeze: there are tiny droplets of water containing viruses forced into the air. Eventually, these droplets will fall upon a surface, but for a while, they float in the air, moved by wind currents. Solid particles formed when a nuclear bomb is detonated can be of vastly different sizes; the large ones fall to the ground fairly quickly (according to the laws of physics—Stokes' law), but as they get smaller, the longer they will remain in the air. If these small particles are pushed up into the upper reaches of the atmosphere, they can remain suspended for years or decades. Eventually, however, even the smallest particle will be pulled upon by gravity and fall to the ground (or other surface). Likewise, bacteria, viruses, and toxins can also be aerosols. A brief discussion of dispersion will help explain how aerosols, whether pathogen or nuclear fallout or chemical droplet, behave.

DISPERSION

When a nuclear weapon detonates or a biological or chemical agent is released, each creates a primary cloud of either solid or liquid aerosols. This cloud then settles to the ground, depending upon characteristics of the aerosol and weather conditions, eventually landing on individuals, plants, equipment, and the ground. The contamination on the ground has a finite lifetime, but can injure people from contact with a secondary cloud of agent that results from being kicked up by people or vehicles. Eventually, the contamination disappears as dispersion reduces the concentration in the cloud, and the agent is destroyed by air, ultraviolet radiation, physical and chemical processes, or weather conditions.

Factors that can increase or decrease the hazard from the primary cloud are related to the local weather conditions. Those factors which tend to diminish the hazard of the primary cloud include: variable wind directions, which causes dilution by redirection of the cloud; presence of wind velocities over six meters (twenty feet) per second, which causes dilution by turbulence; presence of unstable air, which also causes dilution by turbulence; temperatures below 0° C (32° F), which result in less viability or even death of the organisms and freezing of chemicals; temperatures above around 40° C (100° F), which denature proteins and kill organisms; and precipitation, which washes out aerosol particles from the atmosphere. Factors which tend to increase the hazard of the primary cloud are generally the opposites of those previously cited: steady wind direction; wind velocity under 3 meters per second; stable air (atmospheric inversion); temperatures between 20° C (70° F) and 35° C (95° F); and no precipitation. For a military or paramilitary operation, with the goal to cause as much damage as possible, these latter conditions are sought.

Winds vary almost constantly in force and direction, and are affected by every irregularity on the surface. Much work has been done on the effect of these irregularities of motion on the distribution of pollution through the atmosphere, and it is important to determine what can be learned about the probable distribution of chemical and biological agents from this work. The weather scientist, Frank Pasquill, defined six categories of atmospheric turbulence that are generally accepted for making rough estimates for the calculation of the spread of atmospheric pollutants; other scientists have added a few more categories for specialized studies. While his studies involved transport of pollutants, the factors are the same for fallout particles, chemical droplets, and biological organisms.

Of the turbulence classes, Pasquill category A is the most turbulent, and is quite rare. Category F is the most stable atmospheric state for which realistic predictions can be made. Examination of long-term meteorological records have shown that, in most places, classes E, F, and those of greater stability than F are likely to occur at around midnight on about half the nights throughout the year. The longer the nights, the longer will be the period in which these degrees of stability will be observed; in mid-winter, it may even extend for the twelve hours from about 6:00 PM to 6:00 AM. For the purposes of this book, maintenance of this degree of stability for two hours would suffice; this degree of stability occurs in most locations during different times at all periods of the year. Two hours is long enough for a biological or chemical cloud to be effective as a weapon. For this reason, a chemical or biological attack around midnight might be a likely possibility.

Unfortunately, most of the information about the dissemination of aerosols in the atmosphere under different conditions of atmospheric stability has been derived from studies in open countryside; there is much less known about movement in urban areas and in areas of varying altitude. Since most military or terrorist attacks would likely involve population centers, the models are much less accurate. It is usual to predict the likely atmospheric stability from the wind speed, the state of the sky, and the time of the day. Over cities, atmospheric turbulence may be augmented by the extra surface roughness and by heating and resulting thermal air currents from industrial, commercial, and domestic processes.

Pasquill provided estimates of the vertical and horizontal spread of any pollutant, whether industrial or warfare agent, released at ground level after it had traveled different distances over open country. A study carried out by the U.S. National Air Pollution Control Administration in the city of St. Louis (McElroy & Pooler, 1968) provides the only published evidence of the extent of the additional vertical spread that is to be expected over urban areas.

The Pasquill model describes the dispersion, or extent, of a release of particles as a function of wind speed and turbulence in what is called a Gaussian distribution (named after the German mathematician, Frederick Gauss, who first devised the equations). Whenever a solid or liquid particle is released, it is moved by air currents, moving in the direction that the wind is blowing. How much the cloud of many of these particles spreads out is a function of turbulence. In most cases when there is any wind at all, the resulting cloud spreads out vertically and horizontally, with the horizontal spread usually greater than the vertical. This results in a cloud with a cross section that is oval. The ovals become larger as the cloud is moved away by the wind. Envision the vapor trail from a high-flying jet aircraft to picture a Gaussian plume. Eventually, the particles in the cloud begin to fall to the ground by gravity. Figure 1.1 shows a generalized dispersion pattern of a Gaussian plume.

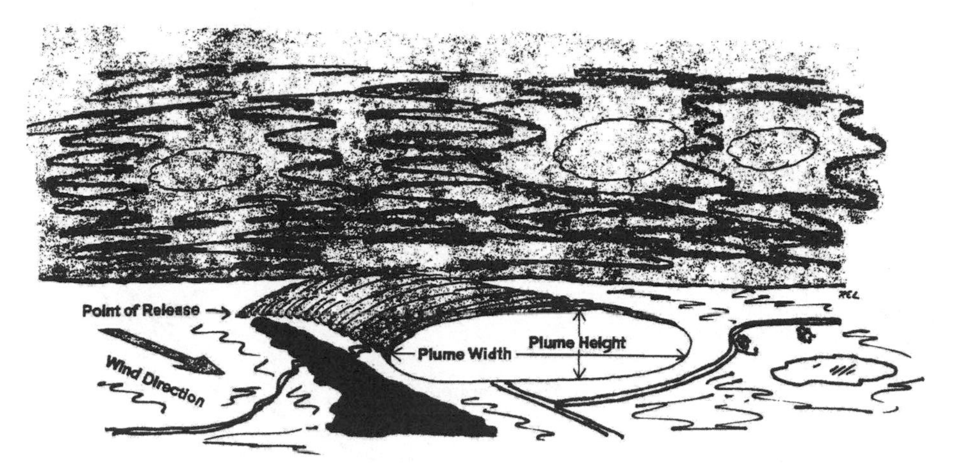

Figure 1.1 General Dispersion of an Aerosol Cloud

Once the primary cloud settles to the ground, those factors that tend to decrease the hazard due to ground contamination include: low ground temperature, which causes decreased viability of biological agents and reduced volatility of chemical agents; high wind velocity, which dilutes the agent or fallout; unstable air, which dilutes the agent or fallout; and heavy precipitation, which dilutes and washes the agent or fallout into the soil. Factors that tend to increase the hazard of ground contamination are related to the stability of the atmosphere. Factors that characterize stable air include low wind velocities and temperature inversions.

Persistency is the term used to describe the duration of an area's toxic contamination. Persistent agents are generally considered to be those for which contamination lasts a day or more. Non-persistent agents are generally considered to be those where contamination dissipates in a matter of minutes or hours. Chemical reactions that affect the persistence of agent in the environment include photochemical reactions from sunlight, especially ultraviolet (UV) radiation, and other reactions with compounds present in the environment. Since UV levels and temperatures are generally lower at night, use of chemical or biological weapons is more likely during periods of darkness (which also allows for stealth and escape).

The calculation of the precise lifetime of a biological or chemical warfare agent in the field can be very complex. This estimation requires knowledge of the parameters of the agent, plus climate information for the specific location, including temperature, humidity, rainfall, and solar flux, among others. Nevertheless, assumptions can be made for average values for humidity, rainfall, and solar flux, yielding approximate lifetimes that indicate the persistency under generalized climatic conditions.

Almost all weather occurs in the lowest portion of the atmosphere, the layer called the troposphere. Higher up is the stratosphere; some nuclear blasts, especially thermonuclear explosions, can penetrate the barrier, called the tropopause, between the troposphere and the stratosphere, sending radioactive debris so high up that it can

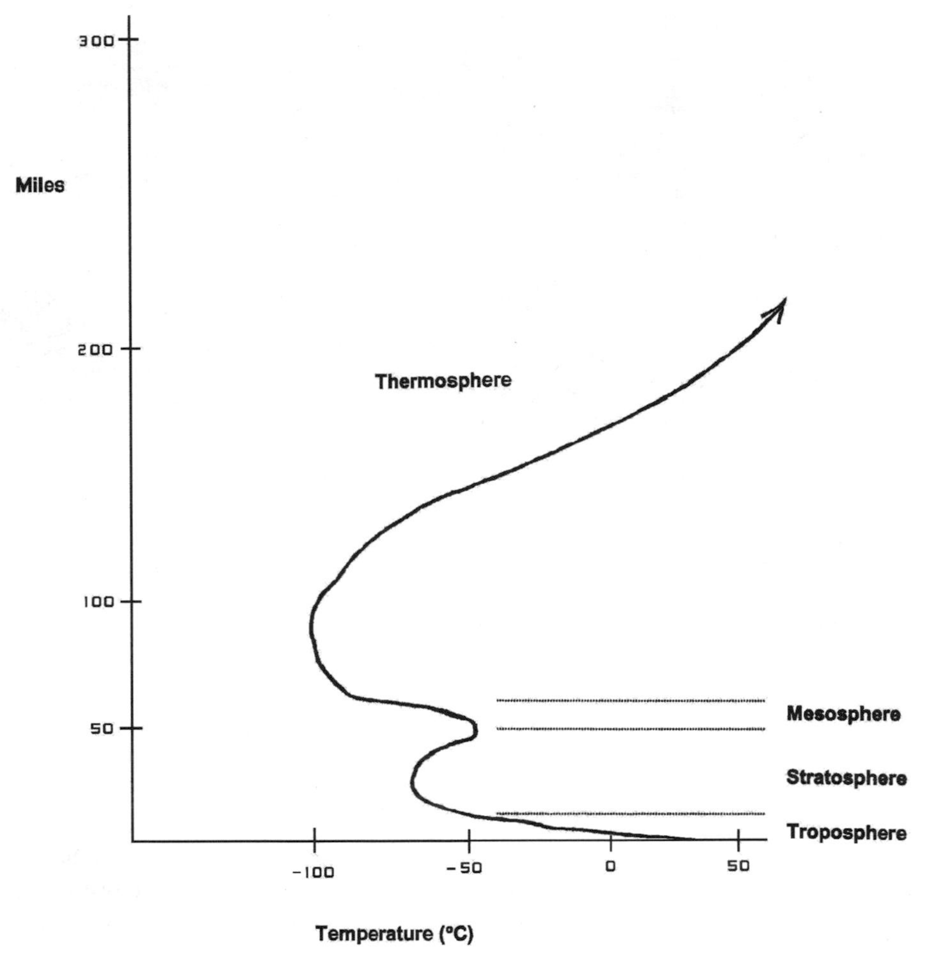

Figure 1.2 Layers of the Atmosphere

take years for the fallout to occur. Fortunately, many of the mixed fission products decay in a short period of time, so the radioactivity may decrease to a lower amount by the time it returns to earth. Figure 1.2 shows the generalized structure of the earth's atmosphere.

EXPOSURE VERSUS DOSE

Most measurements of biological and chemical agents involve the amount of aerosol in the air. If there is a person at that location, this would be the person's *exposure* to the agent. However, the amount actually taken in by the person can be less than the exposure. The chance that a particular *dose* will be received by a person when the air contains a toxic material is proportional to its concentration in the air and to the length of time of the exposure. The product of concentration and time is referred to as the aerosol concentration times the time (C times t, or C × t, or just Ct). If the concentration changes with time, the total dosage received will be proportional to the area under the concentration/time curve when graphed, and the Ct will be equal to this area. Where the concentration is constant, the Ct is simply the product of the concentration and the time of the exposure; this is much easier to calculate. If concentration is expressed in milligrams per cubic meter (mg/m^3) and time is in minutes, then the Ct is expressed in $mg\text{-}min/m^3$. For example, if a person is at a location in which the air contains 10 mg/m^3 of an agent and stays there for 10 minutes, the Ct will be 100 $mg\text{-}min/m^3$. If the concentration is 20 mg/m^3, the person will receive 200 $mg\text{-}min/m^3$, or twice as much, if they stay there for 10 minutes. If the agent is absorbed only through the lungs and not through the skin, for the person at rest breathing an average of 100 liters = 0.1 cubic meters (m^3) of air in the 10 minutes, they will have taken in 1 mg of the agent when the concentration is 10 mg/m^3, but they are unlikely to have absorbed all of it. However, if the concentration had been twice as great or the exposure time twice as long and everything else had been equal, the person would have received twice the dose. The Ct is a measure of the intensity of exposure, and not of the dose that actually penetrates into the body. This concept applies as much to pathogens as to chemical agents. The term LD_{50} (Lethal Dose 50%) is used for the dose that has a 50% probability of producing a specified response to infection, considered in this book as signs or symptoms of disease or death. It is also used to express 50% probability of infecting or killing one individual. LD_{50} can be used for both chemical and biological agents. Solely for infective agents, the term ID_{50} (Infective Dose 50%) is used for the dose that has a 50% probability of producing a specified response to infection, in this case, signs or symptoms of disease or death. This concept will be developed later in this chapter.

In summary, *exposure* is the amount of contaminant in the area, *exposure intensity* is the amount times the length of time exposed, *dose* is the amount taken into the body, and *dose rate* the time relationship of the amount taken in. Exposure would be the concentration of the agent (often in mg/m^3). The exposure intensity would be the concentration times the time (often in $mg\text{-}min/m^3$). The dose would be the amount actually taken into the body, often expressed as weight of the material in milligrams (mg), but sometimes related to the weight of the person as milligram per kilogram (mg/kg), since this is the best expression of the harmful effects. The dose rate is simply the amount taken in the body in a unit of time, say a minute (mg/min).

TOXICITY, DOSE, AND EFFECTS

When a person is exposed to a contaminant, a portion, the *dose*, is taken into the body. The effects from this contaminant being taken in are usually called the *response*. A graph can be drawn of the relationship between these two, and is called

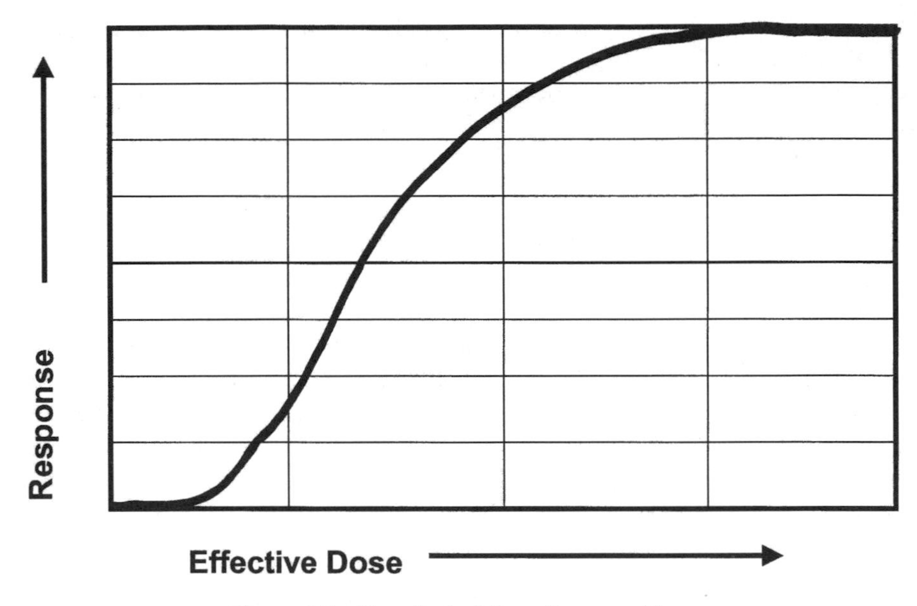

Figure 1.3 Hypothetical Dose-Response Curve

a *dose-response curve*. For most contaminants, including chemical and biological agents as well as some forms of radiation, the shape of the curve is sigmoid (s-shaped) as shown in a representative dose-response curve, Figure 1.3.

What this example graph shows is, at first, there is little or no effect from no or little dose. As the dose increases, the effects start off slowly, then increase more rapidly. Effects are pronounced until adding more dose doesn't change the response (which could be defined as death). Different contaminants will have different rates of increase of effect with dose, so the slope of the graph will be different, but many materials produce a curve similar to that above.

Similarly, a graph can be made of the number of people affected by a given exposure or dose. Such a graph is called a *population response* graph. The same is often similar to that of a dose-response graph. An example of such a graph is shown in Figure 1.4.

As the dose increases, the first effects are small as only the most susceptible people are affected, then most people become injured until essentially all are injured or killed. The point at which the dose crosses the 50% of the population affected, assuming death is the result, is the Lethal Dose 50, mentioned earlier. As a reminder, the LD_{50} (Lethal Dose 50) of a chemical is the dose that will kill 50% of the people receiving it. It may also be defined as the dose that has a 50% probability of killing any particular individual. Correspondingly, the ID_{50} (Infective Dose 50) is the dose that has a 50% probability of producing any defined effect of infection due to biological agents. Likewise, the LD_5 or ID_5 indicates the dose likely to cause an effect in 5% of those exposed and the LD_{95} or ID_{95} is the dose likely to affect 95% of

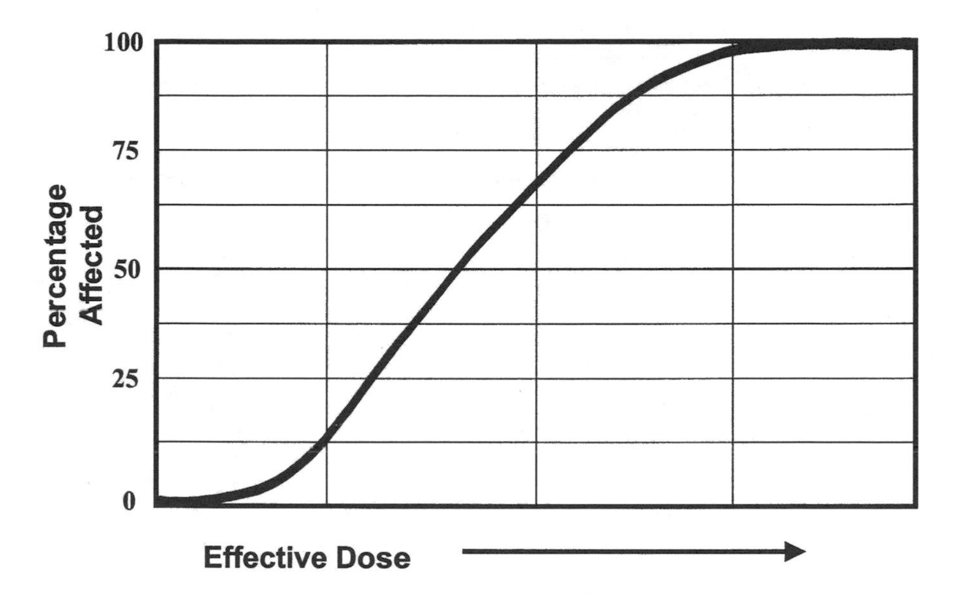

Figure 1.4 Hypothetical Population Response Graph

exposed persons. These are the doses where the curve crosses the 5% or 95% population lines on the graph.

Neither the LD_{50} nor the ID_{50} gives all the information needed to understand the relationship between dose and effect. The additional information needed is provided by the slope of the probit (a statistical representation) line, which can be calculated from the data required for an accurate determination of an LD_{50} or ID_{50} (it can be used for either chemical or biological agents). The slope value is dimensionless, so its meaning is independent of the units of dosage. The steepness of the slope indicates the factor by which the LD_{50} or ID_{50} must be multiplied to give the probability of producing a response to the agent. For many cases, the graph of the dose-response curve looking at the probit slope uses a logarithmic scale for the dose because of the (sometimes) large range of doses required (especially for biological agents); this results in a largely straight line rather than the s-shaped curve and the slope can be better defined. The steeper the slope, the more rapid the probability of an effect increases with increase in dose. For example, if the probit slope is 0.5, the ratio of LD_{95} to LD_5 is very large, some 3.8×10^6; for a slope of 1, the ratio drops to 1.95×10^3; for a slope of 5, the ratio is 4.55; and for a slope of 20, the ratio is only 1.46. Table 1.1 shows selected values for different probit slope factors. For most biological agents, the slope is highly unlikely to be very great; in fact, many probit slopes for biological agents are less than 1. For chemical agents, the slope can be quite large.

At low probit slopes, the probability of effect changes so slowly with dose, even in the neighborhood of the LD_{50}, that it is not even possible to specify the LD_{50} with

TABLE 1.1 Influence of Probit Slope

Slope of the Probit line	Ratio of ID_{95} to ID_5	Ratio of ID_{95} to ID_{50}
0.5	3.8×10^6	1,950
0.6	3.0×10^5	551
0.7	5.0×10^4	224
0.8	1.3×10^4	114
0.9	4,500	67.3
1.0	1950	44.1
1.5	156	12.5
2.0	44.1	6.65
5.0	2.13	2.13
10.0	1.46	1.46
20.0		1.21

any great exactness even when large numbers of animals were used in studies for its estimation. This has a bearing on the comparison of toxicities. Small differences between the LD_{50} values of two compounds will imply real differences in toxicity if both slopes are large, but very large differences in the LD_{50} may have no practical significance if both slopes are small. In the intermediate range of values of the slope, it is essential to consider the slope of the probit line and the LD_{50} together when making comparisons between effectiveness of agents.

As a simple example, Figure 1.5 shows the probit slopes of two chemicals (it could be biological agents as well as in Figure 1.6). There are two agents, A and B, being looked at. Both agents have the same LD_{50}. Agent A causes no effects even as the dose increases until a certain point is reached; this is called the *threshold level.*

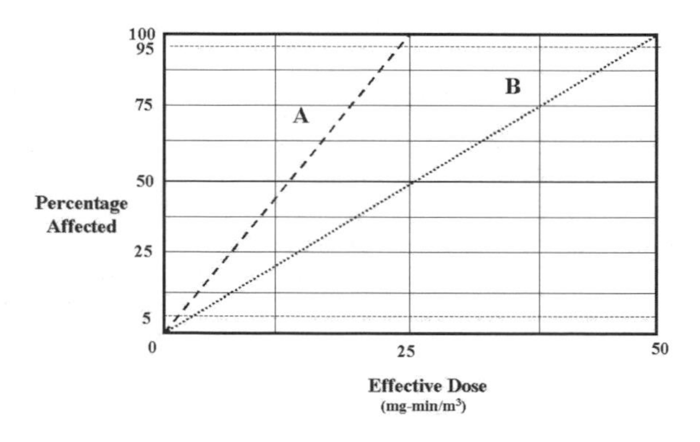

Figure 1.5 Comparison of Effectiveness of Two Chemical Agents

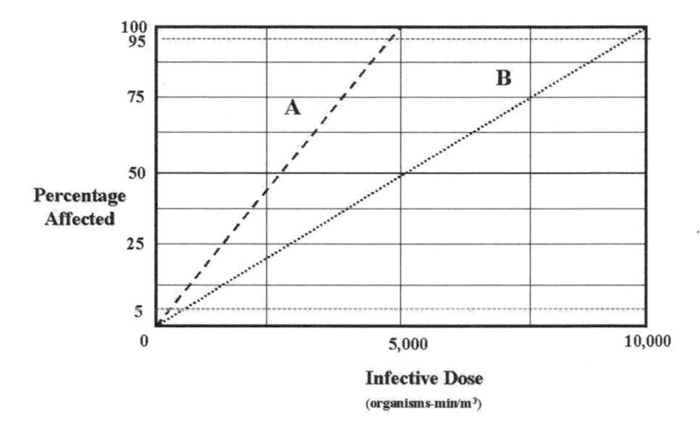

Figure 1.6 Comparison of Effectiveness of Two Biological Agents

Above this level, there are effects from the dose. This is typical of most chemical agents. Agent B has an effect from even the smallest dose. This is typical of cancer-causing chemicals and many biological agents. The same idea follows for comparing two chemical agents or two biological agents.

When the probit slopes are small (*e.g.*, 0.5 and 1.0), larger ratios of LD_{50} values can be misleading. Some very highly toxic chemical agents can have probit slopes between 5 and 10, but most biological agents have slopes less than or near unity (a value less than or close to 1). This means that a biological agent cloud may travel a considerable distance downwind before its dosage drops from an LCt_{50} to an LCt_5, but that exceedingly high concentrations are needed to give an ID_{95}. Conversely, for chemical agents, most are effective at very low doses, but only for a small distance from their point of release.

EFFECTS OF NUCLEAR, BIOLOGICAL, AND CHEMICAL WEAPONS ON PEOPLE

The harmful and possibly fatal effects of nuclear, biological, or chemical weapons on people (and objects) are varied and depend upon the type of weapon, the amount of exposure, the amount of agent taken into the body, climate and weather, and individual susceptibilities. In order to better understand these effects, a brief and simple introduction to human anatomy and physiology is necessary. Details of the effects from each of these weapons will be described in later chapters.

HUMAN ANATOMY AND PHYSIOLOGY

The human body is composed of many specialized and interrelated organ systems. Biological and chemical agents, as well as radiation, may affect one or more of these systems. To understand the effects of nuclear, biological, and chemical weapons, it is necessary to look at how the human body works. The effects of the various weapons on the body will be described in following chapters.

Some of the larger and more important organ systems of the human body are: the skin, the digestive, the respiratory, the circulatory, the lymphatic, the endocrine, the nervous, and the reproductive systems. Any of these systems or organs can be affected by biological and chemical substances and radiation. It is not the purpose of this chapter to make you a biologist or physician; only those technical terms which are necessary will be used, and every attempt will be made to keep the topics as simple as possible.

Simple definitions used in this section are:

Anatomy is the study of the structure of plants and animals.

Physiology is the study of how plants and animals function.

Metabolism is the total of all the chemical reactions taking place in the body.

Exposure refers to ingestion, injection, or inhalation of a chemical, biological, or radioactive substance or particulate or electromagnetic radiation.

Dose refers to the amount actually taken into the body.

Most exposures to radioactive, biological, or chemical agents experienced by victims are by means of only a few organ systems, chiefly the skin or eyes, respiratory system, circulatory system, or digestive system. But, once chemical or biological substances enter the body, they can be moved about and affect other systems such as the nervous and endocrine, far from the location of entry. Radiation like x-rays and gamma rays can affect the body even when external to it. Each major human organ system will be discussed in turn.

The Skin

The skin is the largest organ of the human body, and is the most exposed in the event of a biological or chemical attack or to radioactive fallout. The skin forms a barrier to protect the body from the environment, but it is much more, since it is the location of hair follicles, sweat glands, and oil glands.

The skin consists of an outer layer, the *epidermis*, which is actually two parts: the outermost *corneum*, a tough mixture of flattened dead skin cells, oil, free fatty acids, and dirt, and the *germinativum* just below and made of living skin cells. Skin cells are made there and move upward with time, taking up to 30 days to reach the surface, where they live about two weeks and eventually are removed by rubbing and washing. An injury to the outer skin can take up to a month to heal. The next layer is the *dermis*, consisting of connective tissue, hair roots, sweat glands, oil glands, and fat deposits. The oil glands produce oils that are oxidized to form free fatty acids. These fatty acids tend to make the skin a hostile environment for most bacteria and viruses. The oils and acids also act as a thin chemical barrier for exposure to some

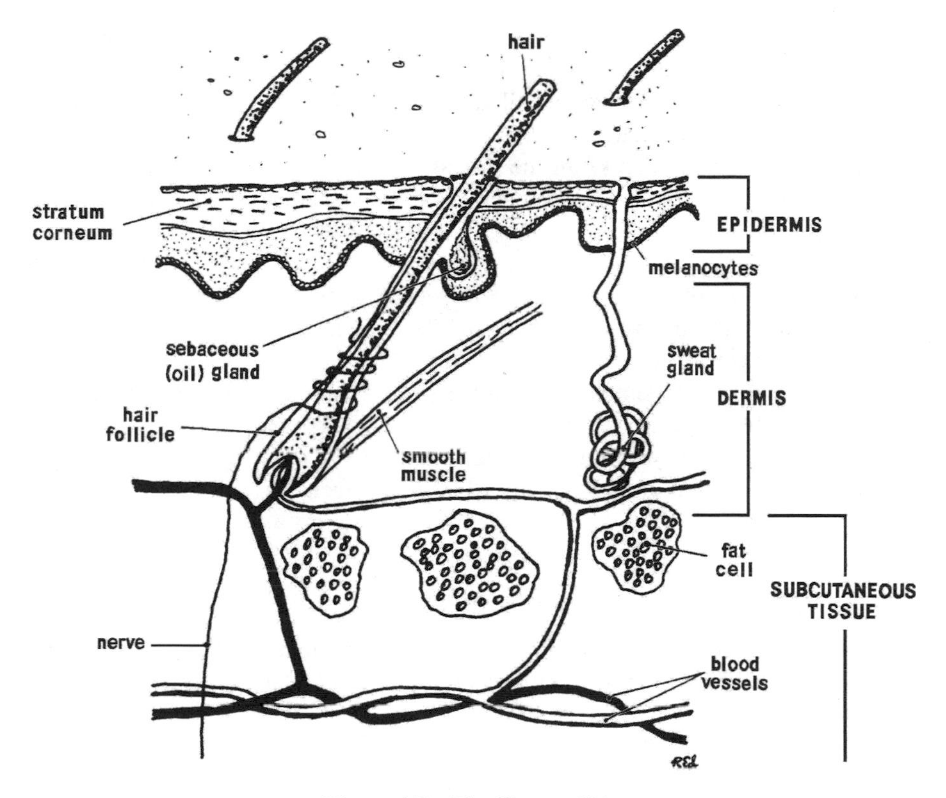

Figure 1.7 The Human Skin

chemical agents. The sweat glands help in regulation of the body's temperature by evaporation of the moisture they produce. Blister agents, like the sulfur mustards, are designed to destroy the underlying skin cells and tissue, making recovery long and painful. An injury to the lower skin may take months or even years to heal. Beta radiation, in particular, can cause severe burns and other destruction to the cells of the skin, since most of the particle energy is dissipated in the first few millimeters of travel through matter. Absorption of a chemical or biological agent through the skin is called *dermal absorption*.

Ocular System

The system which allows for vision is called the *ocular system*. It consists of the *eyes*, the *optic nerves*, and the portion of the brain in which electrical impulses from the eye are converted into vision. The external portion surrounding the eye is called the *conjunctiva*. The eye is composed of an outer layer, called the *sclera*, covering the entire *eyeball*. At the front of each eye is a *lens*, which thickens and thins in order to focus light onto the receptors at the rear of the eye. The lens is covered by the *cornea*, which can be damaged by a number of chemicals. In the disease called *cataracts*, the cornea becomes partially opaque, preventing light from passing through properly. Between the cornea and the lens is a material called the *vitreous humor*. The interior of the eye is filled with a water-like material called the *aqueous humor*. This is the material that gives pressure to the eye and keeps the eyeball round. Excess pressure in the eyeball can be caused by disease as well as chemicals; this is known as *glaucoma* and can cause blindness by damage to the retina. The *retina* is the rear of the eyeball, containing the light sensing cells. *Cone cells* response to different colors while *rod cells* response to intensity of the light. These cells generate electricity when hit by light and are connected to the brain's vision centers by the optic nerve. In the brain, the electrical impulses are converted into sight.

Each of these components can be damaged by chemical and biological agents, especially the blister agents and the nerve agents. The eyes are very sensitive to harm; many of the first symptoms of chemical attack are runny eyes, blurred vision, and tearing. Absorption of a chemical or biological agent through the eye is called *ocular absorption*.

Respiratory System

Humans can exist for many days without food, hours without water, but only minutes without oxygen. In addition, metabolism produces carbon dioxide that can be toxic

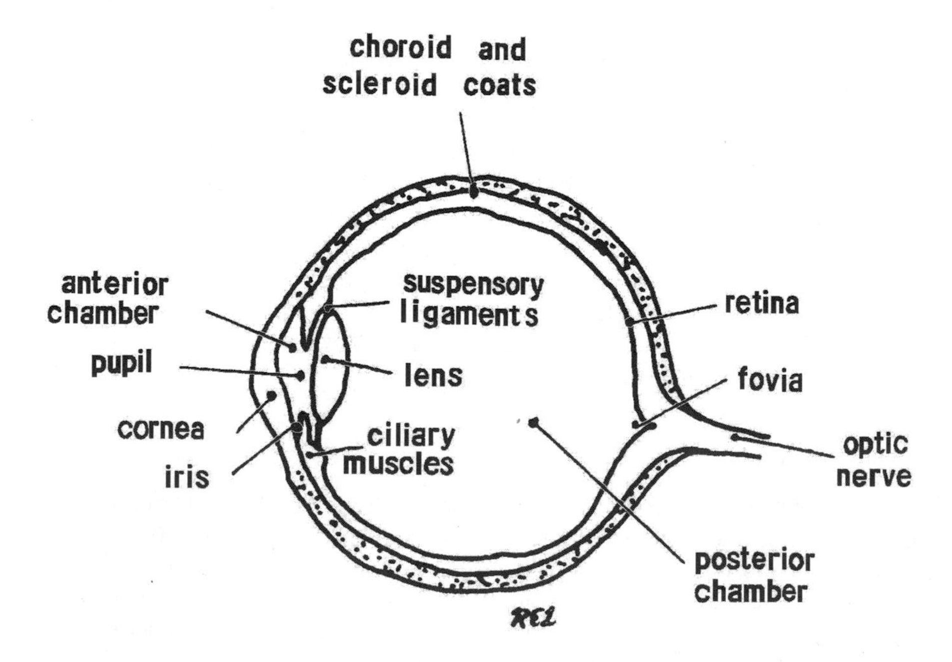

Figure 1.8 The Human Eye

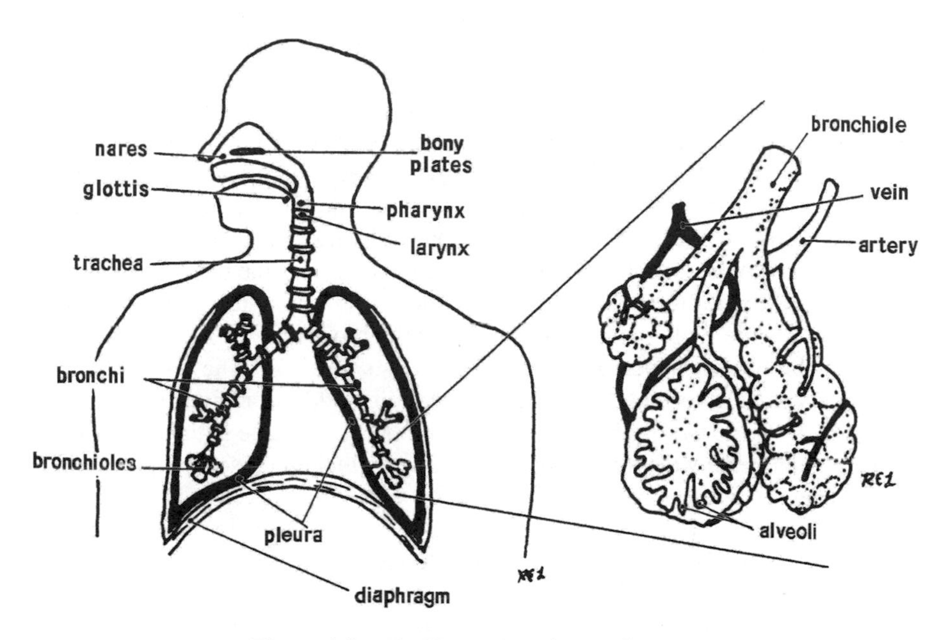

Figure 1.9 The Human Respiratory System

in sufficient concentrations. The respiratory system brings the needed oxygen to the body and removes the carbon dioxide.

Humans get their energy from the "burning" of food (oxidation) in their cells. In this process, oxygen combines with food to produce water and carbon dioxide. The oxygen and carbon dioxide are carried by the blood throughout the body. Any chemical or biological substance that is inhaled enters the body by the respiratory system and may be transported by the circulatory system. Chemical substances being inhaled are carried by the air and are breathed into the lungs by the force of the *diaphragm*, a sheet of muscles located in the body cavity, which pulls on the lungs to force air in and then pushes on the lungs to force air out, much like a bellows.

Air entering the nose is warmed and filtered by a series of bony shelves that are covered with mucous membranes. Larger particles of the chemical substance and some bacteria are partially filtered out by the nasal hairs and mucous plates. Some chemical substances may enter the bloodstream directly through the nose since it contains many blood vessels very close to the surface. The rest of the chemical substance and most biological organisms then pass through the *pharynx* and pass the *glottis*, a flap that prevents food from the mouth from entering the respiratory system. It then passes through the *larynx*, which contains the vocal cords, and into the largest of the respiratory passages, the *trachea*. The trachea is a large tube, about 2 centimeters (1 inch) in diameter, ringed with bands of cartilage that prevent it from collapsing. All these passages are lined with *cilia*, which are small hair-like projections that beat like small brooms to sweep some of the particles up and out of the respiratory tract. Unfortunately, the substance then is swallowed and enters the digestive system.

The trachea branches into two major passages called *bronchi* (singular: *bronchus*), one going to each lung. These, in turn, branch into ever-smaller passages called *bronchioles*. All these contain cartilage to prevent collapse. The very smallest bronchioles, however, do not have cartilage rings and simply end in sacs called *atria* (singular: *atrium*), each of which contain many chambers called *alveoli* (singular: *alveolus*). It is at the alveoli that the exchange of oxygen to the blood's hemoglobin and carbon dioxide from the blood takes place. It is also where much of inhaled chemical or some biological substances enter the bloodstream. The alveoli on the end of the bronchiole appear like bunches of grapes on a vine. The cells of the alveoli are very sensitive to many chemicals, and the blood lies only one cell thickness away so transfer of these substances may easily occur. Damage to the lungs by blister agents can destroy the alveoli, preventing the normal processes from occurring.

The lungs appear like pink, spongy masses for the normal person, but can become black and rigid, especially after some time, for victims exposed to some toxic agents (as well as from smoking and mining). The lungs are surrounded by tissues called *pleura* that are two sheets separated by a fluid allowing motion of the lungs as they expand and contract. Most chemical agents and some biological agents are intended for exposure by inhalation to the respiratory system due to the sensitivity of the cells composing it and the importance of respiration to life. Early symptoms of chemical agent exposure include changes in respiration rate, labored breathing, choking, and gasping for breath.

Digestive System

The digestive, or *gastrointestinal,* system is a long tube from the mouth to the anus. Its function is to pass food through the body, dissolve and digest the food, and pass wastes out. Food or other substances that enter the mouth pass over the *tongue* and the *glottis*, a flap separating the mouth from either the digestive or respiratory systems, as necessary. The material is then moved by ripple-like motion down the *esophagus*, a tube leading to the stomach. Although some digestion of food takes place for starch in the mouth, the first real digestion occurs in the *stomach* where the food is mixed with hydrochloric acid and stomach enzymes that help break down the food. There is usually very little uptake of most chemical or biological substances in the stomach (alcohol and some similar chemicals are exceptions) and most chemical agents are resistant to acidic attack. The stomach is a saclike muscle; it contacts and expands to pulverize and mix food with strong acid. Some pathogens are actually destroyed by the conditions in the stomach, but a few, especially viruses, can survive. Muscles at each end of the stomach close it off from the other parts of the digestive system.

Passing by the muscles at the end of the stomach, food or other materials move into the upper portion of the small intestine called the *duodenum* where various enzymes normally break down food constituents in a highly alkaline environment.

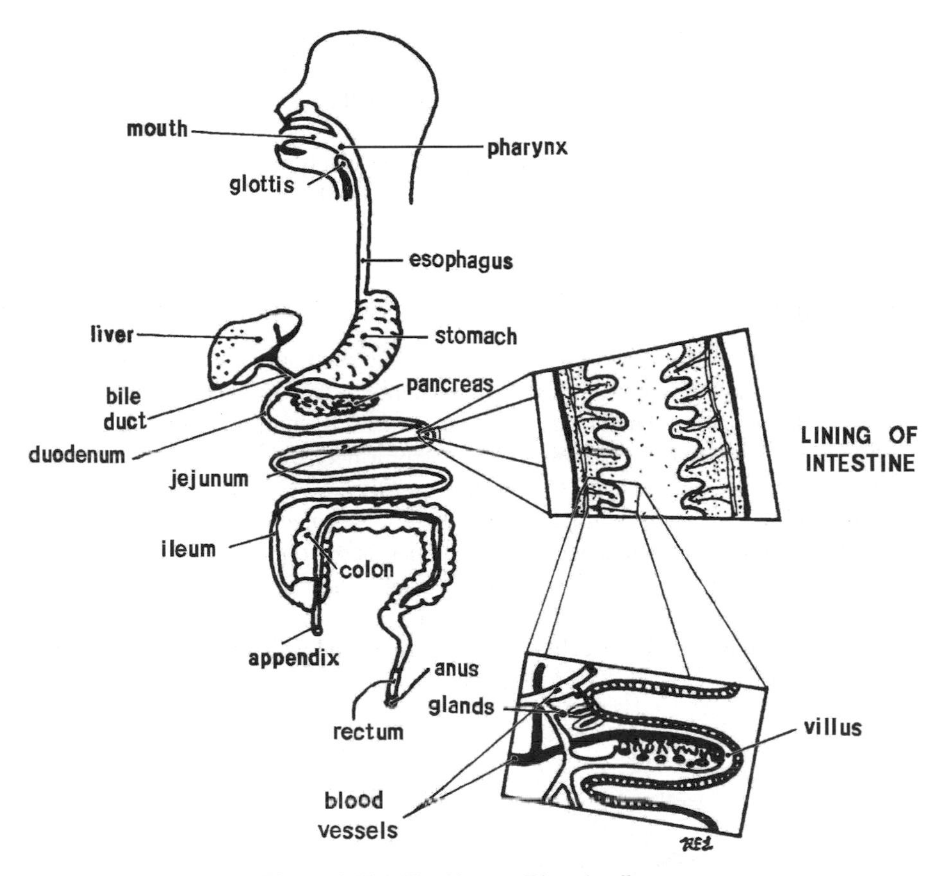

Figure 1.10 The Human Digestive System

The duodenum is the primary location where fatty foods are broken down into simpler compounds and where some substances enter the body. A number of chemical agents can enter the body at this point even though a few are weakened by alkaline conditions. The duodenum is only about 25 centimeters (10 inches) long. The next 3-meter (10-foot) portion is called the *jejunum*, which is very rich in enzymes that further digest the food and can be the site of chemical reactions involving chemical substances. The remaining 4 meters (12 feet) of small intestine is called the *ileum*. It is the location where the food products and much of the ingested chemical substances pass to the blood through hair-like projections into the intestines. These hair-like objects are called *villi* (singular: *villus*). They are very small so they greatly increase the area of absorption. Most chemical agents, and some biological organisms, enter the bloodstream at this point.

The small intestine then joins to the large intestine, collectively called the *colon*, which are chiefly storage locations although there is small amount of water transfer into and out of the colon. The food is mixed with mucous and bacteria and becomes waste matter called feces. When the feces reach and fill the *rectum*, which is normally empty, the desire to eliminate becomes great (usually after about 36 hours after ingestion). When the muscles of the *anus* open, the wastes are passed to the outside of the body, and the process of digestion and elimination is complete. Some chemical and biological substances may pass completely through the body and be eliminated in the feces, but effective ones are designed to remain in the body long enough to adversely affect it.

Nervous System

The human nervous system is very complex, and we shall only briefly discuss it. The nervous system is the target of many of the chemical warfare agents. These agents are called nerve agents. The *brain* is the center of the system. It functions as a processing center with impulses coming from receptor nerves throughout the body and going out to control all the body's functions.

The brain is connected with all parts of the body by nerves; these nerves are not continuous, but have junctions where they almost, but not quite, touch. By chemical processes involving the body chemicals acetylcholinesterase and acetylcholine, electrical impulses can jump these spaces, called *synapses*. As soon as the electrical impulse makes the leap, the circuit is then opened again by chemical presence. Some

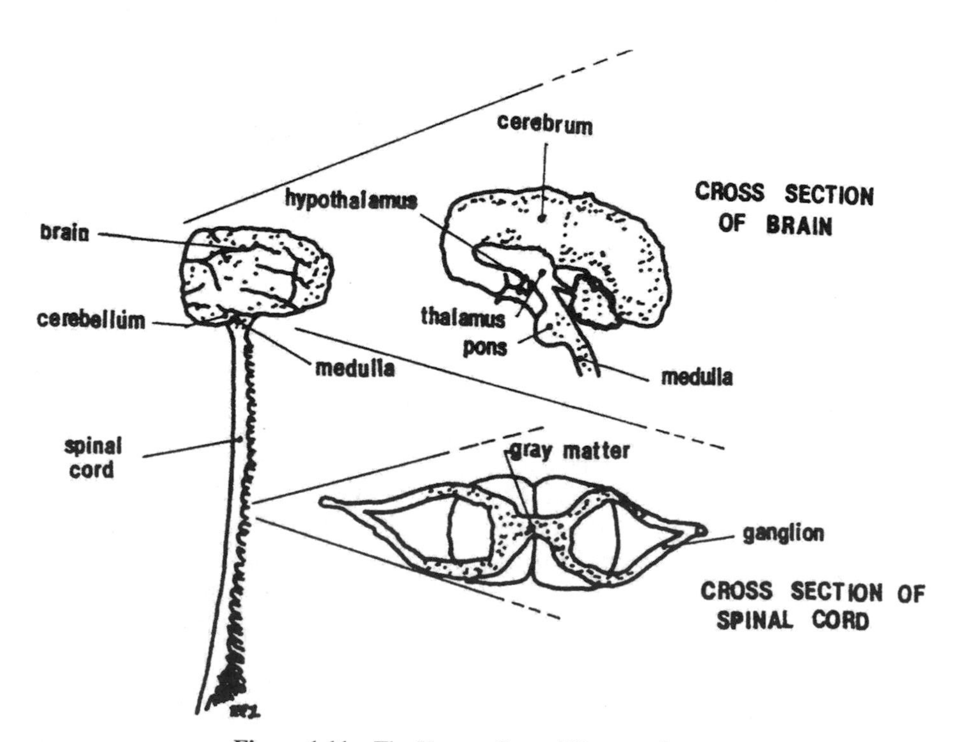

Figure 1.11 The Human Central Nervous System

of the nerve agents are substances that can block the transfer of these electrical pulses. There are other substances that can cause, in effect, a short circuit by allowing continuous impulses to occur. In both cases, normal nerve activity cannot take place, and death can result.

The nervous system is usually divided into two parts: the *central nervous system* (often abbreviated CNS) consisting of the brain and spinal cord, which is a bundle of nerve fibers running through holes in the protective vertebrae of the backbone; and the *peripheral nervous system* consisting of nerves that connect the spinal cord with sensory receptors and glands and muscles of the body. Some agents are designed to affect one, the other, or both.

The brain has several parts, each with a different function. The chief parts are the *cerebrum*, *cerebellum*, and the *medulla oblongata*. The cerebrum is the seat of thought, memory, and reasoning; it is the target of those chemical agents designed to remove the user from reality (like Agent BZ or lysergic acid diethylamide—LSD). It is also affected by acute high doses of radiation. The cerebellum is the seat of muscular coordination; it is the target of those chemical agents designed to cause muscular paralysis. The medulla oblongata controls automatic functions of the body such as breathing, heartbeat, *etc.*; it is the target of agents designed to render unconsciousness or to kill.

Reproductive System

Bacteria and other lower forms of life reproduce by a process known as *mitosis*, which simply involves duplication of cellular material and splitting into two identical cells. This is also the process by which individual cells of the human body duplicate. One cell becomes two; the two cells become four; and so on. This is an asexual form of reproduction. Many chemical agents, some biological agents, and some effects of radiation can affect the normal cellular duplication, leading to injury, death, or abnormal cellular growth called cancer.

Humans, like all higher animals, reproduce the organism by a process called *meiosis* or sexual reproduction. Most simply, this process is the adding together of cellular material from reproductive cells of both parents. All the cells of the human body contain within the nucleus certain chains of chemicals called *chromosomes*. The chemical making up these carriers of genetic information is *deoxyribonucleic acid* (DNA). The ways that this chemical can arrange itself determine all the characteristics of the offspring (hair color, eye color, height, *etc.*) as well as in the cells of the person themselves. The DNA is like a template which transfers information from one generation to the next and as the individual ages.

Nucleic acids are made of polynucleotide chains formed by many nucleotides bonded together. Each nucleotide contains three parts: a phosphate, a sugar, and a nitrogenous base. There are two different kinds of sugars in a nucleotide, deoxyribose and ribose. If the polynucleotide chain forms DNA, the sugars in its nucleotides are deoxyribose; if the nucleotides contain ribose as its sugar, RNA results. *Ribonucleic acid* (RNA) is involved in telling the DNA when and how to replicate, and may supply material to the DNA. There are five different bases in a nucleotide. These chemical bases are adenine (A), cytocine (C), guanine (G), thymine (T), and uracil (U). Uracil is only found in RNA, while thymine is only found in DNA. The two strands of DNA are connected at each base, like the rungs of a ladder. Each base will only bond with one other base: adenine (A) will only bond with thymine (T), and guanine (G) will only bond with cytosine (C); they are called *base pairs*. The pattern of base pair linkages that is formed in the DNA conveys the genetic information for reproduction and heredity.

Some chemicals and most forms of radiation can damage the DNA, so that the structure can fall apart or fail to copy properly. If only one side of the ladder is broken, the DNA might be able to repair the damage. If both strands are broken, repair is much less likely. The amount of damage depends upon the amount of energy released; alpha radiation releases all its energy in a short distance, so can be the most damaging form of internal radiation. In cases of DNA damage, there can be miscommunication of cellular reproduction information to succeeding generations, resulting in cancer to the organism or even birth defects of offspring. Similarly, damage to RNA, which serves several functions in the cell from passing messages to the DNA to supplying some of the raw material needed for DNA replication, can result in cellular damage. This damage is most commonly to the existing cells and less to succeeding generations. Therefore, the possibilities are: the cell can die; the cell can repair the damage (especially if only one of the two DNA strands is broken); the cell can exhibit DNA damage by becoming cancerous; or the cell (especially eggs and sperm cells) can transmit faulty information, resulting in genetic effects like birth defects.

Some chemical substances can cause damage to the DNA and the chromosomes. All the cells of the human body, except the sex cells, contain the same number of chromosomes, 46 in 23 pairs. The sex cells contain only one-half this number or 23 chromosomes. This is so because a union of the two sex cells leads to the original 46. This process is the basis of heredity. Which chemical substances can cause heredity changes (that is, effects on future generations) is a matter of speculation at this time and, while more research is needed, the possibility remains that the chemical warfare substances could affect the heredity code and cause damage to future generations. A number of chemicals, including lysergic acid diethylamide (LSD), have already been implicated in breaking or otherwise damaging chromosomes.

Male sex cells, called *sperm*, are formed in the male gonads, the *testes*. The process of sperm formation is called spermatogenesis. These cells begin their development from precursor cells, which are formed at the edges of the testes. These precursor cells contain 46 chromosomes. They divide by a process called mitosis or cell division by which identical daughter cells result from a splitting of the parent cell. As the population of cells increases, the cells move to the center of the testes where they become mature sperm following the last two stages of meiosis and end with only 23 chromosomes. These sperm look like small tadpoles, having a rounded or arrow-shaped head and a long tail. There are a number of chemical substances that can lead to reduction in numbers and viability of the sperm.

Female sex cells are called eggs or *ova* (singular: *ovum*) and are formed in the female gonads, the *ovaries*. The eggs are formed by a process initially quite similar to that for sperm. The immature cell divides by mitosis but, unlike the sperm, into two unequal sized cells with one receiving almost all the cellular material. A second division of the larger cells produces another unequal division with the larger body becoming the egg and the other, like the first smaller cell, not involved in reproduction. Some chemical substances can damage the ova and make them less viable.

We now have an egg cell, which is much larger, and a sperm cell, both with 23 chromosomes carrying the hereditary coding in their sequence of genes. The union of the two cells is called *fertilization* with the gene sequence of the new DNA decided by chance from the two parents' DNA. The details of coding and heredity are too complex to cover in this text.

The single fertilized cell then divides by mitosis repeatedly in the process of making a child. Usually this process continues in a normal fashion until birth; but, sometimes there are changes because of chance, a physical cause, or a chemical cause.

The process of formation and development of an embryo is actually a programmed sequence of cell proliferation followed by differentiation (or specialization) and cell migration leading to organ formation. The most critical period of human embryo formation is during the first three months (trimester) of pregnancy since it is at this time that the most rapid cell proliferation takes place and upon which all subsequent development is based. A well-known fact of ionizing radiation is that rapidly reproducing cells are more vulnerable to the effects. During embryo development, even very small amounts of radiation can cause birth defects, cancer, or death. The very first two weeks, however, is generally not a period in which chemical agents can cause birth defects since prenatal death followed by natural abortion usually occurs if severe damage takes place to the embryo.

Physical deformities and mental defects are also possible if the mother was exposed to chemical or biological agents as well as radiation the very early weeks of pregnancy. The baby may suffer from low birth weight and slow physical and mental development if it survives. The long-term effects of most chemical agents, especially on future generations, is largely unknown. Knowledge of the effects of radiation on the developing embryo is somewhat better and will be discussed in a later chapter.

Endocrine System

The endocrine system consists of those glands of the body that do not have ducts. While this might seem a strange definition, it really is not. There are many glands in the human body that produce chemicals, called *enzymes*, which promote chemical reactions. Many of the glands in the body pass their enzymes to other parts of the body by ducts. For example, the salivary glands of the mouth pass saliva directly into the mouth to digest starches, and the liver passes bile to digest fats into the duodenum through a duct. Other glands, however, pass their chemicals,

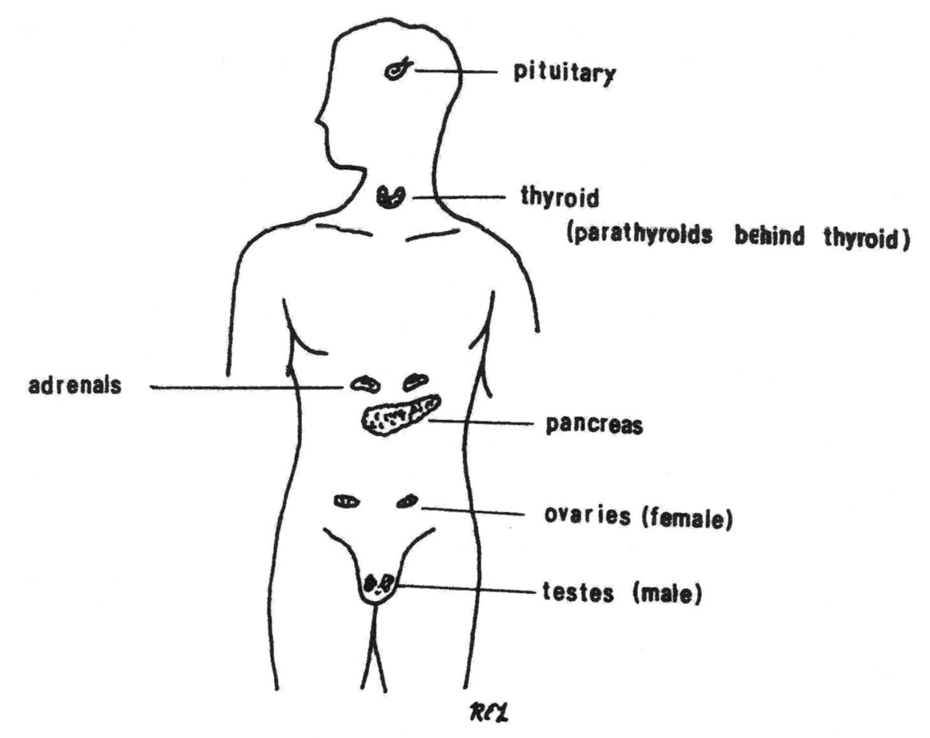

Figure 1.12 The Location of the Human Endocrine Glands

called *hormones*, to other parts of the body through the blood. A few glands are both ducted and ductless, passing different chemicals for different functions. Since enzymes are chemicals, there are chemical and biological agents which can affect the normal functioning of the body.

There are seven ductless glands in the body, several of which can be affected by chemical substances: the *gonads*, which produce hormones to promote the development of the sex cells; the *pancreas*, which produces a hormone used in the control of the amount of blood sugar used as food by the cells; the *thyroid*, which produces a hormone used in control of metabolism; the *parathyroids*, which produce a hormone that maintains calcium and phosphorus levels; the *pituitary*, which is the most complex endocrine gland and produces growth hormones and also plays a role in control of the other endocrine glands; and the *adrenals*, which produce several hormones that function in the conversion of proteins to carbohydrates and also to stimulate the body to respond in emergencies.

At this time, few, if any, chemical or biological weapons directly target the endocrine system; however, there are chemicals, called *endocrine disrupters*, which mimic certain hormones and may have both short-term and long-term adverse effects.

Blood and Circulatory System

It is believed that life began in ancient seas, and so, life even today needs to be surrounded by a mild salt solution. Every cell of your body needs to live in such an environment. Since the cells of the human body may be far removed from sources of food, there is needed some way for food materials to be brought to the cells and for wastes to be removed. The blood performs this function, but can also move chemical substances around the body. The *heart* is the center of the circulatory system. It is fundamentally a pump, pulling blood from parts of the body and pumping it under pressure to other parts. The heart is almost totally cardiac muscle. The parts of the heart are such that they act as if all the cells were really one. Since the blood flows in a continuous circuit, we can start at any point; but, let's begin at the point in the heart where blood is pumped to the lungs.

Blood is pumped to the lungs by the right (as viewed from the front) chamber called the *ventricle*. The blood is pumped through the lungs where oxygen is absorbed by hemoglobin and waste gases are expelled. The now-oxygen-rich blood is pumped back to the heart's left *atrium* (sometimes also called auricle). The bicuspid value between the atrium and ventricle opens and the blood is pumped into the left ventricle. From here, it is pumped to all the parts of the body. The blood vessel first passed through is the *aorta*, the largest artery of the body. From the aorta, the blood is pumped through ever-narrowing vessels until, at last, the very tiny *capillaries* are reached. It is at the capillaries that the gaseous exchange occurs so that oxygen is given to every cell and the cell's wastes are transferred to the blood. It is also

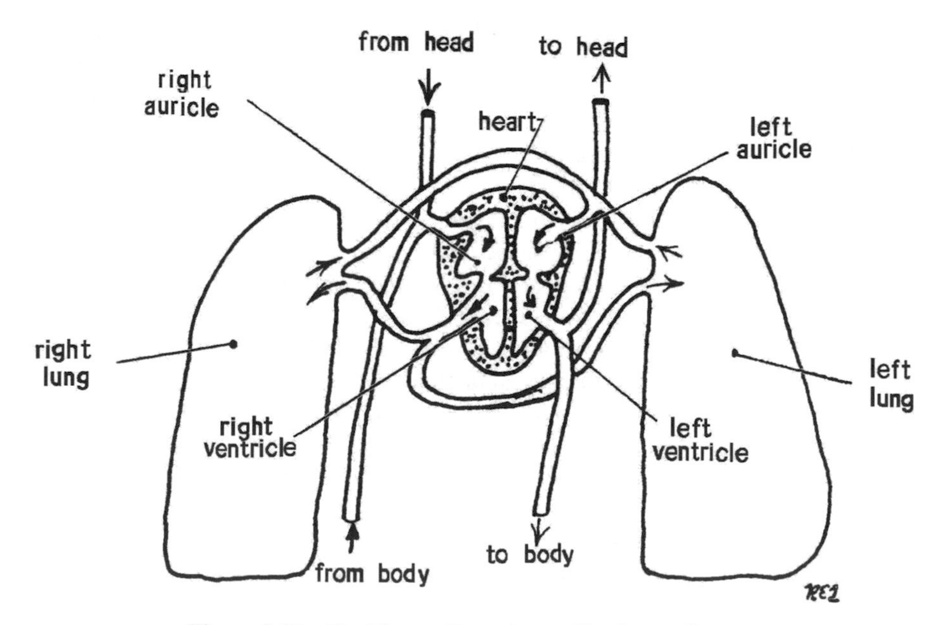

Figure 1.13 The Human Heart-Lung Circulatory System

where chemicals can be transferred from the blood to the cells. From the heart to the capillaries, all the vessels are called *arteries*. As the blood leaves the capillaries, the return system is used. All the vessels from the capillaries back to the heart are called *veins*. Thus, the blood in the arteries is rich in oxygen and low in wastes, but the blood in the veins is low in oxygen and rich in wastes. The blood moves into ever-enlarging vessels until it returns to the right atrium through the *vena cava*, the largest vein of the body, from which it is pumped to the right ventricle. The cycle then repeats itself.

Many factors influence the rate of the heart's pumping and the pressure of the blood in the vessels. There are many chemical agents that can affect either or both; a number of chemicals similar to tranquilizers and beta blockers reduce blood pressure as well as slowing heart rate, even to the point of coma and death.

The blood is a multi-purpose fluid. It carries the red blood cells, the white blood cells, the platelets, and a large number of antibodies that attack specific biological organisms or chemicals, as well as carrying dissolved gases and chemicals. Red blood cells, essential for carrying oxygen to the cells of the body, are produced by the long bones of the arms and legs in the bone marrow. White blood cells, such as lymphocytes and granulocytes, are important in the body fighting off infection. These can be damaged by radiation and some chemicals. Radiation is particularly damaging to the blood, first for red cell formation effects and later with white blood cell and platelet destruction. Radiation causes a reduction in the body's ability to make more red blood cells. The process of making red blood cells is called the *hematopoetic process*. The next cells affected are the lymphocytes, formed in lymphoid tissues, critical in the body's immune response to pathogens. The next cells affected by radiation are the platelets, which assist in the formation of blood clots; without platelets, bleeding cannot be controlled. Finally, the granulocytes, formed in the bone marrow, are reduced in number; these white blood cells are important in fighting infection. Blood agents interfere with the transfer of oxygen and/or carbon dioxide, leading to cell damage and, often, death. Nerve agents are carried by the blood to the entire body.

The combined effects of a radiation attack, weakening the immune system, and a biological agent attack, causing disease, or a chemical agent attack, causing alteration of physiology, would compound the effects of either alone. It is well known that it takes longer for many injuries, especially burns, to heal or diseases to end if the victim has also been exposed to radiation. It is possible that a "dirty bomb" might be used in conjunction with a biological or chemical agent to compound the contamination as well as multiply the effects. Fortunately, the amount of radiation in most "dirty bomb" scenarios is not sufficient to produce the hematological effects to a degree where the immune system of many victims would be impaired.

Lymphatic System

There is another system, sometimes called the body's second circulatory system: the lymphatic system. This represents another route by which fluids can flow from the cells into the blood. Larger molecules, such as proteins, cannot pass from the cells into the blood in the capillaries. However, there must be a way for them to return to the blood. This is one of the functions of this system. The fluid carried by this system is called *lymph*, and is produced throughout the body, but chiefly by the liver and intestines.

Protein molecules are carried by the lymph from the cells back to the blood. The vessels of the lymphatic system are similar to those of the blood circulatory system, being larger at the points where they enter the bloodstream and tiny capillaries where they contact the cells. These capillaries are different from the blood capillaries since there are minute spaces into them which can allow molecules such as proteins to enter. The lymph is returned to the blood at several locations. One-way valves allow the flow to pass into the blood but not for the blood to enter the lymphatic system. Another important role of this system is the carrying of fats and some other nutrients from the small intestines to the blood rather than by way of the villi.

Finally, very large particles such as bacteria and other pathogen agents can enter the system. As the lymph passes through the *lymph nodes*, some bacteria are removed and destroyed. Sometimes, though, the lymph nodes can be temporarily overwhelmed, as during a cold or a deliberate attack using high concentrations of pathogens or toxins. The lymph nodes can become swollen and painful. Locations of lymph nodes under the neck and armpits are commonly swollen during infections. It is infection of the lymph nodes by plague bacteria which forms *buboes*, swollen nodes, and gives the name *bubonic plague*.

Urinary System

Liquid wastes are removed from the body by means of the urinary system. Many chemical agents or the products of their metabolism within the body (sometimes as toxic as the chemical agents themselves) are removed from the body via the urine.

The use of urinalysis, the study of components in the urine, for detection of agent exposure is the most common method to determine if chemical substances have been taken into the body. As blood passes through the *kidneys*, many waste substances are removed. These wastes, including chemical substances and chemical substance metabolic products, are carried to the *bladder*, which is a storage sac, until discharged from the body. The liquid waste is called *urine*, and is generally a light yellow color from the breakdown products of protein metabolism.

The two kidneys are able to remove many wastes from the blood and concentrate them for removal. The kidneys are very rich in blood vessels from which they remove the products of metabolism that would be quickly fatal if not removed. The concentration of a chemical agent is intended, in part, to overwhelm this detoxification mechanism. The wastes pass from the kidneys through tubes called *ureters* to the urinary bladder. Muscles at the bottom of the bladder hold the urine until it is ready to be passed. When this happens, the urine goes through the tube called the *urethra* to the outside of the body. In the female, this is a separate system; however,

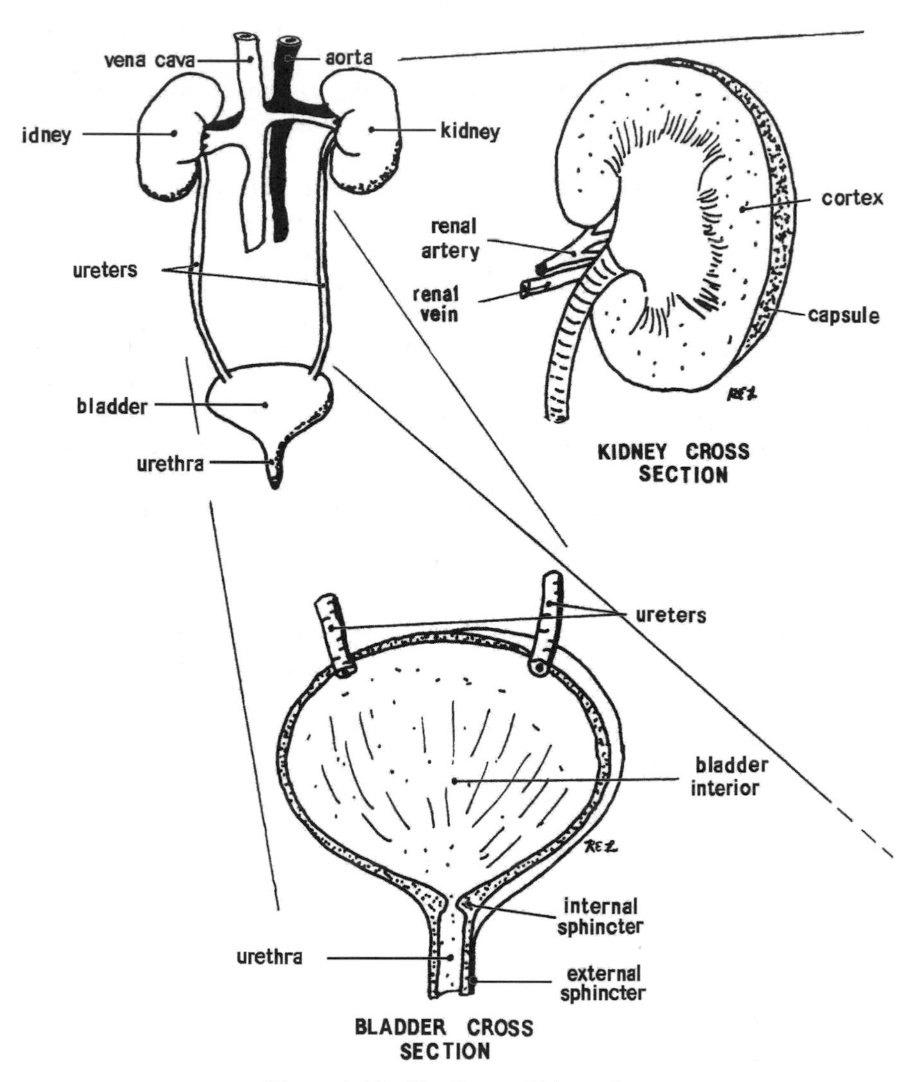

Figure 1.14 The Human Urinary System

in the male, the urine is routed through the penis with constricting muscles blocking flow to the reproductive system.

The Liver

One of the most amazing organs of the human body is the *liver*. It acts as a chemical filter and processing plant to remove toxic materials from the body and detoxifies many chemicals into less harmful substances. In addition, the liver also produces an enzyme, called *bile*, which aids in the digestion of fats and produces much of the lymph fluid.

The job of removing toxins from the blood is a very large order, especially for the victim of a chemical or biological agent attack. The various types of materials that may enter the bloodstream can each require a different reaction to be rendered less harmful. The liver is able to detoxify many different types of chemicals; however, it can be harmed itself by some substances, either by direct action on the liver or by overwhelming the liver's capacity to detoxify. Several types of cancerous tumors can be promoted in the liver by chemicals. Long-term exposure to certain common chemical substances, especially alcohol, can reduce the detoxifying ability of the liver or even destroy it. The liver is the target organ for hepatitis viruses.

In the absence of improper chemical substances or disease, the liver can usually continue its function, filtering hundreds of gallons of blood daily, producing enzymes and lymph, and detoxifying harmful chemicals throughout a person's life. This organ is one of the most critical for life and is the primary means by which chemical substances are removed by the body. Agents, either chemical or biological (like the hepatitis viruses), which weaken or destroy the liver would be very potentially harmful over the long term.

SUMMARY

Weapons of mass destruction, whether nuclear, biological, or chemical, or even in combination, are designed to produce fear in addition to their physical health effects. Each of these weapons can harm the human body. These topics will be further discussed in the following chapters on the nature of the weapons, the effects on buildings and people, the status of these weapons in the world today, the delivery methods of these weapons, and the ways to protect from exposure and ameliorate the effects.

Each of the following chapters will describe in some detail the history, employment, effects, and protection from each of nuclear, biological, and chemical weapons which either have been used or are likely to be used.

Part 1

RADIOLOGICAL WEAPONS

2

NUCLEAR RADIATION

ATOMS AND ELEMENTS

All matter (solid, liquid, or gas) is composed of very small particles called *atoms* (from the Greek, *atomos*, meaning "indivisible"). To better understand how nuclear weapons work, it is necessary to know a little about the structure of atoms and how we define characteristics of atoms. The simplest vision of an atom is a central sphere, the *nucleus*, made up of positively-charged particles called *protons* and uncharged particles of about the same size and weight called *neutrons*. Circling around this core at relatively great distances are much smaller and lighter negatively-charged *electrons*, like planets circling the sun. These electrons travel in defined paths or orbits, depending upon their relative energy. If an electron gains energy, it can move to a higher orbit; if it loses energy, to a lower orbit. Such electron transitions result in absorption or emission of energy as visible light, x-rays, or ultraviolet radiation. There is no net electrical charge on the atom because there are the same number of negative and positive charges. If one or more of the electrons are lost by some means, the atom has an unbalanced charge and is then called a *positive ion*. If more electrons are attracted to the atom, it becomes a *negative ion*. Figure 2.1 shows a simplified representation of atoms held together by this electrostatic attraction of opposite charges (*ionic* bonding). Figures 2.2, 2.3, and 2.4 show representations of the three types of hydrogen molecules ("normal" hydrogen or H, deuterium or D, and tritium or T) held together by a sharing of electrons (*covalent* bonding).

The neutrons in the center, or nucleus, of a single atom can be simplistically thought of as being made up of protons (p) and neutrons (n) with encircling electrons

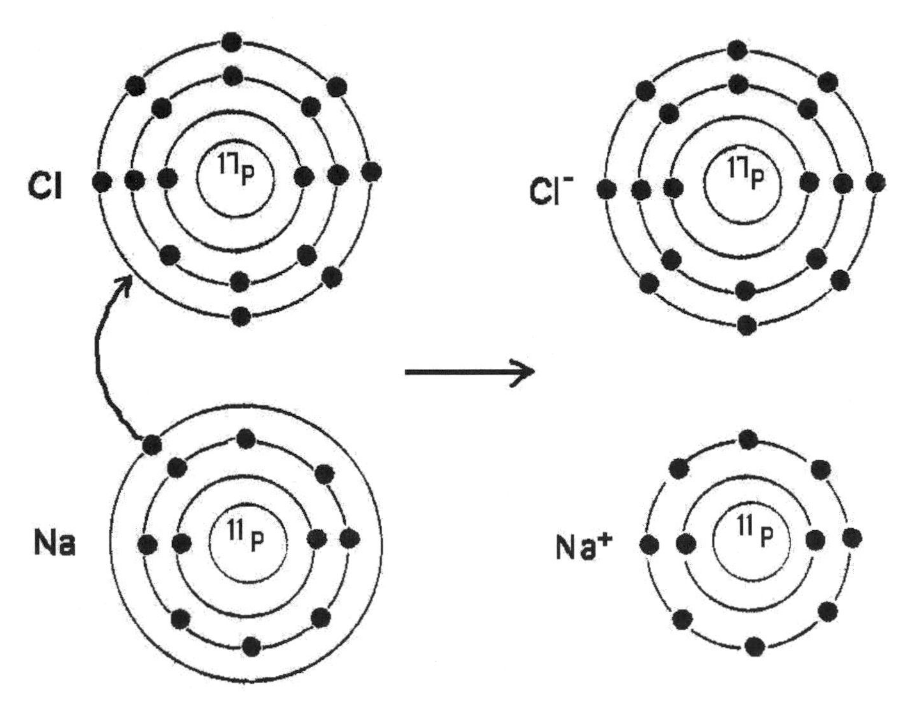

Figure 2.1 Representation of Ionic Bonding

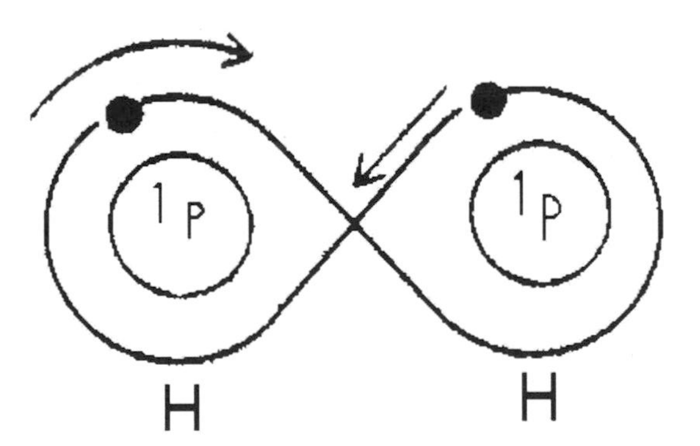

Figure 2.2 Figure Representation of Hydrogen Molecule

(e). There is the same number of protons as electrons; thus, there is no net charge. While this is not technically correct and is a simple explanation, it is sufficient for our needs. Most of the time, the atom's nucleus stays just that way, but rarely, if acted upon by internal or external energy, a neutron can split into parts, ejecting an electron as a beta particle (and another very small particle called a neutrino) and becoming a proton. This process is known as beta decay. By this beta process, called *transmutation* or *radioactive transformation*, the chemical nature of the atom has changed since it is now an atom of the element of one greater atomic number.

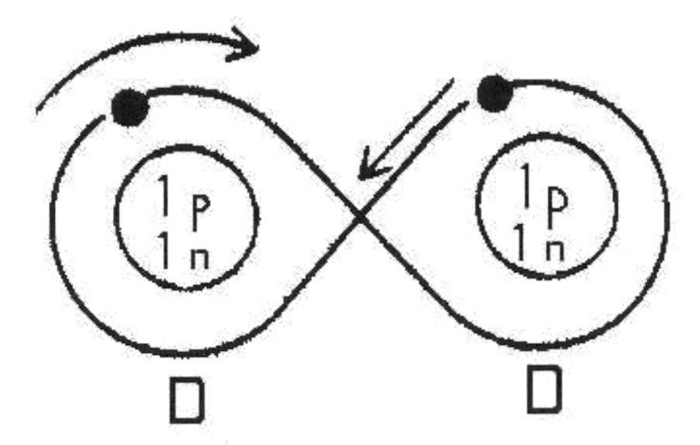

Figure 2.3 Representation of Deuterium Molecule

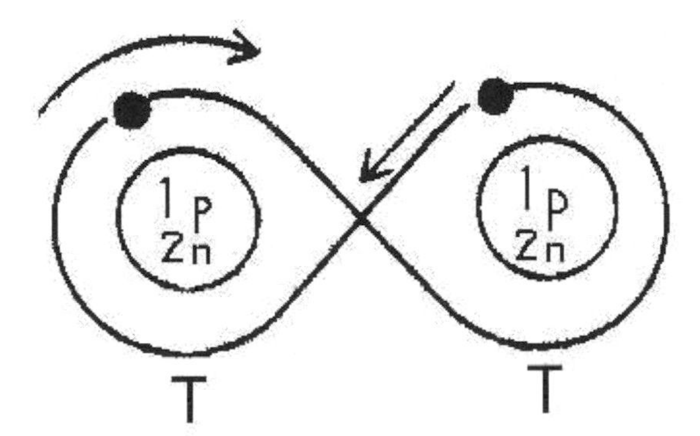

Figure 2.4 Representation of Tritium Molecule

Neutrons (and other subatomic particles, like pi-mesons) act like the "glue" to hold the nucleus together since the positively-charged protons would otherwise repel each other. The force they exert to hold the nucleus together is known as the *nuclear force*. This is the source of nuclear energy, whether in a controlled reactor generating electricity or uncontrolled in an atomic weapon.

Around the atomic nucleus circle the negatively-charged electrons. While electrons play several roles in nuclear physics, they are chiefly responsible for chemical processes. We shall spend most of our time considering atomic nuclei rather than electron structure in this book.

Atoms are often described by their *atomic number* and *atomic mass*. The atomic number is the number of protons in the atom's nucleus. All atoms of a chemical element have the same number of protons and, thus, what is called the "atomic number." Atoms are sometimes defined by their atomic number; other times, they are defined by their mass (or weight) based on the *atomic mass unit* (amu). This unit is the weight, or mass, of $1/12^{th}$ that of one carbon-12 atom, and is equal to 1.66×10^{-24} gram. Protons have a mass of slightly more than 1 atomic mass unit (amu) (1.6726

$\times 10^{-24}$ gram), but is so close that is often considered as 1 amu. A neutron has a mass very slightly heavier (1.67492×10^{-24} gram), but can also be considered as about 1 amu. Electrons are so very light (only 9.0186×10^{-28} gram at rest) that they can be ignored in weight calculations. Thus, the atomic mass is the sum of the mass of protons and neutrons in the atom's nucleus; however, this is usually simplified to simply adding the number of protons and neutrons since this is essentially the same value. For example, for uranium-235 (U-235 or ^{235}U), each atom of this isotope of uranium has 92 protons (the atomic number of uranium) and 143 neutrons. Thus, the atomic number of uranium is 92, and the atomic mass of U-235 is 235 (92 + 143). The more naturally abundant U-238 differs from U-235 in that each atom has 146 neutrons rather than 143.

You may know that an atom of the simplest element, hydrogen, has one proton and no neutrons in its nucleus, so the atomic number is 1 and the atomic mass is 1 amu. It is symbolically ^{1}H or H-1, where the "1" is the atomic mass. There are two other *isotopes* (atoms of the same element with different atomic masses) of hydrogen: deuterium and tritium. Both of these are important in nuclear weapons. Deuterium atoms have one proton (so it is an isotope of hydrogen) and one neutron, for an atomic mass of 2 amu. It is given symbolically as ^{2}D or ^{2}H or just D. Deuterium is also called hydrogen-2 or H-2. There is a third isotope of hydrogen, tritium, having one proton and two neutrons for an atomic mass of 3 amu. It is given symbolically as ^{3}T or ^{3}H or just T. Tritium is also known as hydrogen-3 (H-3 or ^{3}H). Tritium is itself radioactive. Both of these heavier isotopes of hydrogen are important in thermonuclear weapons. Some of the material in this book is simplified, but accurately describes the processes involved in nuclear weapons. Although there are more than 30 subatomic particles (which themselves may be made of even smaller particles called quarks), the role or function of most of these is unknown. For the purpose of this book, an atomic nucleus may be considered as being made up of protons and neutrons. While this simplified picture of an atom ignores other nuclear particles and quantum concepts, it is sufficient for our needs.

Atoms may combine to form molecules. Pure materials, called *elements*, are composed of only one kind of atom. *Compounds* are composed of two or more elements, and are made up of molecules. The atoms making up molecules may be held together by electrostatic attraction of oppositely-charged *ions* when atoms have lost and gained orbital electrons (ionic bonding), or by a sharing of electrons (covalent bonding). The study of atoms, molecules, elements, and compounds as well as the transformations which take place is called *chemistry*. There are some 90 or so naturally occurring elements, often depicted on the *periodic chart*, having recurring similar chemical and physical characteristics. A version of the period chart of the elements, including those not naturally occurring, appears as Figure 2.5.

Each element is represented by one or two letters of the Roman alphabet. Most are related to the name, but a few seem strange (for a couple of examples, Au for gold is from the Latin "*aurum*" and K for potassium, also from the Latin "*kalium*"). On the periodic chart, the atomic number (the number of protons) is a whole number and consecutive from 1 to 118 (actually, only some 109 or so have been discovered), and the atomic mass (or weight) is a fraction due to this being a weighted average of

1.01 H 1																															4.00 He 2
6.94 Li 3	9.01 Be 4																									10.81 B 5	12.01 C 6	14.01 N 7	16.00 O 8	19.00 F 9	20.18 Ne 10
22.99 Na 11	24.31 Mg 12																									26.98 Al 13	28.09 Si 14	30.97 P 15	32.06 S 16	35.45 Cl 17	39.95 Ar 18
39.10 K 19	40.08 Ca 20	44.96 Sc 21															47.90 Ti 22	50.94 V 23	52.00 Cr 24	54.94 Mn 25	55.85 Fe 26	58.93 Co 27	58.71 Ni 28	63.55 Cu 29	65.37 Zn 30	69.72 Ga 31	72.59 Ge 32	74.92 As 33	78.96 Se 34	79.90 Br 35	83.80 Kr 36
85.47 Rb 37	87.62 Sr 38	88.91 Y 39															91.22 Zr 40	92.91 Nb 41	95.94 Mo 42	(99) Tc 43	101.07 Ru 44	102.91 Rh 45	106.4 Pd 46	107.87 Ag 47	112.40 Cd 48	114.82 In 49	118.69 Sn 50	121.75 Sb 51	78.96 Te 52	126.91 I 53	131.30 Xe 54
132.91 Cs 55	137.34 Ba 56	138.91 La 57	140.12 Ce 58	140.91 Pr 59	144.24 Nd 60	(145) Pm 61	150.4 Sm 62	151.96 Eu 63	157.25 Gd 64	158.93 Tb 65	162.50 Dy 66	164.93 Ho 67	167.26 Er 68	168.93 Tm 69	173.04 Yb 70	174.97 Lu 71	178.49 Hf 72	180.95 Ta 73	183.85 W 74	186.2 Re 75	190.2 Os 76	192.22 Ir 77	195.09 Pt 78	196.97 Au 79	200.59 Hg 80	204.37 Tl 81	207.12 Pb 82	208.98 Bi 83	(209) Po 84	(210) At 85	(222) Rn 86
(223) Fr 87	(226) Ra 88	(227) Ac 89	232.04 Th 90	(231) Pa 91	238.03 U 92	(237) Np 93	(244) Pu 94	(243) Am 95	(247) Cm 96	(247) Bk 97	(251) Cf 98	(254) Es 99	(253) Fm 100	(256) Md 101	(253) No 102	(257) Lw 103	(257) Rf 104	(262) Db 105	(263) Sg 106	(262) Bh 107	(265) Hs 108	(266) Mt 109	110	111	112	113	114	115	116	117	118

Figure 2.5 Periodic Chart of the Elements

the various isotopes (the number of protons plus neutrons). The shape of the chart is related to the electron configurations of the atoms making up the elements, and isn't needed for the purpose of this book.

INTRODUCTION TO NUCLEAR RADIATION

Radioactivity is the process by which unstable atomic nuclei release energy in order to arrive at a more stable configuration. Just as water flows downhill, atoms tend toward their lowest energy state. Radioactivity is sometimes known as decay, since the original radioactive atom changes into another kind of atom or into an atom with less energy. There is a fairly large number of naturally occurring radioactive materials. Most naturally occurring radioactive materials are members of three natural decay chains, each beginning with an isotope of uranium or thorium and ending with an isotope of lead. These three chains are: the *actinium series*, beginning with uranium-235 (^{235}U) and decaying through a number of steps to become stable (non-radioactive) lead-207 (^{207}Pb); the *thorium series*, beginning with thorium-232 (^{232}Th) and decaying though a number of steps to become stable lead-208 (^{208}Pb); and the *uranium series*, beginning with uranium-238 (^{238}U) and decaying in steps ultimately to become stable lead-206 (^{206}Pb). Almost all the naturally occurring radioactive materials are members of one of these chains or another (there are a few exceptions).

The elements of the uranium series and the radioactive decay processes involved are shown in Figure 2.6 (darker arrows represent the more likely decay processes).

There are also a large number of radioactive materials not present in nature, but only produced in nuclear reactors or by colliding particles in devices such as

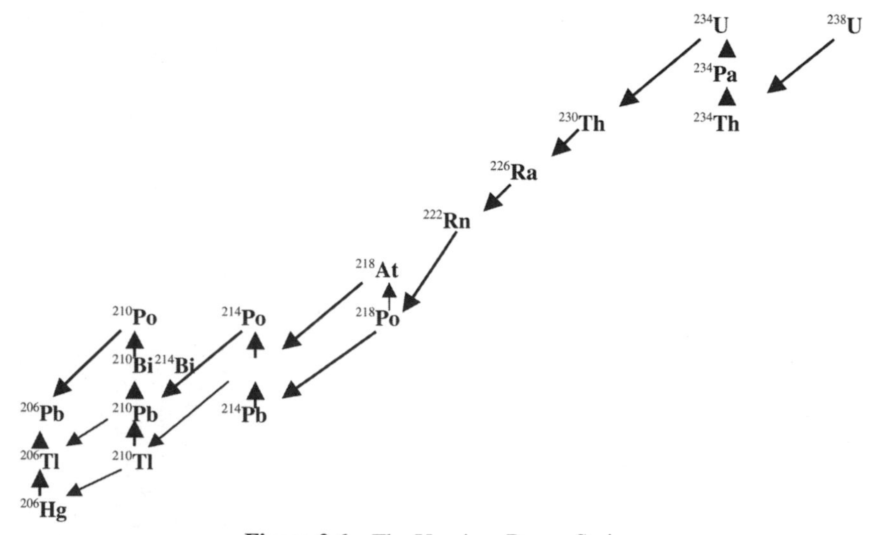

Figure 2.6 The Uranium Decay Series

cyclotrons. Most are produced by irradiation of non-radioactive atoms. There is one more decay chain, beginning with artificially produced plutonium-241 (^{241}Pu) and decaying through steps to stable bismuth-209 (^{209}Bi). While not occurring in nature, this chain of radioactive materials, called the *neptunium series*, can be produced in an atomic bomb; in fact, that is how it was discovered. Many of the deleterious effects of a plutonium bomb and resulting fallout result from members of this series.

Since it is not possible to determine exactly when a particular radioactive atom will undergo the release of energy, a concept called "*half-life*" has been developed. Some will be decaying at any given moment; given a large enough number of decaying nuclei, an average number can be estimated. From this, how many of the parent nuclei will remain after a given time can be determined. Simply, the time it takes for one-half of a given number of radioactive nuclei to decay is the half-life. After that period of time, one-half of the original atoms will have decayed; after two time periods, three-fourths; after three time periods, seven-eighths; and so on. For practical purposes, radioactivity can be considered as being gone after some ten half-lives, at which time only 0.1% of the radioactivity remains. Half-lives can range from microseconds to billions of years, depending upon the rate at which decay is taking place, so these ten half-lives can range from seconds to tens of billions of years. It is probably clear that if nuclei are emitting radiation quickly, this material is said to be highly radioactive. The reverse is also true. In each of the naturally occurring series, the first member has a very long half-life, on the order of millions or billions of years; if not, they would have disappeared long ago.

The energy of a nuclear process is usually given in electron volts (eV). An *electron volt* (eV) is a measure of energy equal to 1.6×10^{-19} joules (actually a very small amount, approximately equivalent to the energy expended by a fly in beating its wings once or twice, since it takes 2.62×10^{16} eV to equal just one thermal calorie). Multiples of the eV are the kiloelectron volt or keV (10^3 eV); million electron volt or MeV (10^6 eV); and billion electron volt or BeV (10^9 eV). An electron volt (eV) is a very small amount of energy, but if there are billions and billions of nuclear reactions, each yielding 2 to 100 MeV, the resulting energy can be tremendous.

It may seem strange that at least some of the materials involved in nuclear weapons are actually not very radioactive. Uranium-235 has a half-life of 713 million years, and releases a fairly weak alpha particle (4.39 MeV) and gamma ray (0.18 MeV). On the other hand, plutonium-241 has a short half-life of only 13.2 years, so almost all is gone in just a century. For this reason, the Pu-241 in nuclear weapons must be regularly replaced. Even Pu-241 isn't highly dangerous unless taken into the body or placed on the skin; it releases a weak 0.02 MeV beta particle and a very weak x-ray.

PROCESS OF ALPHA RADIATION

Alpha radiation is particulate radiation. The *alpha particle* consists of two protons and two neutrons, essentially a helium nucleus or helium ion, with an atomic number of 2. The atomic mass of the particle is 4 amu. Since there are no electrons to counter the two positive charges of the protons, the alpha particle possesses an electric charge of +2. The symbol is ^4He.

When an atom releases an alpha particle, the weight is reduced by 4 amu and the atomic number by 2. As an example, when an atom of polonium-210 (atomic number 84; symbol Po) emits an alpha particle, the remaining atom becomes lead-206 (atomic number 82; symbol Pb). The energy of the alpha particle being released is characteristic of the particular transmutation. This energy can be thought of as the difference between the energy of the emitting (parent) atomic nucleus and that of the resulting (progeny) nucleus. The alpha particle resulting from the Po-210 to Pb-206 transformation has an energy of 5.3 MeV initially. Alpha radiation, consisting of relatively massive particles, cannot travel very far without bumping into another atom, and thereby losing energy. This energy is imparted to whatever medium the alpha particle is in, whether the air, a piece of paper, or living tissue.

PROCESS OF BETA RADIATION

Beta radiation is also particulate radiation, consisting of free (not in atoms) electrons, called a *beta particles*. Ions are formed when orbital electrons are gained or lost while circling nuclei, but beta particles are emitted from within the nucleus of an atom as a neutron becomes a proton. An electron has a very small mass, some 1,800 times less than that of a proton. A beta particle possesses an electric charge of -1. The symbol is ^0e (since it's mass is so small as to be insignificant).

In beta emission, there is an increase in atomic number since the neutron changed into a proton and beta particle (plus an even smaller particle called a neutrino), leaving one more proton than before the release of the beta particle. There is essentially no change in the atomic mass. For example, when thorium-234 (atomic number 90) undergoes beta emission, the result is an ejected electron and a remaining atom of protactinium-234 (atomic number 91).

The energy possessed by the departing electron is not a constant like with the alpha particle, but can vary from the difference in energy of the atoms before and after the emission down to almost zero. Usually, the beta particle energy is about $1/3^{rd}$ of the total energy released with the rest going to the neutrino. If there are a large number of atoms undergoing beta emission, the result is a spread of beta particle energies over a spectrum. Beta particles are capable of passing a little deeper into matter than alpha particles, depending upon their energy.

There is a second form of radiation of electrons; this is called secondary or delta rays. It results from a beta particle speeding past an atom possessing orbital electrons. Since both particles have a negative charge, they tend to repel each other. If this coulomb force is great enough, the orbital electron can be repelled to move to a higher energy level in the atom (called excitation) or even ejected from the atom, forming an ion and a free electron (called a secondary electron). Secondary electrons typically have very short paths.

A third form of radiation involving electrons is *bremsstrahlung*, or braking, radiation. It results from the fast-moving beta particle coming near an atom's nucleus.

The nucleus causes the beta particle to change direction of travel. In accordance with laws of physics, when a moving particle changes path, there is a loss of energy that reappears as electromagnetic radiation. This bremsstrahlung radiation can cause damage beyond the maximum range of the beta particle. Bremsstrahlung radiation is greater in metals than in many other materials; for this reason, shielding of beta sources often does not use heavy metals, like lead, common to shields for other kinds of radiation. This will be discussed in more detail in the section on shielding.

PROCESS OF GAMMA RADIATION

Gamma radiation is not particulate but *electromagnetic* in nature, like visible light or radio waves. Gamma radiation consists of electromagnetic waves travelling at the speed of light. In many ways, gamma radiation is similar to *x-rays*; it can pass through human tissue and even some thickness of metals. The major difference between x-rays and gamma radiation is the source of the electromagnetic energy. If the radiation is generated by the relocation of electrons from one energy level (orbit) to another in the atom, the radiation is x-radiation. If the radiation is generated by nuclear transformations within the nucleus of an atom, the radiation is gamma. Gamma radiation has no mass.

When a nucleus releases an alpha or beta particle, there is often a gamma ray also emitted. This gamma ray usually results from a difference between an excited state of the progeny (daughter) nucleus and its lowest energy, or ground, state. Unlike the spectrum of energies seen in beta decay, the gamma energy is characteristic of the particular nuclear transformation, just like that of the alpha emission.

Gamma radiation can result from particulate radiation, from other nuclear processes, and from bremsstrahlung radiation. Gamma radiation, being electromagnetic in nature, can pass through matter just like x-rays. And, also like x-rays, it can give up some of its energy to the matter, whether inanimate object or living tissue.

PROCESS OF NEUTRON RADIATION

Neutrons have a mass of about 1 and no electric charge. The symbol is ^{1}n. No radioisotopes, other than several rare fission fragments with very short half-lives, naturally emit neutron radiation. However, californium-252 (Cf-252 or ^{252}Cf), which is an alpha emitter, does undergo some nuclear fission spontaneously, at a rate of one fission for every 31 alpha particles emitted. Effectively, Cf-252 might be considered as a neutron emitter, giving off some 2,300,000 neutrons per second per microgram. The neutrons are emitted with a wide range of energies, but have an average of about 2.3 million electron volts (MeV). The release of neutrons from Cf-252 is important in functioning of the nuclear weapon initiator.

All other neutron sources result from nuclear reactions. For example, beryllium-9 (Be-9 or ^{9}Be) can be impacted by a stream of high-energy deuterons in a cyclotron to produce boron-10 (B-10 or ^{10}B) and a neutron from each reaction. The same Be-9 can also be bombarded by alpha particles to form carbon-12 (C-12 or ^{12}C) and a neutron; this source can be made by grinding together an alpha-emitter like radium or polonium with beryllium. This design has been used to initiate nuclear weapons. The resulting neutrons have a wide range of energies.

Neutrons can be called thermal, slow, and fast, depending upon their energy. *Thermal neutrons* have the lowest energy (usually about 0.025 eV). A thermal neutron moves at about 2200 meters per second. *Fast neutrons* are those having energies greater than 0.1 MeV. Neutrons between these values are called *slow neutrons*. The energy (and resulting speed) of a neutron is important in nuclear fission. Much of the early research in nuclear energy used slow or thermal neutrons to initiate fission; in fact, commercial nuclear power reactors use slow neutrons. Later, it would found that U-235 can be fissioned using slow or fast neutrons, and that fast neutrons were better in an atomic bomb.

LINEAR ENERGY TRANSFER (LET), STOPPING POWER, AND EFFECTS

The energy given up by either particulate or electromagnetic radiation depends upon the type of radiation, the energy of the particles or rays, and the material they are passing through and interacting with. The energy can be described as a transfer of energy from the radiation to the matter. It is this transfer that causes the damage from radiation, whether to structural materials or to living organisms. The predominant result of radiation energy transfer is ionization. When a charged particle, alpha or beta, passes through a material, atoms along the path of the radiation become ionized.

Ionization can then result in damage to deoxyribonucleic acid (DNA), the basis of cellular reproduction. As mentioned in Chapter One, DNA is an acronym for deoxyribonucleic acid, a type of nucleic acid. When the radiation produces ionization of the atoms of the DNA, the structure can fall apart. If only one side of the ladder is broken, the DNA might be able to repair the damage. If both strands are broken, repair is much less likely. The amount of damage depends upon the amount of energy released; alpha radiation releases all its energy in a short distance, so can be the most damaging form of internal radiation. In cases of DNA damage, there can be miscommunication of cellular reproduction information to succeeding generations, resulting in cancer to the organism or even birth defects of offspring. Similarly, damage to RNA, which serves several functions in the cell from passing messages to the DNA to supplying some of the raw material needed for DNA replication, can result in cellular damage. This damage is most commonly to the existing cells and less to succeeding generations. Therefore, the possibilities are: the cell can die; the cell can repair the damage (especially if only one of the two DNA strands is broken); the cell can exhibit DNA damage by becoming cancerous; or the cell (especially eggs and sperm cells) can transmit faulty information, resulting in genetic effects like birth defects.

Interestingly, ultraviolet radiation (UV) can also damage DNA even more effectively than gamma or particulate radiation. This is because UV directly affects DNA, unlike the indirect effects of ionization from other radiations produced by nuclear weapons. However, UV cannot penetrate beyond the upper layers of the skin.

The concept of *linear energy transfer* (LET) is that, as energy is transferred by the radiation, the amount transferred depends upon how far the radiation has traveled. LET is defined as the average energy released per unit length of path, and is usually given in units of energy per distance traveled or keV per micrometer (thousands of electron volts per micrometer (mm). Typical LETs can range from 0.3 keV/mm for a soft 3 MeV x-ray to 3.0 keV/mm for a 0.6 keV beta particle from tritium to 110 keV/mm for a 5.3 MeV alpha particle from polonium.

Stopping power is simply the amount of average energy absorbed by a particular material per unit path. It is typically given in MeV per centimeter (MeV/cm). Stopping power depends upon the chemical and physical characteristics of the material, the type of radiation, and the energy of the radiation. Some materials, usually heavy metals and other dense substances, absorb more energy than others, like water or living tissue. It should be remembered that stopping power is a function of the material to a particular type of radiation. The effects of all forms of radiation therefore depend upon the type of radiation, the energy of the radiation, and the nature of the material the radiation is passing into or through.

FISSION FRAGMENTS

When a nuclear weapon is detonated, or—to a lesser extent—even the processes within a nuclear reactor occur, fission results in many fragments having high initial velocities and high energies on the order of 65 to 100 MeV per fragment. They are also highly charged since most of the uranium's electrons have been stripped off by the fission. Each fragment can have charges as high as +20 so can cause extreme ionization in matter. Because the path length of most fission products is very short, the resulting ionizations are very intense. Some fission fragments can have LETs up to 9,000 keV/mm.

PHYSIOLOGICAL EFFECTS OF ALPHA RADIATION

The large alpha particles cause intense ionization over a short path. Because of the relative large size and double charge, an alpha particle is capable of very intense ionization effects as it passes through matter. However, also due to its size and charge, an alpha particle cannot travel very far. Just as a large car is harder to get moving, requiring more energy than a small car, the large alpha particle requires a lot of energy to get it moving. Like a large car hitting a wall, the alpha particle gives up a lot of energy very quickly. This energy can cause intense ionization in atoms passed or hit by the particle. Since it gives up its energy quickly, it doesn't travel very far. Even a thin piece of paper can stop an alpha particle. The large size also increases the possibility that the alpha particle will hit both strands of a DNA molecule. There are cellular repair mechanisms to fix broken single strands of DNA, but these mechanisms are largely ineffective at repair of both. Internally in the lungs or stomach or intestines, an alpha particle can cause severe damage due to releasing a lot of energy quickly and breaking both strands of the DNA.

Several effects are possible: the cell can be killed by the radiation; the DNA can be altered so as to cause or induce cancerous growth of the cell; or the damage to the DNA may cause birth defects or other heredity changes in the offspring. Repair of the DNA or other cellular materials is much less likely with alpha radiation than with the other types. However, recall that the alpha particle doesn't travel very far, so only a few cells might be affected unless there are many alpha particles.

PHYSIOLOGICAL EFFECTS OF BETA RADIATION

Beta particles give up their energy over a longer path than alpha particles. Due to the small size, they are much less likely to cause double strand breaks in DNA; cellular repair mechanisms are more likely to fix the damage done by beta particles. The primary biological injury from beta particles is a skin burn. Heavy metals, such as lead, do not make good shields for beta particles because of the effect known as Bremsstrahlung or secondary radiation. On the other hand, light metals, like aluminum, make good shields for beta radiation. Beta particles give up less energy in a given length of path since they travel farther than alpha particles.

Like with alpha radiation, several effects are possible: the cell can be killed by the radiation; the DNA can be altered so as to cause or induce cancerous growth of the cell; or the damage to the DNA may cause birth defects or other heredity changes in the offspring. Repair of the DNA or other cellular materials is more likely with beta radiation than with the other types. The skin burns can completely heal. Genetic effects, unless the radiation was received directly by the gonads, are less likely than from alpha radiation.

PHYSIOLOGICAL EFFECTS OF GAMMA RADIATION

The interaction of electromagnetic rays with matter are by three main mechanisms: the photoelectric effect in which the ray hits an orbital electron giving it all the ray's energy to be ejected as a photoelectron; Compton scattering in which the ray hits an orbital electron giving it some of the ray's energy to be ejected as a recoil electron while the rays continues with lower energy; and pair production in which the ray passes by a nucleus, especially of materials of high mass number, instantaneously disappearing with an electron and positron being formed. A positron, a form of antimatter, is similar to an electron, but has positive rather than negative charge. This process requires at least 1.02 MeV of energy, so is more likely with high-energy rays. Any energy of the ray above this will simply impart kinetic energy to the two new particles. The electron can then cause ionization like the beta particle. The positron, being anti-matter, cannot exist for long, but can cause ionization as it passes atoms. It will eventually come to rest with an electron to which it is attracted by opposite charge; both will disappear and two photons will be formed. These photons, although generally weak, can go on to produce other ionizing effects.

Gamma radiation, being electromagnetic in nature, can pass through matter rather easily, depending upon its energy. The primary biological effect of gamma radiation is to cause ionizations in the cells. Damage to DNA molecules is, most likely, a single strand break, but very energetic gamma rays can cause double strand breaks. The single breaks can be repaired some of the time by the cell. One of the effects of radiation is to cause damage to the genetic material, the DNA, of the cell with the result that a mutation may appear in subsequent generations. Like alpha and beta radiation, gamma (and x-) radiation can thus injure the victim who is exposed to the radiation by causing or inducing cancer as well as causing hereditary effects seen as birth defects.

PHYSIOLOGICAL EFFECTS OF NEUTRON RADIATION

More of the neutron's energy is imparted to nuclei having low atomic weight than to heavier nuclei because of the similarity of masses. For this reason, materials of low atomic mass, especially those possessing many hydrogen atoms, are very effective at absorbing neutron energy, thus slowing or moderating the neutrons. This is why heavy water, ordinary water, paraffin, and many plastics are very efficient neutron moderators. The hydrogen nucleus, which has absorbed energy, can be pushed out of the molecule. The result is a moving hydrogen ion, possessing a charge and capable of causing ionization in the material it passes through. The weight (mass) of the neutron is about one amu. Because the colliding neutron can have almost any energy, there is a spectrum of energies of the resulting hydrogen ions. The largest component of the cells, and thus the human body, is water, so the potential for neutron interactions is very great.

Neutron radiation has a potential for cellular damage similar to alpha radiation. Double strand breaks in DNA are possible with less chance of repair than from beta or gamma radiation. While neutrons have no charge, secondary radiation is possible with subsequent effects similar to beta radiation. However, neutrons have other characteristics; they can hit atoms with sudden transfer of kinetic energy, and they can be absorbed into the nucleus of an atom, forming an unstable radioactive material. This latter process is called neutron activation. The subsequently radioactive atom can emit any of the forms of radiation.

TIME-DISTANCE SHIELDING

In order to protect a person from nuclear radiation, there are three variables: the length of time exposed; the distance from the source of the radiation; and the presence of any shield material between the source and the exposed individual. To reduce the exposure, the duration of time in the area should be the least possible; the distance away should be as far as possible; and the proper kind of shielding should be employed.

RADIATION SHIELDING

The barrier, or shielding, to protect from exposure to nuclear radiation depends upon the type and energy of the radiation. However, it must be remembered that when radiation is stopped in matter, the energy of the particle or wave is transferred to the material. Alpha radiation can be shielded by very thin materials, even a piece of paper. Beta particles can travel somewhat farther, depending upon their energy, but can be stopped by thin sheets of lower atomic number metals like aluminum. Gamma and x-rays can travel great distances, even through feet of steel or lead, if they have enough energy. Neutrons generally travel a distance less than gamma rays but more than alpha particles; this depends upon the density and nature of the absorbing material. Substances with a large number of hydrogen atoms such as heavy water, plastics, and even regular water are effective neutron shields; graphite is also effective as a neutron shield.

SUMMARY

Nuclear radiation may consist of alpha particles, beta particles, neutrons, and energetic electromagnetic radiation. Alpha particles consist of two protons and two neutrons without any electrons; it is essentially a helium nucleus stripped of electrons. Since there are two positive charges, the alpha particle has a mass (weight) of 4 amu. They can only travel a very short distance in matter. A beta particle is essentially an electron travelling freely, not going around an atomic nucleus. The electron and its positively-charged antimatter cousin, the positron, are very small, some 1,800 times less massive than a proton. They can travel farther than an alpha particle, but are still stopped by thin pieces of matter. Electromagnetic radiation, in quantum theory, can be considered as particles called photons moving at the speed of light. These photons have no weight and possess energy related to the frequency of the radiation. Depending upon the energy of the ray, it can pass through thick pieces of matter. An energetic x-ray or gamma ray can pass through even inches of heavy metals like lead. Neutrons, being uncharged, cannot cause ionization directly, but they can pass near nuclei. They can be captured by a nucleus or scattered in a collision with the nucleus. In a collision, some of the neutron's energy is transferred to the atomic nucleus as kinetic energy, like collisions between billiard balls or a bowling ball and the pins.

In order to protect from exposure to nuclear radiation, there are three factors: the length of time exposed, the distance from the source of the radiation, and the presence of shielding. To reduce the exposure, the duration of time in the area should be the least possible; the distance away should be as far as possible; and the proper kind of shielding should be employed.

3

BRIEF HISTORY OF NUCLEAR WEAPONS

INTRODUCTION

The story of nuclear weapons reached fruition during World War Two, chiefly by the United States. These horrible weapons were the result of almost half a century of research into the structure of matter and atoms. Toward the end of the Nineteenth Century, a little was becoming known about atoms. It was known that there were some ninety unique forms of matter called elements, made of atoms with relative weights between 1 and about 240. Much of the research in physics from the mid 1800s on was into the mysteries of atomic structure.

Many scientists, including Marie and Pierre Curie, James Chadwick, J.J. Thompson, Frederick Soddy, Hans Geiger, Ernest Marsden, Hantaro Nagaoka, Niels Bohr, Enrico Fermi, Frédéric Joliot, Wolfgang Pauli, and others, were involved in early studies of atomic structure and behavior. With the discovery of the atomic nucleus, knowledge of atomic structure took a giant leap forward. In order for positively-charged subatomic particles (protons) to stay together in a very small volume the size of the nucleus, enormous energy must be available, acting like a glue to hold the nucleus together. The release of this so-called binding or atomic energy is the basis of nuclear weapons.

PRE-WORLD WAR TWO

Following the discovery of *natural radioactivity* by Henri Becquerel in 1897, there was a realization that some form of energy was involved. It was known that this energy could fog photographic plates even in the dark and, wrapped in layers of paper, cause certain minerals to fluoresce, and emit radiations of various energies, but even the leading scientists of the day were not sure just how this mysterious energy could be utilized or even how much there might be. Although some noted scientists, such as Max Planck, first denied the existence of *atoms*, the concept and internal structure of atoms were discerned in the early years of the Twentieth Century. It was suspected that some form of energy was possessed by these atoms, even from early studies. With the discovery of the *atomic nucleus* by Ernest Rutherford, physics had to be rewritten because it was thought impossible for particles of the same electrical charge to exist in extremely close proximity.

Two German researchers, Otto Hahn and his assistant Fritz Strassman, continued the work of the Austrian scientist Lise Meitner when she was forced to leave Germany because her family had been Jewish. In December 1938, they found that the element *uranium* could be broken into smaller pieces by irradiation with slow uncharged subatomic particles called *neutrons*. Somehow the neutron affected the center, or *nucleus*, of the uranium atom, causing it to split in fragments; this process, called *nuclear fission*, would be the beginning on the road to atomic weapons. Fission was named by Meitner's nephew and physicist Otto Frisch, who compared it to the biological process by which a bacterium divides in two in mitosis.

Earlier, in 1933, the Hungarian physicist Leo Szilard had theorized that if an atom could be split, very large amounts of energy might be released in the process. Szilard, a friend and collaborator of Albert Einstein, at the time felt that here might be a source of sufficient energy to bring peace to the world when nations wouldn't have to fight over coal and oil. It is interesting that his first thoughts were of peaceful uses of atomic energy.

Evidence was appearing of both the effects of radiation on living organisms and the energy released during nuclear fission. A number of physicists had skin burns and cancers from their work with radioactivity. Marie Curie died July 4, 1934 of cancer, possibly as a result of her research on radium and polonium. It was also known that radiation could induce anemia.

In the world of late 1939, with Hitler's armies in Poland and poised to attack both east and west, Einstein wrote a carefully-worded letter to President Franklin Roosevelt, edited by Szilard, Eugene Wigner, and Edward Teller. In it, Einstein briefly explained the concept of a new weapon—an *atomic fission bomb*. This bomb, using one of uranium's three natural isotopes, U-235, could release unimaginable amounts of energy in the form of light, radiation, and heat. The physicists and mathematicians involved were fearful that the Germans were working on just such a bomb.

Werner Heisenberg, later to become chief of the Nazi atomic research project, had attended a meeting in Berlin in September 1939 which suggested that the Allies, especially the United States and England, were studying nuclear fission and suggested that Germany consider such a weapon. In that same year, Heisenberg came to the conclusion that either heavy water or graphite could be used as a neutron moderator (a material used to slow down rapidly-moving neutrons "*fast neutrons,*" so they could be better controlled) in a *self-sustaining chain reaction*. He also knew that the newly-discovered element *plutonium* might be an even better fuel than uranium. The Germans understood the theory very well; development of a weapon would depend upon other factors such as engineering and chemical separation as well as government funding. Soviet nuclear research was generally limited to laboratory work until information supplied by spies leaked details on the U.S. atomic bomb project. Igor Kurchatov and his research group had earlier identified spontaneous fission in uranium. When they published their work in 1940, Kurchatov was surprised that there was no response at all from his counterparts in the U.S. This led him to correctly realize that the Americans were secretly working on fission. Earlier, in 1939, he had alerted Soviet authorities to the potential for a fission bomb and his fear that the Germans were working on the development of one. The Russians had a number of spies inside the Manhattan Project, the U.S. efforts to product an atomic bomb; they fed critical information throughout the research. The first atomic bomb tested by the Russians was an exact copy of an U.S. bomb down to every dimension, but with the exception that it had a different atomic core.

Even the Japanese were conducting studies into the possibility of a fission bomb as early as 1940, under the direction of Tatsusaburo Suzuki. The Japanese held a physics colloquium in 1943 in Tokyo; the physicists agreed that an atomic bomb was theoretically possible and could be designed and built, but would take at least two or three years at great financial costs. They overestimated the time (and costs) needed to separate uranium isotopes. They had also underestimated the dedication of the American and Allied scientists as well as U.S. industrial capabilities. When the first atomic bomb was dropped on Japan, these Japanese physicists knew what had happened, even as the Imperial government first claimed an earthquake had hit Hiroshima. In 1939, British Member of Parliament Winston Churchill had studied Niels Bohr's reports about the energy within the atom, and had warned the British Secretary of State for Air that a bomb might be possible, but he felt that development of a weapon was many years off. The military in nations on both sides of the impending conflict were aware of the potential for a new and frightening weapon using the basic building blocks of nature.

Einstein and his associates knew that they had to get their letter to U.S. President Roosevelt quickly, but they had few political contacts. Einstein first thought about sending the letter via his personal friend Queen Elizabeth of Belgium, especially since Belgium was mining much of the world's uranium in the Congo. Later, it was decided to use either the biologist-economist Alexander Sachs of the Lehman Corporation or the famous aviator Charles Lindbergh. Knowing that President Roosevelt did not care for Lindbergh and was a friend of Sachs, they chose Sachs. When Sachs presented the letter to President Roosevelt, the president quickly realized the importance of such a weapon as well as felt fear that the Germans were already at work on one. An informal committee, the Advisory Committee on Uranium, was set up. It consisted of Szilard, Teller, Sachs, Wigner, Lyman Briggs of the Bureau of Standards, Merle Tuve of the Carnegie Institution, and Army and Navy representatives. After long and sometimes friction-filled discussion, the committee produced a report recommending that research begin with fifty tons of uranium oxide and four tons of graphite. However, due to more pressing matters, the president did not act upon these recommendations until the next year. Research did continue, but in American universities and laboratories. It seems there was never a formal document signed by Roosevelt to begin work on the atomic bomb. All that exists is a brief note from Roosevelt to his advisor Vannevar Bush with a handwritten "OK" accompanying a technical description of an atomic bomb.

Since many of the world's scientists knowledgeable about the atom and nuclear processes were in Germany or German-occupied Europe, it was very likely that the Nazis could indeed produce the first atomic bomb. That they did not is an interesting tale of science, politics, human personalities, and luck.

WORLD WAR TWO EFFORTS

With the war underway in Europe and China, the United States slowly understood that the nation would eventually become embroiled in the conflict. The Manhattan Project, so named because of the involvement of Columbia University of New York City in its early days and the non-existent and secret Manhattan Engineer District of the U.S. Army Corps of Engineers, was the code name for the U.S. atomic bomb project; however, the scientists and engineers worked for what was known as the Metallurgical Laboratory (Met Lab), a ruse to confuse any German spies. Early funding was provided through Columbia University for work on studying if a chain reaction in uranium could be sustained long enough to function as a weapon.

Whether it was possible to build a bomb based upon a nuclear self-sustaining chain reaction was still not clear. Many experiments would be needed to decide for sure. Leo Szilard, having failed to obtain a faculty position at Columbia University in New York, nonetheless obtained permission to use their laboratory facilities for a few months. During that time, he reasoned that a mixture of uranium and water might be used to cause a chain reaction, but earlier research by Enrico Fermi had shown problems with the use of water to slow down neutrons. Szilard and Fermi collaboratively decided that, since the use of either ordinary water or heavy water as a neutron moderator was not practical, the use of graphite was. Graphite might be placed in layers alternating with uranium oxide with a resulting chain reaction. Several reactors were built, but more neutrons were lost than needed to sustain the reaction, due to impurities in the carbon (like atoms of boron) or in the uranium itself.

The Japanese attack on Pearl Harbor in December 1941 brought the U.S. actively into the war. The need for speed in making an atomic bomb was more evident than ever before. Although small reactors had been designed and tested at Columbia

University, all the scientists knew that they would have to build a functioning nuclear reactor to prove their theories. The desire was to make the reactor spherical; this required supports of wood to hold up the alternating layers of uranium oxide and graphite. Fermi called the device a "*pile*" since it looked like a heap of building material. Where to build the nuclear pile was a point of contention. A number of locations were considered, ranging from an aircraft hanger on Long Island to the Polo Grounds to university campuses such as Princeton, Berkeley, Columbia, and Chicago. Although most of the early work of the Metallurgical Laboratory was at Columbia, in January 1942, Arthur Compton, head of the project, made the decision to build it at the University of Chicago. The university made an area under Stagg Field, the doubles squash courts, available for the scientists.

Finally, in late 1942, Fermi's pile, named CP-1, consisting of some 80,590 pounds of uranium oxide, 12,400 pounds of metallic uranium, and 771,000 pounds of graphite, and costing a million dollars, had the last cadmium control rod (used to stop the neutrons) pulled out. The world's first self-sustaining nuclear reactor produced about ½ watt of power for 4½ minutes, until the control rods were reinserted. Earlier, a few nuclear reactors had been used to make radioisotopes, but here was one that could be used to produce energy. It also proved that uranium could be used in a continuing chain reaction. Fermi later built larger piles to produce radioactive materials, both for the bomb project and for medical uses.

To produce a functional atomic bomb would require ultra-pure *fissile material* (material who atoms were capable of being easily split), and a *design* that would bring together pieces to a *critical mass* (since a certain amount of fissile material is needed before a chain reaction can occur) very quickly. A massive project was begun at Oak Ridge, a new town in Tennessee, and Hanford, a military reservation in the State of Washington. Oak Ridge would be the site of the uranium production, and Hanford the site of the plutonium production. Oak Ridge was isolated and surrounded by hills, had abundant electrical energy from the Tennessee Valley Project, and had sufficient water from the Clinch River. Hanford, likewise, had electricity from the dams along the Colorado River, was in a remote part of the high desert, and had the water from the river for cooling. Many, if not most, of the workers at Oak Ridge and Hanford were unaware of the goal of the work they were performing. Another location was selected in the hills just outside Santa Fe, New Mexico, at a former school called Los Alamos, for design and testing of the weapon. The small cadre of scientists there were knowledgeable of the work they were doing, but most of the support workers were not. Soviet spies, most often in low-level jobs, however, were successful in sending secrets to the U.S.S.R. throughout the war.

It was known that, unlike almost every other atom, U-235 was fissile; that is, could be split into pieces by absorbing a neutron. There was now a search for an even better fissile material. Plutonium, first temporarily called eka-osmium, had been discovered at the University of California at Berkley by Glenn Seaborg in 1942 from neutron irradiation of uranium nitrate in a cyclotron. Plutonium possesses a larger *cross section* (a probability) for neutron absorption than uranium. The larger the cross section for absorption, the more likely the neutron is to hit a target nucleus.

Thus, plutonium is a better fissile material than uranium. The neutron impact would produce neptunium and, subsequently, plutonium. In later years, due to its use in weapons and general bad reputation, Seaborg said he regretted naming it after the Roman god of the darkness and the underworld, but had been guided by the names of the solar planets in sequence. The planet following Uranus (uranium) is Neptune (neptunium), then Pluto (plutonium). The planet Pluto had been discovered in the late 1930s, just before the element. Louis Werner and Burris Cunningham chemically isolated the first sample of pure plutonium in 1942. It seems that plutonium can exist in several *oxidation states* (the difference between the number of protons and electrons in a atom in a molecule). The +4 state can be co-precipitated with a rare earth halide like lanthanum fluoride as a carrier, while the +6 state remains in solution and the carrier crystallizes. It was found, by Glenn Seaborg, that alternating oxidation and reduction cycles could separate plutonium from uranium and the other rare metals. The plutonium was separated from the uranium and other fission products of a reactor by first treating it with a strong reducing agent. The plutonium and carrier as crystals would be separated from the other metals. A strong oxidizing agent, like nitric acid, would be used to change the plutonium to the +6 state, and it will dissolve. The solution then contains enriched plutonium. Adding a base, like sodium hydroxide, causes the plutonium to come out of solution as a solid. This process was continued a number of times to produce almost pure plutonium.

Heavy water, or deuterium oxide, is a form of water that has atoms of deuterium rather than simple hydrogen combined with oxygen. The atomic mass, or weight, of ordinary water is about 18, and that of heavy water is 20, so *heavy water* is about 11% heavier than ordinary water. Heavy water, because of the presence of deuterium atoms (^2H) or "*deuterons*," is a very good moderator of fast neutrons; that is, it slows down fast neutrons but doesn't absorb the resulting slow neutrons. The cross section, that is, the likelihood, for absorbing slow neutrons is much less for deuterons than for normal hydrogen atoms. Ordinary water, containing normal hydrogen atoms (^1H), which also slows neutrons, is not practical since it absorbs the resulting slow neutrons. Graphite is also a good neutron moderator, and was used at the first nuclear reactor at the University of Chicago, where Enrico Fermi was able to develop the first self-sustaining chain reaction. The German atomic research project, under Heisenberg, decided to use heavy water as their moderator. Most of the U.S. researchers decided on the use of graphite, a crystalline form of carbon; nuclear resonance effects were reduced more by graphite than by heavy water (this is important in keeping the reaction going). The decision to use heavy water put the Germans at a disadvantage of being dependent upon a very scarce and expensive material. Most of the heavy water in the world was being produced by the Norsk Hydro heavy water plant at Vermork, Norway. It was able to make a ton and a half of heavy water annually in the late 1930s. One of the events which slowed German development in nuclear physics was the damage to that plant by Allied bombing using units of the British Royal Air Force (RAF), thus limiting German access to heavy water for their nuclear program. Adolph Hitler decided that German nuclear research would largely be limited to its potential use as a power source for ground vehicles, and not as a weapon.

The fact that Hitler did not understand the concept of the atomic bomb and, therefore, did not fund the project sufficiently meant that the U.S., British, and Russian fears of being beaten to the bomb were unfounded. After their capture by the Allies, the German nuclear scientists, including Heisenberg, were allowed to hear the radio reports of the new bomb detonated in Japan. The scientists were amazed; they argued that it could not have been an atomic bomb because they had could not have produced one, so how could the "inferior" Allied scientists have done so. Well after the war, Heisenberg intimated that he had not pushed for making a bomb because of his opposition to Hitler, but his real motives and ability to make a functional bomb will probably never be known. Some historians believe that if Heisenberg had been less dictatorial and more open to differing ideas, such as moving beyond the design of flat uranium sheets, and if Hitler had fully funded the project, the Germans might have produced a bomb near the end of the war, possibly before the U.S.

There were a number of competing designs for isotope separation. Some possible methods of producing U-235 were considered: centrifugation, gaseous barrier diffusion, liquid thermal diffusion, and electromagnetic separation. Magnetic separation was clearly possible as evidenced by the work of Alfred O.C. Nier who had separated a sample of uranium into its isotopes, including the sought-after U-235, in a mass spectrometer. However, this process required enormous electromagnets in order to separate the two fairly close (in atomic mass) isotopes (U-235 from the much more common U-238). The first Oak Ridge facility for magnetic separation was called the Alpha I *calutron* (named for the *U*niversity of *Cal*ifornia) and called the racetrack (since it was oval like a horse track). It used pure silver-wound magnets, and was located in facility Y-12. Enormous vacuum pumps were needed to reduce the air pressure inside the calutrons to a fraction of atmospheric pressure. The magnets had to be cooled by circulating oil to remove excess heat. The size of the electromagnetic separation facility at Oak Ridge was huge, as was the need for electrical power. The output of the Alpha calutrons would feed smaller so-called Beta units used to further refine the uranium. In fact, the magnets used for the uranium separation in the Manhattan Project used all the silver at the West Point Depository, some 13,540 tons; it was loaned for the duration of the war to the project to make the wire for the windings of the magnets. After the war, it was melted down and returned to the bank.

Another possibility was to use the slight difference in mass between the two isotopes in a mixture of gaseous uranium hexafluoride (UF_6). Molecules of different mass move a slightly different speeds; the heavier the molecule, the slower it is. The atomic mass of one UF_6 molecule (with U-235 uranium) is 348.99 and the other (with U-238 uranium) is 351.99, a difference of some 0.85%. This is a small difference in weight, but is enough to allow for different diffusion rates given a long enough path. It is like two runners, one able to run at 10 miles per hour and another at 10.0085 miles per hour. Over a long enough running path, there will be quite a distance formed between them. Eugene Booth and John Dunning decided in 1940 to consider this gaseous diffusion process, using UF_6 to separate U-235 from natural uranium by diffusion through a nickel mesh. In order that the gaseous diffusion path was long enough, a very large building was needed to house the diffusion chambers. The K-25 gaseous diffusion plant was built at Oak Ridge, and was 42.6 acres under one roof. It contained five million barrier tubes made of nickel. The liquid thermal diffusion process was similar, with the heavier gas moving slower. Steam was circulated through an upper pipe and cooling water through an outer pipe, causing U-235 to diffuse inward and circulate upward faster than U-238. The S-50 thermal diffusion plant at Oak Ridge experienced a number of maintenance problems, and produced less U-235 than K-25; however, it did produce some of the U-235 for the gun tube bomb, with its output fed to the Alpha calutrons for improved refinement.

The third method, using a cascade of high speed tubular centrifuges to separate materials based upon their density, was also used and was more energy efficient than the gaseous diffusion method, but did not produce as much uranium. The uranium used in the first atomic bomb, Little Boy, was largely produced by electromagnetic separation at Y-12 and less by gaseous barrier diffusion at K-25. Later, isotope separation using centrifuges became very common, largely due to the fact that as little as 4% of the energy is required to separate uranium isotopes by tubular centrifuges than by gaseous diffusion.

There were two ways to produce plutonium: in a nuclear reactor from uranium and by bombarding uranium atoms with neutrons in a cyclotron. In the former method, a small amount of fissile plutonium-239 (Pu-239 or ^{239}Pu) would be produced from the neutron irradiation of uranium at a concentration of some 250 parts per million (ppm). In the second, uranium nitrate hexahydrate was bombarded in a cyclotron with neutrons to produce the plutonium. The method used to produce the plutonium used in the second atomic bomb that exploded over Nagasaki was the large production nuclear reactors at Hanford, Washington rather than the cyclotrons in California. The problem with both methods is not making the plutonium, but separating it from the uranium or uranium nitrate. The process involves alterations of the oxidation state of the plutonium and uranium, forming insoluble precipitates of the uranium compounds, resolubilizing the plutonium compound, and followed by reduction of the plutonium compound to metallic plutonium. The chemistry involved in separating plutonium from uranium and mixed fission products (which become nuclear waste) in the reactor waste is very tedious and sensitive to pH, temperature, and other factors. The Manhattan Project had decided to locate the plutonium production facility far from Oak Ridge, at a site on the Colorado River in Washington, the Hanford Reservation, for a number of reasons. Here was plentiful water for cooling, electricity for operations, and isolation from prying eyes as well as geographic separation from Oak Ridge so spies might not make a connection between the two operations. A number of reactors were built, and plutonium was produced at Hanford for shipping to Los Alamos. The plutonium production reactions weighed as much as 1200 tons. At first, there were three reactors, separated by six-mile spacing for safety. Four chemical separation plants were built at two locations. Chemical separation at Hanford led to the generation of large quantities of nitric acid and other radioactively-contaminated wastes, a problem to be corrected only many years later and at great cost.

There were also competing designs for the mechanism of the bomb itself. Design of the bomb itself would be done at Los Alamos, New Mexico. There, experiments would be conducted in development of shape charges, detonators, criticality, and yield measurements. The director of the laboratory was the brilliant but controversial Robert Oppenheimer, who worked for General Leslie Groves, Director of the Manhattan Engineer Division. One early and simple idea was to push two hemispheres of uranium together using springs, but such a mechanical method might be too slow. Another design was having two hemispheres pushed together perpendicular to their equators by an explosive charge, but there was concern that the explosive shock wave might shatter or distort the hemispheres. The design that was settled upon used a bullet made of U-235 fired, by the propellant cordite, down a gun tube into three rings of U-235. This design became known as the "*gun tube*" design. "Little Boy" was the code name for the bomb made using this design since the bomb was thin and much smaller than the other design under consideration. Little Boy was the first bomb dropped on Hiroshima, since it was of simpler design and the military did not want to risk a failure. A drawing of "Little Boy" appears as Figure 3.1.

Another, more ambitious design, involved forming either an empty spherical shell or subcritical hemispherical masses of fissile material, either uranium or plutonium, around which would be placed high explosives, called explosive lenses. When the explosives were detonated, the uranium or plutonium would be forced into a much smaller volume of more than critical mass. This design was known as the "implosion" design. Since Pu-239 was a much better fissile material than uranium, it was decided to use this radionuclide in the bomb. "Fat Man" was the code name for the bomb that used this design because it required a spherical shell of high explosives around the fissile material, and was almost spherical. Fat Man was the bomb dropped on Nagasaki. A drawing of Fat Man appears as Figure 3.2.

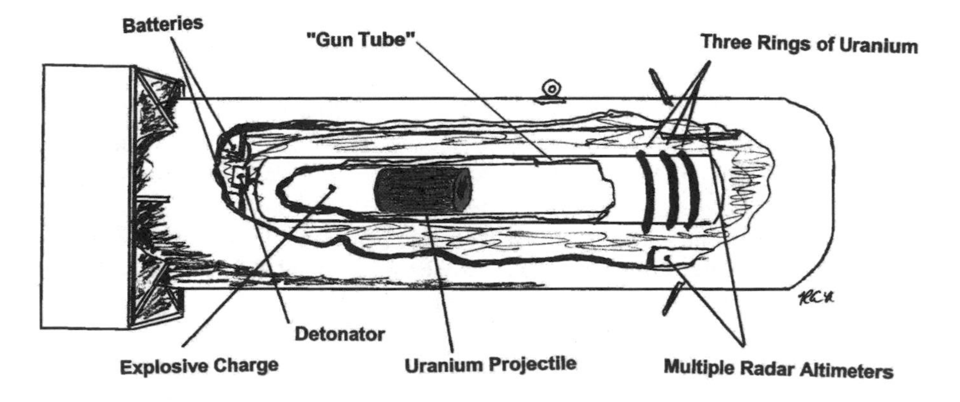

Figure 3.1 Schematic Diagram of "Little Boy" Gun Tube Bomb

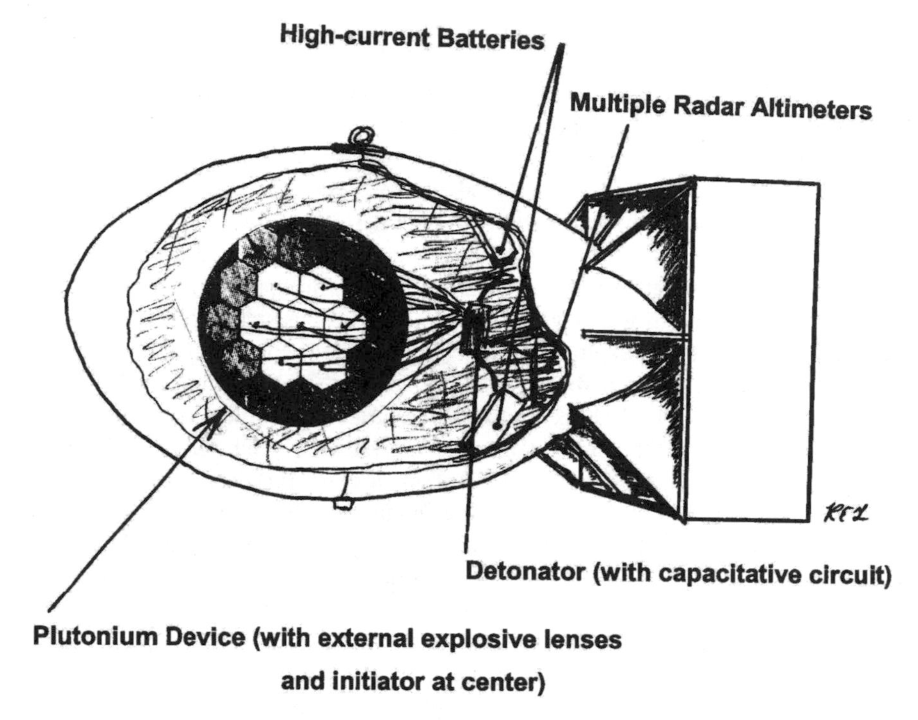

Figure 3.2 Schematic Drawing of "Fat Man" Implosion Bomb

The tedious process of separating U-235 from uranium produced sufficient material for the Hiroshima bomb and more, but with the potential to produce larger quantities of plutonium at Hanford, nuclear weapons development moved from uranium to plutonium as the primary fissile material. All U.S. nuclear weapons, except for Little Boy, have used plutonium with a uranium tamper.

There was so much confidence in the gun tube design that the decision was made not to test it. The implosion design, on the other hand, was much more complex, and it was decided that a test was needed. The test, code named "Trinity," was conducted at Los Alamos on July 16, 1945. The device itself was code named "Gadget." It was the first human-made nuclear detonation in history. Within two weeks, weapons, as opposed to a test bomb, would be ready for delivery by B-29 bombers to targets in Japan.

At *Hiroshima*, a city on southwest Honshu, whose name in Japanese means "broad island," the first atomic weapon, Little Boy, exploded at 8:16 AM on August 6, 1945. Hiroshima was a port city with extensive military facilities and industries, but had been spared much of the damage of the war. The bomb was partially assembled and the firing mechanism engaged in flight; red inert cables switched for the green firing ones, the radar units connected, and the internal batteries activated. The bomb was dropped from a B-29 called the *Enola Gay*, named after the pilot's mother, and fell for 43 seconds before exploding some 1,900 feet above Shima Hospital with an explosive yield equal to some 12–18 kilotons of TNT, although is was probably closer to 16 kilotons. The bomb was detonated by a set of radar units, set for 2,000 feet above the city and aimed for the Aioi Bridge over the Ōta River (there is some evidence it actually exploded somewhat higher). The population at the time of the bombing was about 285,000 civilians plus 43,000 soldiers at the Japanese 2nd Army headquarters. In addition to the Japanese civilians and military, there were many Korean forced laborers living at Hiroshima. Of the 76,000 buildings, some 70,000 were destroyed or damaged with 48,000 totally destroyed. The number who died immediately is not known for certain, but estimates range from 70,000 to 100,000. By the end of 1945, some 140,000 had died from the bombing, mostly from burns and traumatic injuries; by 1950, almost 200,000, the later deaths largely due to radiation effects. Survivors of the explosion reported a "black rain" which began falling soon after the detonation, due to upwelling of humid air into the cooler stratosphere. This had not been seen in the arid Nevada desert at the Trinity explosion. The point on the surface above which the detonation took place is known as the *hypocenter*. Most of the fallout from the atomic bomb at Hiroshima fell on the districts of Koi and Takasu, located some 3 kilometers west of the hypocenter. The fallout largely consisted of unreacted uranium-235, cobalt-90, cesium-137, and a number of other fission products. The presence of uranium-235 in the Hiroshima fallout is relatively unique since all other nuclear weapons exploded in the atmosphere have used plutonium as the fissile material.

Fat Man, Weapon Unit 31, was the second implosion bomb, identical to the Trinity Gadget, and was being made ready on the night of August 7, 1945 for takeoff the next morning when it was realized that the cables between the bomb mechanism and the radar units in its tail were wired in backwards. At 10:00 PM on August 8th, after a hasty repair, it was loaded into a B-29 called *Bock's Car*, named after the regular pilot who wasn't on this particular mission. The batteries for detonating the explosive lenses had required a minimum of 24 hours to fully charge; they could only hold their charge for less than a day. The target was Kokura Arsenal on the northern coast of Kyushu when the plane took off at 3:47 AM on the 9th. The bomb was fully armed, except for connections to the radar units, during the takeoff and flight due to the complex nature of the implosion device that made assembly onboard too difficult. Arriving at Kokura after flying 1,500 miles, the crew found the target completely obscured by rain, clouds, and fog. After two attempts to see clearly enough to drop the bomb, the pilot decided to attack the alternate target, a city named *Nagasaki*. It was the location of the Mitsubishi torpedo factory, which employed a large number of Korean forced laborers, but was also the home of Japan's largest Christian community. *Bock's Car* was running out of fuel and only had enough available to get to Nagasaki and return to Tinian. Even though Nagasaki was also obscured by clouds, the decision was made to drop the bomb using radar. At the last moment, a brief opening in the clouds allowed the bombardier to see the ground through his Norden sight. Fat Man fell from the B-29 and exploded 1,650 feet over the city at 11:02 AM. The yield was estimated at 22 kilotons of TNT.

Because the clouds were difficult to see through, the bomb actually detonated over the hills that surround the city and not the city center. The steep hills tended to confine the explosion, and less damage and loss of life was experienced than at Hiroshima. Still, over 70,000 people died either because of the blast or from radiation by the end of 1945, and some 140,000 by 1950. It is estimated that about 50,000 people died in the initial explosion.

The third plutonium implosion bomb was ready in New Mexico, and would be delivered to Tinian Island in the Pacific on August 12th or 13th with scheduled dropping on August 17th or 18th; the target was proposed to be Tokyo. On August 15, 1945, the Emperor of Japan notified the Allies that he was willing to surrender under the terms of the Potsdam Accord, worked out among President Roosevelt, Secretary Josef Stalin, and Prime Minister Churchill. The third Fat Man was never delivered.

BRITISH AND COMMONWEALTH WORLD WAR TWO DEVELOPMENTS

Between January and April 1940, the British investigated the possibility of an atomic bomb using fast, or unmoderated, neutrons. After brief experimentation, they concluded that an atomic bomb using uranium would be impossible using fast neutrons since too many neutrons would get lost or be captured by the more-prevalent uranium-238 atoms. In March, 1940, Otto Frisch and Fritz Peierls, two refugee German scientists in England, developed a new concept. They discovered that an atomic bomb could be built and detonated using only a few kilograms of uranium-235, the lighter isotope of uranium, fissioned wholly by fast neutrons. Frisch and Peierls observed that using only uranium-235, there is no need to slow down the neutrons, and no moderator would be required.

In their briefing paper, they warned that no man-made structure could withstand the resulting explosion; that no defense against such a weapon would be possible; that the intensely radioactive fission products would remain deadly for many years after the explosion; and that the Nazis were probably developing such a bomb. Based upon their report, in April 1940, the British government set up a top-secret Committee of Experts, later known as the MAUD Committee (a code name chosen from the first name of one member's nanny), to investigate the feasibility of an atomic bomb, and James Chadwick, British scientist who discovered the neutron, was asked by the Government of the United Kingdom (U.K.) to verify, in utmost secrecy, whether the Frisch-Peierls concept for an atomic bomb could work.

Several research teams swung into action at four British universities to explore the idea of an atomic bomb using uranium-235, and a nuclear research team from France was invited to continue slow neutron research at Cambridge; but the project was given a low priority since it was not expected to produce a bomb. The most formidable task facing the MAUD Committee and the researchers was how to enrich uranium by separating out the unwanted uranium-238. Using uranium supplied by McGill University in Canada, the highly corrosive uranium hexafluoride was produced; this vaporizes to a gas when heated. This gas can be diffused through a very fine membrane with the lighter U-235 atoms passing through somewhat more easily than the heavier U-238. Multiplied tens of thousands of times, this process can be used to increase the concentration of U-235. This is the same process as one of those that the U.S. scientists decided upon. But passive diffusion is a slow, expensive, and technologically sophisticated method, requiring huge energy inputs and a very large physical plant. Thus the concept of uranium enrichment through gaseous diffusion, the technology that produced the explosive material used in the Hiroshima bomb, was born in England.

A delegation, called the Tizard Mission, was sent to North America in August 1940 to explore the possibility of relocating British military nuclear research facilities across the Atlantic Ocean, out of reach of the German Luftwaffe. The British delegation visited, among others, Enrico Fermi, then at Columbia University; he was pursuing slow neutron research using very pure graphite as a moderator. They informed Fermi of the Frisch-Peierls concept for an atomic bomb, but he was highly skeptical as his research was geared toward using nuclear energy to heat water to steam for electricity generation, not making atomic bombs. In Canada, the delegation met George Laurence of the Canadian National Research Council (CNRC), who had secretly built a slow neutron experiment using graphite as a moderator.

When the delegation returned to Britain, they reported that the slow neutron researches being conducted at Cambridge, Columbia, and in Canada, were irrelevant to the effort of making a bomb. But since nuclear reactors for power generation could have post-war value, they arranged that money be sent to support the Canadian fission experiments. In November 1940, two British researchers, Professors Norman Feather and Egon Bretscher, discovered an unexpected link between nuclear reactors and atomic bombs. When a U-238 atom captures a neutron, it is transmuted into Pu-239: a man-made element, never seen in nature. On purely theoretical grounds, the two men predicted that plutonium-239 could be used as a nuclear explosive due to its large neutron cross section, or probability of catching a neutron. Since plutonium has different chemical properties than uranium, it should be possible to obtain pure plutonium from pure uranium mixture. Since uranium enrichment would not be required, a chemical separation process would be all that would be needed. It is immediately obvious to Feather and Bretscher that a nuclear reactor fueled with uranium would produce substantial amounts of Pu-239 as a by-product. However, since at that time no one had achieved a self-sustaining nuclear chain reaction, no one had seen any plutonium. Since their abstract predictions about Pu-239 could not be tested, their idea of a plutonium bomb was not taken seriously. But, unknown to the British, the American physicist Glen Seaborg showed that Pu-239 fissions even more readily than U-235 with either fast or slow neutrons, using small amounts of Pu-239 produced in a new University of California cyclotron, the world's first particle accelerator.

In May 1941, the American government ordered some 8 tons of uranium from Eldorado, a private radium-mining company in Canada; the purpose of this uranium was to demonstrate the nuclear reactor concept. The uranium was a discarded byproduct from processing radium ore mined at Port Radium in the Northwest

Territories. The British realized that the U.S. was very interested in nuclear research. The MAUD Committee made its report to the British government in the Summer of 1941. The Committee reached three conclusions: the Frisch-Peierls scheme for producing a uranium bomb was feasible; work toward building such a bomb should receive high priority; and close cooperation with the U.S. was of great importance. The option of using plutonium received only a short a mention in the MAUD Report.

By the fall, Prime Minister Winston Churchill gave the atomic bomb project his approval. The British, in secrecy, ordered two tons of uranium from Eldorado.

By October, the Americans received a copy of the MAUD Report and were impressed. Quickly, they changed their minds about the feasibility of an atomic bomb. The U.S. suggested a cooperative effort with Britain, but Britain was noncommittal. No actions were taken.

One week before the Japanese attack on Pearl Harbor, President Roosevelt committed the U.S. to a unilateral effort to build an atomic bomb. The Americans elected to pursue both the uranium option using enrichment and the plutonium option using nuclear reactors. The U.S. ordered 60 tons of refined uranium oxide from Canada; this required restarting the radium refinery and reopening the Port Radium Mine.

A British delegation went to America and was astounded at the speed that the atomic bomb project had quickly assumed. It was clear that the Americans would soon outpace the British in gaseous diffusion, nuclear reactors, and bomb design. Fermi was then at the University of Chicago building a pilot plant for a large graphite-moderated uranium reactor to produce plutonium. Seaborg and others from the University of California had relocated to the University of Chicago to investigate the complicated chemical properties of plutonium and its separation from uranium.

Slow neutron research, which the British had largely ignored, suddenly acquired military significance. It seemed imperative that the heavy water research team be relocated to Chicago, but the Americans had become very security-conscious. Of the six senior scientists in the Cambridge heavy water group, only one was British, and the Americans were in no mood to trust atomic secrets to non-British foreigners. Sensing the Americans' edginess, the British decide to back off. Earlier, the MAUD Committee had suggested sending the heavy water team to Canada. It seemed a good compromise with the fission scientists close to Chicago, but they were still in the British Commonwealth, and safe from German bombing.

Canadian Prime Minister at the time, Mackenzie King, received a delegation from Britain which told him about Britain's top-secret military project, code-named "Tube Alloys," for building atomic bombs. The British pointed out to him that a privately-owned Canadian company had a mine in the Northwest Territories which was rich in uranium. The British wanted to establish control over that uranium. The Canadian government bought up shares of Eldorado, and soon had a controlling interest. Ownership of the mine was split three ways, with Canada a junior partner to the U.S. and the U.K.

Malcolm Macdonald, the British High Commissioner in Ottawa, proposed moving the heavy water research group from Cambridge to Canada, creating a joint French-British-Canadian nuclear research team. Over the next few months, all the necessary arrangements were made, with the British paying the salaries of the people they sent over and the Canadians paying all other expenses. Although the Nuclear Research Commission (NRC) was Ottawa-based, it was decided to house the nuclear research team in Montreal, where lab space and accommodations were easier to obtain. A portion of the Medical Wing of the Université de Montreal, on the slopes of Mount Royal, was leased and refurbished for this purpose. The heavy water team started arriving in late 1942.

In December 1942, the U.S. Army ordered an additional 500 tons of Canadian uranium from Eldorado, even though delivery of the earlier 350-ton order was not complete. Eldorado was subsequently told to interrupt all deliveries of Canadian uranium to the U.S. in order to refine some 1200 tons of rich uranium concentrates, originally obtained from the Belgian Congo, when the material was discovered in a warehouse on Staten Island, New York, where it had been stored since 1939.

On January 2, 1943, a blunt letter was sent to the Canadians by the Americans, followed by a memorandum ten days later, representing a harsh turnabout on the part of the Americans. The U.S. was embarking on an intensive effort to produce plutonium using heavy water so it was unable to assist the British-Canadian heavy water team in obtaining additional supplies of heavy water from the U.S.-owned heavy water plant in Trail, British Columbia, as well as technical information about plutonium. For security reasons, the U.S. would no longer share any information on heavy water production, the manufacture of uranium hexafluoride, the method of electromagnetic separation, the physical or chemical properties of plutonium, the details of bomb design, or the facts about fast neutron reactions. In addition, the U.S. would only supply heavy water to the Montreal group if it agreed to direct its research along lines suggested by the U.S. This blunt communication reflected the fact that the U.S. Army had taken over the atomic bomb project, under the command of General Leslie Groves. He regarded the Montreal group, with so many different nationalities, as a serious security risk. These actions put the Montreal effort on hold just as it is getting started. Despite the American policy of non-cooperation, important work was conducted at the Montreal Laboratory. Neutron measurements involving heavy water and graphite were carried out. They developed a number of design concepts for a nuclear reactor using heavy water as the moderator.

In February 1943, Bertrand Goldschmidt, later a leader in developing the French nuclear weapons program, plutonium industry, and nuclear power industry, visited Chicago and brought back with him to Montreal a sample of fission products. Apparently the Chicago scientists didn't know about the new restrictions. From this sample, the Montreal radiochemists separated out three micrograms of pure plutonium. Further experiments were then done to evaluate different methods of separating plutonium from uranium.

The British discovered that the U.S. had completely tied up Eldorado's mining and refining capacity for years to come, despite prior assurances that the British would have joint control over Canadian uranium resources. In July 1943, General Groves visited Canada. He was told that heavy water was the best choice for producing plutonium, and therefore the Montreal project should be given high priority. The

General concurred, but had many reservations. At about the same time, in London, senior U.S. officials were engaged in frank discussions with the British to clear up some misunderstandings about British motives. Churchill drafted a series of agreements on nuclear cooperation which the Americans promised to relay to the President. This resulted in the 1943 Quebec Agreement, signed by Roosevelt and Churchill at Quebec City on August 19. Operation Overlord, the Allied invasion of Europe through Normandy, was also finalized at the same meeting. The Quebec Agreement required that Britain and the U.S. share resources "to bring the Tube Alloys (Atomic Bomb) project to fruition at the earliest moment."

In December 1943, a Combined Policy Committee ruled that British scientists could join in three aspects of the U.S. effort: the gaseous diffusion project at Oak Ridge, the electromagnetic project for separating uranium isotopes at Oak Ridge, and the bomb development work at Los Alamos. This led to a mass exodus of British scientists to North America, but nothing was mentioned about the Montreal Lab. Another agenda item was a Combined Development Trust to control and allocate all of the world's supplies of radioactive ores. As Canada has not signed the Quebec Agreement, she did not sign the Declaration establishing the Combined Development Trust either. Nevertheless, one of the six trustees had to be a Canadian. Still nothing was mentioned about the Montreal Lab.

In early 1944, it was announced in the House of Commons that the Eldorado mine and refinery had become national assets. After some delay, the Combined Policy Committee decided that a large-scale nuclear reactor would be built in Canada, using heavy water as a moderator, with the U.S. providing the necessary materials. However, no information about the chemical properties or even the biomedical hazards of fission products or plutonium was transmitted to the Canadians. The Montreal team had to figure it out for themselves. The Americans agreed to donate a few irradiated fuel rods to the Montreal group. The Americans delivered spent fuel rods of natural uranium (containing Pu-239) and of thorium (containing U-233).

The Montreal team knew little about the U.S. method for separating plutonium, except that it was based on precipitation. Precipitation has a big disadvantage; it can only be done in batches, and use of batches could cause a criticality event. The Montreal team wanted a process that could run continuously, mass-producing plutonium for bombs. Over two hundred different solvents were studied to separate plutonium away from the mixed fission products by creating two liquid fractions which would not mix. If this could be done, the plutonium-bearing fraction could be separated mechanically and continuously.

The sixteen months between the Combined Policy Committee's decision and the dropping of the atomic bombs on Japan were productive times at the Montreal Laboratory. The basis was laid for three post-war nuclear programs: the Canadian, the British, and the French. At Montreal and later, at Chalk River, the British made detailed plans for their post-war nuclear industry, including civilian uses. Incidentally, the word "crud" was derived from laboratory environmental samples labeled "*C*halk *R*iver *u*nidentified *d*eposits."

With no heavy water available in England, the British settled on graphite as the moderator for their nuclear program. Accordingly, a Graphite Group was formed at Montreal to work out the details. By the end of the war, the basic design work had been completed for Britain's first major experimental reactor at Harwell, called BEP0 (British Experimental Pile 0). All the graphite used in the first few British reactors came from Ontario.

One month after the Japanese atomic bombings, on September 6, the new ZEEP (Zero Energy Experimental Pile) reactor started operation at Chalk River. A much larger NRX (Nuclear Reactor eXperimental) reactor was also in early stages of construction. The next day, September 7, Igor Gouzenko, a cipher clerk at the U.S.S.R. embassy in Ottawa, revealed the existence of a large Soviet spy ring in Canada. One of its missions was to obtain information about the atomic bomb project. Two of the British scientists associated with the Canadian team were identified as spies.

Suddenly, the war was over. Harry Truman was the new U.S. President, and Clement Atlee was the newly elected British Prime Minister. Neither of them knew anything about the atomic bomb before coming to office. Fearful of the potential consequences of a nuclear war, the heads of three governments, the U.K., the U.S., and Canada, met in Washington and issued a frank statement in October 1945, recognizing that atomic bombs represented a new level of destruction and that only peaceful uses of nuclear power should be investigated.

BRITISH AND COMMONWEALTH POST-WAR DEVELOPMENTS

The Canadian NRX reactor went into operation in 1948. The British, eager to build a large scale plutonium separation plant in England, built a small-scale pilot reprocessing plant at Chalk River, drawing on the experience of the Montreal Laboratory.

Following the Russians by some four years, Britain exploded its first atomic bomb in Australia, incorporating some of the plutonium that was produced and separated at Chalk River. This first British nuclear weapon was detonated on October 3, 1952. It was a fission implosion device very similar to the Trinity Gadget and Fat Man. Unlike the U.S. and U.S.S.R., which has tested only on land, the British set off their first test underwater, some 30 meters (90 feet) off the northwest coast of Australia under an obsolete frigate. The yield was twenty-five kilotons, and the ship disappeared. It was explained that the British feared the Soviets might smuggle a weapon aboard a ship into a harbor, so they wished to see the effects of underwater detonation. Underwater detonations have been found to produce a large amount of fallout.

An accident at Chalk River blew the roof off and destroyed the core of the NRX reactor. Hundreds of American and Canadian soldiers were brought in to help during the radioactive clean-up operation. One of these was Jimmy Carter, the future U.S. President. The damaged NRX core was buried on the Chalk River site. In 1953, an explosion at the Chalk River reprocessing operation, separate weapons-grade plutonium from irradiated NRX fuel, killed one man and injured three others. Another accident, this time in 1958, at Chalk River's NRU (Nuclear Reactor

Uranium) reactor, resulted in a metallic uranium fuel rod catching fire, spreading intense radioactive contamination throughout the reactor building and into the atmosphere. Over 600 soldiers were called in to help with the radioactive clean-up.

In 1961, Canada sold a version their 125 megawatt CANDU (Canadian Deuterium-Uranium) power reactor to Pakistan; it was called KANUPP. In 1963, Canada sold India a 200 megawatt CANDU power reactor called RAPP-1. In 1967, Canada sold India a second 200 megawatt CANDU power reactor called RAPP-2. In May 1976, after India detonated an atomic bomb, Canada formally ended its nuclear relationship with India, but Atomic Energy of Canada Limited quietly restored cooperation with India later. In December 1976, Canada formally ended its nuclear relationship with Pakistan, but Atomic Energy of Canada Limited later restored cooperation with Pakistan.

RUSSIAN NUCLEAR WEAPONS

At the end of World War Two, only the U.S. possessed nuclear weapons. The U.S.S.R., fearing world domination by the U.S., quickly decided that they would have to build an atomic bomb. Information from numerous spies within the U.S. atomic project gave them all they needed to build their first weapon, code named JOE I by the U.S. (apparently in reference to Josef Stalin). It was identical in all ways to Trinity and Fat Man down to every physical dimension, with the exception that it was fueled with U-235 rather than Pu-239 since the Soviets did not possess plutonium in sufficient quantity to fuel a bomb. Shortly afterwards, the Soviets detonated a similar design using uranium-235, but of much smaller size and greater yield, called JOE II. Most of the tests in the former Soviet Union took place at Semipalatinsk, Kazakhstan.

By 1952, the U.S.S.R. had built a gaseous diffusion plant and a heavy water reactor for tritium production. Tritium is a heavier isotope of hydrogen than deuterium, with an atomic mass of 3 (H-3, ^{3}H, or T); it is important in hydrogen bombs. Both the U.S. and U.S.S.R. continued research into weapons design; each side produced weapons with greater yield and smaller size. While the scientists of the former Soviet Union were aware of the physics behind a thermonuclear, or hydrogen, weapon, there was less emphasis in its development than in the U.S. The first Soviet design for a thermonuclear weapon was one that had been considered, but rejected, by the U.S. researchers. Called a "layer cake" design and using lithium-6 deuteride (^{6}LiD) obtained by an electromagnetic separation technique, it eventually suffered from limitations in explosive yield. The first Soviet test, code named JOE 4 by the U.S., was detonated from a tower and used a layered design of alternative layers of lithium deuteride and uranium. There was extensive fallout in the region with widespread contamination. Although not weaponized (for delivery), it yielded about 400 kT. The design, using conventional high explosives, was ultimately limited to no more than about twice that yield. The Soviets had not learned how to use radiation for compression rather than chemical explosives. They would. The former Soviet weapon designers were apparently most interested in increasing yields, and soon discovered how to make a radiation-compressed design, similar to the U.S. design. They built larger and larger bombs; it is possible that the former U.S.S.R. developed the largest nuclear weapon ever made, a 50 megaton (MT) device; whether it was weaponized and could be deployed is not public knowledge. There is a drawback to very large weapons: as the yield doubles, the area damaged or destroyed is not doubled, but is only something like 40% greater. The 50 MT bomb would only do damage some seven times greater than one of 1 MT. Use of expensive, and short-lived, tritium as well as plutonium in making such a superbomb is less practical than making some 50 smaller ones.

Most Soviet, now Russian, strategic nuclear weapons fall generally in the following sizes: 50 kilotons of TNT (kT) for tactical use, 200 kT, 1 megatons of TNT (MT), and 10 MT. For comparison, the bombs dropped at Hiroshima and Nagasaki were both less than 20 kT, equivalent to the force of detonating 20,000 tons or 40,000,000 pounds of TNT. In general, the strategic nuclear weapons of the former Soviet Union are of larger yield than those of the U.S. Tactical weapons can have yields of only a few thousand pounds of TNT equivalent up to some five kT.

The inheritor of the U.S.S.R. military, Russia, is currently estimated to have about 5,000 strategic nuclear warheads, plus another 3,400 tactical nuclear weapons. It should be noted, however, that estimates of Russia's tactical nuclear arsenal vary widely, ranging upwards to 10,000–15,000 when estimates include weapons awaiting dismantlement.

Although Russia has made dramatic reductions in its nuclear forces since the end of the Cold War, a limiting factor in the pace of reductions has been the funding to destroy systems. Russia has taken control of all nuclear weapons stationed in the former Soviet republics, particularly the strategic weapons formerly deployed in Kazakhstan, Ukraine, and Belarus. For economic reasons, Russia's strategic nuclear arsenal is likely to decline to fewer than 2,000 warheads by 2015, according to U.S. intelligence estimates. Declines have also been particularly dramatic in Russia's delivery systems. In 1990 Russia had 62 ballistic missile submarines; today there are 17 operational submarines, and some of these are aging. Though there were plans to deploy several new Borey-class submarines, construction has been suspended since 1998.

Russia continues to conduct test launches of its intercontinental ballistic missiles and to replace some missiles. The SS-25 Topol missile, with a new variant sometimes called the SS-27, mobile single warhead is currently being deployed. It is believed that the total of Russian Intercontinental Ballistic Missiles (ICBMs) will continue to decline. Russia has three classes of bombers with a nuclear mission: some twenty-nine Tu-95 MS6s (Bear H6s), thirty-four Tu-95 MS16s (Bear H16s), and fifteen Tu-160s (Blackjacks).

The Strategic Arms Reduction Talks (START) II treaty limits Russia and the United States to 3,500 strategic deployed warheads. In the Treaty of Moscow signed May 24, 2002, both the United States and Russia agreed to reduce their strategic nuclear arsenal to between 1,500 and 2,200 weapons by 2012. However, Russia would likely have made this reduction regardless of the treaty or U.S. cuts due to fiscal necessity. This reduction may be imperiled by the U.S. plan to move most of the nuclear weapons taken out of the active stockpile into a reserve stockpile, where they could easily be rearmed. Both the START and Moscow treaties do not restrict tactical or reserve weapons. Russia will likely retain approximately 3,000 tactical warheads, in addition to an unknown number of reserve weapons.

FRENCH, CHINESE, INDIAN, PAKISTANI, ISRAELI, AND SOUTH AFRICAN NUCLEAR WEAPONS

France

French nuclear research began well before World War Two. In the period between the two world wars, nuclear physics was at an advanced stage in France due to the work of Pierre and Marie Curie, Frederic Joliot-Curie, and Irene Joliot-Curie. General Charles de Gaulle was informed by various scientists of the progress made in American research in these matters and of its military implications by 1945, prior to the Trinity test. In Autumn 1945, after the Hiroshima and Nagasaki explosions, he made a decision to create a French Atomic Energy Commissariat (AEC).

The instability of the Fourth Republic in France after the war and the lack of financial means held back French nuclear research. A five-year plan for the development of atomic energy was prepared by Felix Gaillard, under the Pinay government which was in power between March 1952 and January 1953; it was intended to find a remedy for the French electrical energy deficit while also producing nuclear weapons. The plan was to produce 50 kilograms of plutonium each year that would allow six to eight nuclear bombs to be produced. In December 1956, a Committee for the Military Applications of Atomic Energy was secretly created. This committee was to provide coordination between the Atomic Energy Commissariat and senior military officials. Earlier in 1956, there had been a directive for the establishment of a program to build delivery vehicles. Finally, a program was outlined in December 1956 for a future strategic nuclear bomb. The return to power of General de Gaulle in 1958 marked the end of any French indecision in these matters. At the meeting of the French Defense Council, he announced the date of the first French nuclear explosion and decided to accelerate the French nuclear program.

The first French bomb was detonated in the Sahara Desert in 1960. In parallel, France developed the systems needed to deliver a bomb: planes initially, then land-based missiles, and, finally, missiles launched from submarines. In all, 41 tests in the atmosphere and 134 tests in boreholes in Pacific atolls were conducted by France between 1960 and 1991, chiefly on Mururoa and Fangataufa. Added to those in the Sahara, France conducted a total of 192 tests up to 1992. In April 1992, President Francois Mitterrand, through his Prime Minister, announced the suspension of all French nuclear testing for that year. This began a unilateral French moratorium on nuclear tests which was renewed several times, finally to be suspended by the new

French President, Jacques Chirac, in 1995. As President of the French Republic, Mr. Chirac announced the resumption of nuclear tests by France to be a final series of eight tests between September 1995 and May 1996. The French President announced that France would carry out this final campaign of nuclear testing in the Pacific, and then sign a universal and verifiable Comprehensive Nuclear Test Ban Treaty (CTBT). The last French nuclear test took place on 26 January 1996.

As soon as the Cold War was over, France renounced a development programs for land-based S45 strategic missiles, and accelerated the withdrawal of two systems, Pluton missiles and AN-52 bombs. It began scaling back its nuclear programs for sea-launched systems (a new generation of ballistic missile submarines for air-launched systems using the Mirage 2000N and ASMP missiles) and for a new ground-launched system of Hadès missiles. In 1996, France further reduced its nuclear forces by scaling down its fleet ballistic missile submarines by one to a total of four, canceling the Plateau d'Albion test site for land-based missiles and withdrawing the obsolete Mirage IVP bombers from service. Since the end of the Cold War, France is the only nuclear weapon state to have dismantled all its test facilities and given independent international experts access to its nuclear test sites. By ratifying the Protocols to the Treaty of Rarotonga in 1996 and the Comprehensive Nuclear Test-Ban Treaty in 1998, France enshrined in legally binding international instruments the decision announced by the President of the Republic to cease all nuclear testing. In addition, France is the only nuclear power having announced and started the dismantling of its fissile material production facilities. Since 1992, France has not produced weapon-grade plutonium. At the end of 1997, it closed the Marcoule reprocessing plant where this plutonium was produced and the Pierrelatte enrichment plant where highly enriched weapon-grade uranium was produced. France has ceased all production of fissile material for nuclear weapons.

France is presently modernizing its sea-based deterrent to include two new Triomphant class submarines, which carry a new ballistic missile, the M-45. The controversial nuclear testing at Mururoa Atoll in 1995 and 1996 was reportedly done to perfect the warhead design for the M-45. Two more Triomphant class submarines will be deployed by 2008. The French are pressing forward with an advanced submarine-launched ballistic missile, the M-51, complete with a stealthy, maneuvering warhead called the TN-76.

The means of air delivery will remain potent, though the last French nuclear gravity bombs have been retired. The Mirage 2000N and carrier-based Super Etendard fighter-bombers are available to deliver short-range nuclear ASMP missiles. A follow-on to the current ASMP missile, dubbed the ASMP+, is under development and is slated to enter service in 2007. The new French multi-role aircraft, the Rafale D which will have a nuclear mission, should be ready by then as well.

However, today, the French nuclear arsenal, although reduced in number, remains very significant, the fourth largest in the world. The French concept is of showing the will and of having the capacity to deter an adversary by inflicting damage that is out of proportion with the stake of a conflict. France retains a nuclear capability. This includes an underwater component equipped with four submarine launchers (SNLE) carrying M45 missiles and three squadrons of Mirage 2000N aircraft as well as a number of Super Etandard planes, carried on an aircraft carrier, which can carry a nuclear-tipped missile. France is believed to have roughly 350 nuclear warheads, all bomber and submarine based. France plans to deploy two new nuclear-powered ballistic missile submarines (SSBMs) by 2008, bringing its total number of SSBMs to eight.

People's Republic of China

The Chinese atomic bomb projects have been the most secretive, even after the fact. Almost nothing is known for sure about nuclear weapons development in China. It is known that, in 1964, China exploded its first atomic bomb and, in 1967, its first thermonuclear bomb in western China. When Sino-Soviet relations cooled in the late 1950s and early 1960s, the Soviet Union withheld assistance to China on atomic bombs, abrogated the agreement on transferring defense technology, and began the withdrawal of Soviet advisers in 1960. Despite the termination of Soviet assistance, China committed itself to continue nuclear weapons development to break what they called "the superpowers' monopoly on nuclear weapons" to ensure Chinese security against the Soviet and United States threats and to increase Chinese prestige and power internationally.

When China decided in 1955 to develop atomic bombs, it faced a number of technological choices as to the most appropriate route to follow. At that time, China could only work on one path for economic reasons, and had to choose between producing Pu-239 from a reactor or developing a method of producing U-235 through isotope separation. The uranium path offered two alternatives, either chemical separation or physical separation. For the plutonium path, chemical separation of Pu-235 from the mixed system of U-235 and U-238 would have been easier than physical separation, but the separation of plutonium and uranium is difficult due to the high radioactivity of the system and the severe chemical toxicity of plutonium. Therefore, the Chinese chosen path was the physical separation of U-235 and U-238 isotopes, much like the early U.S. effects at Oak Ridge. The implosion method of detonating an atomic bomb was considered more technically advanced, although there were questions as to whether China was capable of producing a uranium bomb detonated by the implosion method.

China made remarkable progress in the 1960s in developing nuclear weapons. In a 32-month period, China successfully exploded its first atomic bomb on October 16, 1964, launched its first nuclear missile on October 25, 1966, and detonated its first hydrogen bomb on June 14, 1967. The first Chinese nuclear test was conducted at Lop Nor and was codenamed CHIC 1. It was a tower shot involving a fission device with a yield of about 25 kilotons. Uranium-235 was used as the nuclear fuel, which indicated Beijing's choice of the path of creating high-yield nuclear weapons as quickly as possible. Of the ten test shots that followed by September 1969, six are believed to have been related to thermonuclear development. The others had as their goals the adaptation of CHIC 1 for bomber delivery and test of a missile warhead in a test code named CHIC 4. The third nuclear test was conducted on September 9, 1966 dropped from a Russian Tu-16 bomber. In addition to uranium-235, this nuclear device with a yield around 100 kT contained lithium 6 (^{6}Li or Li-6), which attested to China's eagerness to test a thermonuclear weapon, since lithium is important in hydrogen bombs. CHIC 6, an airdrop test on June 17, 1967, was the first Chinese full-yield, two-stage thermonuclear test.

The so-called Cultural Revolution disrupted the strategic weapons program less than other scientific and educational sectors in China, but there was some slowdown in succeeding years. The success achieved in nuclear research and experimental design work permitted China to begin series production of nuclear and thermonuclear devices small enough to be used as missile warheads.

Later Chinese nuclear tests, such as CHIC 12 and CHIC 13, were suggestive of a new phase of the Chinese nuclear test programs. Both were deliberately low yield weapons. It appeared possible that CHIC 13 was delivered by an Chinese F-9 fighter aircraft, and may have been a proof test of a small weapon. One of the objectives of the final series of Chinese nuclear tests was to miniaturize China's nuclear warheads, dropping their weight from 2200 kg to about 700 kg in order to accommodate the next generation of solid-fueled missile systems.

In addition to the development of a sea-based nuclear force, China began considering the development of tactical nuclear weapons. Chinese Army exercises featured the simulated use of tactical nuclear weapons in both offensive and defensive scenarios, beginning in 1982. Reports of Chinese possession of tactical nuclear weapons remained unconfirmed until 1987. In 1988, Chinese specialists tested a 1 to 5 kT nuclear device with an enhanced neutron radiation yield, advancing the country's development of a very low yield neutron weapon and laying the foundation for the creation of nuclear artillery.

There is considerable uncertainty in published estimates of the size of the Chinese nuclear weapons stockpile. In the late 1980s, it was generally held that China was the world's third-largest nuclear power, possessing a nuclear deterrent force of some 225 to 300 nuclear weapons. Other estimates of the country's production capacities suggested that, by the end of 1970, China had fabricated around 200 nuclear weapons, a number which could have increased to 875 by 1980. With an average annual production of 75 nuclear weapons during the 1980s, some estimates suggest that, by the mid-1990s, the Chinese nuclear industry had produced around 2000 nuclear weapons for ballistic missiles, bombers, artillery projectiles, and landmines.

China is seeking to increase the credibility of its nuclear retaliatory capability by dispersing and concealing its nuclear forces in difficult terrain, improving their mobility, and hardening its missile silos. China's nuclear arsenal is believed to be undergoing a rapid modernization program begun in the mid-1980s. By increasing the size, accuracy, range, and survivability of their nuclear arsenal, Chinese leaders aim to strengthen Beijing's deterrent. China hopes to mimic the United States and Russia in deploying its nuclear weapons in a sea-, air-, and land-based triad. U.S. intelligence and defense agencies predict that China may increase the number of warheads aimed at U.S. targets from some twenty to more than a hundred. In the near future, China will likely place emphasis on the development of improved ballistic missiles. Development efforts have stressed increasing the number of mobile, solid-fuel, intercontinental missiles. Currently, China is thought to have a host of nuclear missiles at its disposal. These include: 20 liquid-fueled intermediate range (IRBM) Dong Feng-4s (DF-4s); 48 medium range solid-fueled DF-21s, which are launched from mobile platforms; and 20 silo-based intercontinental (ICBM) DF-5s, which can reach the United States. The older DF-3 ICBMs have being retired. Other solid-fueled short-range missiles, the DF-11 and DF-15, called the M-11 and M-9 when exported to her clients, may have nuclear capability. The DF-21A missiles, of which there are about 50, also have the same warhead with a yield of 250 kT. Two new mobile solid-fuel intercontinental ballistic missiles (ICBMs) are under development. China may have a Xia submarine-launched program, using the Julang missile with a 1,000-kilometer range and a nuclear yield of around 250 kT. Chinese aircraft capable of carrying nuclear weapons include the Hong-6, with a 3,000-kilometer range, and the Qian-5, with a limited 400-kilometer range.

U.S. intelligence services have estimated that China has long had the ability to develop multiple reentry vehicles (MRVs) for its missiles. Should China choose to further develop these systems, ICBMs could be retrofitted within a few years. U.S. deployment of a missile defense system could precipitate such action.

The weakness of the Chinese Air Force had led Beijing toward dependence on Russian aid. Today, China relies primarily on two types of aircraft for its nuclear force, about 100 Soviet-based medium-range bombers—the Hong-6—and 30 shorter-range Qian-5. A supersonic fighter-bomber, the JH-7, has been under development for more than a decade, but is not currently outfitted to carry nuclear bombs. Owing to technical problems, few have been deployed. China has purchased around eighty Russian SU-30 multi-role aircraft. Additionally, the Russians have sold China fifty-eight SU-27s, along with production rights and engineering assistance, which should allow China to produce another two hundred SU-27s by 2015. While both the SU-30s and SU-27s could be modified to fulfill a nuclear mission, there is little indication that such modifications are underway.

Efforts have been made to upgrade China's ballistic missile submarine fleet, but technical difficulties have limited progress. China is believed to have 12 Julang I submarine-launched ballistic missiles stored at Jianggezhuang Submarine Base, where its one nuclear-powered ballistic missile submarine, the Xia, is housed. It is not clear whether this sub is operational. China has a long-term plan to build four to six new submarines, each of which will carry sixteen Julang II missiles of intercontinental range. The new subs are not likely to be deployed for many years. China may also have another 120 non-strategic nuclear weapons, for a total of about 400.

India

Knowing of the U.S. and U.S.S.R. advances in nuclear weapons, in 1948, India felt it was forced to establish an Atomic Energy Commission (AEC) to explore for uranium ore. It was well-known that some monazite sands in India contain relatively high concentrations of thorium, so it seems likely that uranium might also be present. The United Nations (U.N.) created an agency for the control of radiation, the International Atomic Energy Agency (IAEA), and proposed strict guidelines for transfer of, protection from, and uses of nuclear energy. In 1954, the Chairman of India's AEC rejected safeguards and oversight by the IAEA. By 1956, India had completed negotiations with Canada to build a 40 megawatt research reactor. It was a copy of the Canadian NRX reactor and was called the "Canadian-Indian Reactor, U.S." or CIRUS. The U.S. had supplied the heavy water. CIRUS was not designed for electricity production, but as a very efficient producer of man-made isotopes, including weapons-grade plutonium, which the Canadians had sold to the Americans during the 1940s and 1950s to help defray the cost of nuclear research.

In 1958, India began designing and acquiring equipment for its own Trombay plutonium reprocessing facility, giving the nation a dual-use capability that could lead to atomic weapons. The U.S. trained Indian scientists in reprocessing, and handling plutonium. Two purchases were made in 1963, two U.S. 210-megawatt boiling-water reactors for the Tarapur Atomic Power Station, made by General Electric, and a Canadian 200 megawatt CANDU power reactor called RAPP-1, modeled after the Douglas Point power plant in Ontario. The U.S. and India agreed that any plutonium from India's reactors would not be used for research into nuclear weapons or for any military purpose. The next year, India built their first plutonium reprocessing plant at Trombay. The Chairman of India's AEC proposed a subterranean nuclear explosion project in 1965. The U.S. withdrew military aid from India after the India-Pakistan War in that year. Soon afterwards, India declared it could produce nuclear weapons within 18 months. When the Nuclear Non-Proliferation Treaty was completed, India refused to sign.

In 1969, France agreed to help India develop breeder reactors. In May, 1974, India tested their first nuclear device with a yield of about 15 kilotons, and called the test a "peaceful nuclear explosion" using weapons-grade plutonium produced in the Canadian-supplied CIRUS reactor. The explosion took place at the Pokhran site in the Rajasthan desert, near the border with Pakistan. The name of the bomb test was "Smiling Buddha." Canada suspended nuclear cooperation while the U.S. allowed continued supply of nuclear fuel. By 1976, the U.S.S.R. had assumed the role of India's main supplier of heavy water.

By the early 1980s, India had acquired centrifuge technology and built uranium enrichment plants at Trombay and Mysore. India signed an agreement with Pakistan in 1991 prohibiting attacks on each other's nuclear installations in order to ease tensions between the states. In 1992, the Rare Metals Plant at Mysore began producing enriched uranium. The next year, India announced independent development of supercomputing technology that could be used to test nuclear weapon designs. Fuel reprocessing plant at Kalpakkam, a large-scale plutonium separation facility, completed the last phase of pre-operating trials. In 1998, India announced plans to sign a deal with Russia for two 1,000 megawatt nuclear reactors. That same year, in May,

India conducted five underground nuclear tests, and declared itself a nuclear state. It is believed that India has about 60 nuclear warheads using Intermediate-Range Ballistic Missiles (IRBMs).

Pakistan

In 1959, Canada sold a 125 megawatt CANDU power reactor to Pakistan; it was called KANUPP. In 1972, following its third war with India, Pakistan secretly decided to begin a nuclear weapons program to match India's developing capability. Canada supplied the reactor for the Karachi Nuclear Power Plant, as well as a heavy-water production facility. Due to instability in the region after India's first test of a nuclear device, by 1974, Western suppliers embargoed nuclear exports to Pakistan. The next year, purchase of components and technology for the Kahuta uranium-enrichment centrifuge facility began after return to Pakistan of Dr. Abdul Qadeer Khan, a German-trained metallurgist who took over the nuclear program. Canada stopped supplying nuclear fuel for Karachi. In 1977, a German seller provided vacuum pumps and other equipment needed for uranium enrichment. Britain sold Pakistan 30 high-frequency inverters for controlling centrifuge speeds. The U.S. halted economic and military aid because of Pakistan's nuclear-weapons program. France had agreed to build a plutonium reprocessing plant at Chasma, but soon cancelled the contract.

In 1979, the U.S. imposed economic sanctions after Pakistan was caught importing equipment for an uranium enrichment plant at Kahuta. In 1981, a smuggler was arrested at a U.S. airport while attempting to ship two tons of zirconium, which could be used in fabricating nuclear weapons, to Pakistan. In spite of this, the Reagan administration lifted economic sanctions and restarted military and financial aid because of Pakistani help to Afghan rebels battling the Soviets. In 1983, it was rumored that China began to supply Pakistan with nuclear weapon design. U.S. intelligence discovered a Pakistani centrifuge program intended to produce material for nuclear weapons. In 1985, the U.S. Congress passed the Pressler Amendment which required economic sanctions unless the White House annually certified that Pakistan was not embarked on a nuclear weapons program. Islamabad was certified every year until 1990.

The next year, Pakistan and China signed a pact on the peaceful use of nuclear energy, including design, construction, and operation of reactors. Pakistan acquired a tritium purification and production facility from West Germany, and, in 1989, a 27-kilowatt research reactor was built with Chinese help, but Pakistan agreed to international monitoring. By 1990, fearing a new war with India, Pakistan made the cores for several nuclear weapons. The U.S. administration, under the Pressler Amendment, imposed economic and military sanctions against Pakistan. The next year, Pakistan put a ceiling on the size of its weapons-grade uranium stockpile, and entered into an agreement with India that prohibited the two states from attacking each other's nuclear installations. A 1993 study by the Stockholm International Peace and Research Institute reported about 14,000 uranium-enrichment centrifuges had been installed in Pakistan. German customs officials seized another 1,000 gas centrifuges bound for Pakistan.

In 1996, Pakistan bought 5,000 ring magnets from China to be used in gas centrifuges for uranium enrichment. China told the U.S. government it would stop helping Pakistan's unsafeguarded nuclear facilities. Islamabad completed a 40-megawatt heavy-water reactor that, once operational, could provide the first source of plutonium-bearing spent fuel free from international inspections. Reacting to fresh nuclear testing by India, Pakistan conducted its own atomic explosions in 1998. The Pakistani nuclear arsenal is believed to be between 15 and 25 warheads. Delivery systems are largely limited to Intermediate Range Ballistic Missiles (IRBMs).

Israel

The secrecy with which Israel shrouds its nuclear arsenal renders estimates of its size highly unreliable, possibly even more so than the Chinese. China has at least announced that they possess nuclear weapons. Israel developed nuclear weapons with French help in the 1950s and 1960s, and enjoyed the tacit approval of the U.S. since the Nixon administration. Despite refusals to comment on the issue by the Israeli government, the Israelis clearly have a sizeable nuclear arsenal. There are two interesting loopholes in Israel's repeated pledge never to be the first to introduce nuclear weapons into the region. The U.S. "introduced" weapons in the region in the 1950s when nuclear bombs were briefly stored at Dharan, Saudi Arabia, and at sea aboard ships of the Mediterranean Sixth Fleet. It is believed that Israel might not keep its nuclear weapons fully assembled. The well-equipped Israeli Air Defense Force could deliver nuclear weapons, particularly with U.S.-supplied aircraft such as the F-4E and F-16. However, Israel also produces ballistic missiles. The Jericho I can strike Syria, and the Jericho II brings the entire Middle East within Israel's range, particularly Iran and Iraq. Israel may also have some tactical nuclear weapons. The Shavit space launch booster could be adapted for a long-range nuclear delivery role, whereby Israel would be able to quickly develop an intercontinental ballistic missile. The Israeli arsenal will likely remain stable in number for years to come. Israel signed the Comprehensive Test Ban Treaty, and is not known to have ever tested any of its weapons. It is believed that Israel is unlikely to reduce or eliminate its nuclear arsenal, estimated at between 100 and 200 weapons. Delivery is likely largely by Intermediate Range Ballistic Missiles (IRBMs) and, possibly, in tactical use, from aircraft.

South Africa

To date, South Africa remains the only independent nation to voluntarily destroy their entire nuclear weapon stockpile. South Africa, with help from the French, obtained sufficient plutonium to produce at least nine weapons. These weapons, upon the fall of the apartheid regime, were declared to the International Atomic Energy Agency (IAEA), transferred to the IAEA, and were destroyed.

POST-WAR U.S. DEVELOPMENTS

There is another fissionable isotope of uranium, U-233, which was investigated as bomb material. The U.S. tested a number of U-233 bombs, but the presence of U-232 in the fissile material was a problem since this could lead to possible pre-detonation.

Although Fat Man had contained a sphere of solid plutonium metal, research at the very end of World War Two had indicated a better design using a thin outer sphere of uranium with an internal void of air and a small central core of plutonium. This design is called the composite, or levitated, core. With the levitated design, much smaller and lighter weapons were possible. Where Fat Man had used 6.2 kilograms, as little as 3.2 kilograms can now be used.

Thermonuclear bombs, often called "hydrogen bombs" or simply "H-bombs," function entirely differently than fission bombs, but require a fission bomb to initiate the thermonuclear process. When the first thermonuclear bomb was tested, some of the scientists worried that the nitrogen in the air might be ignited with a resulting widespread catastrophe. Fortunately, their fears were not realized. Thermonuclear bombs produce energy from the process of forming helium from hydrogen atoms under great pressure and heat; this process is known as "fusion" since the hydrogen nuclei are fused together to make a helium nucleus. Many people think that the thermonuclear bomb was a follow-on to the nuclear bomb; in fact, work on the hydrogen bomb began as early as 1942, even before there was an atomic bomb. Because an atomic bomb destroys the fissile material when it detonates, a theoretical upper limit to the yield of a fission bomb is about one megaton; more explosive power would require a different design. Dr. Edward Teller realized that deuterium nuclei could be fused to form heavier elements while releasing binding energy. Deuterium is much easier and cheaper to separate from heavy water than U-235 is from uranium, but tritium was difficult to obtain until later reactors were built to produce tritium. A one megaton bomb could be made using only about 25 pounds of liquid deuterium. But, liquid deuterium must be kept at cryogenic temperatures under pressure. Later designs used much less mass of lithium deuteride in a "dry bomb" which did not need cryogenic cooling. However, a nuclear fission bomb is still needed to initiate the thermonuclear reactions.

The theory behind the hydrogen bomb is simple, but the design was not. Edward Teller, who had desired such a weapon for a long time, had a conversation in early 1951 with Stanislaw Ulam, then a mathematician at Los Alamos. The problem lay in being about an implosion, not by high explosives, but much more rapidly. Together, they came up with the idea, known as radiation implosion, to build a large cylindrical casing that would hold an atomic bomb and hydrogen fuel at opposite ends. The flash of the exploding bomb would hit the case, causing it to glow and flood the interior of the casing with radiation pressure sufficient to compress and ignite the hydrogen fuel. The first design was made by Dick Garwin, a researcher in Fermi's laboratory at the University of Chicago. Other scientists who worked out the details of the design were Marshall Rosenbluth and Conrad Longmire.

The first designs for thermonuclear weapons were called Alarm Clock by the U.S. and Sloika ("Layer Cake") by the U.S.S.R. In these weapons, a few grams of a deuterium/tritium gas mixture were included in the center of a fission core. The first boosted weapon test was the U.S. Greenhouse Item, which yielded 45.5 kT on 24 May 1951, an oralloy (named after Oak Ridge + alloy) design exploded on island Janet at Enewetak Atoll. This experimental device used cryogenic liquid deuterium-tritium instead of gaseous forms. The boosting approximately doubled the yield over the expected unboosted value. Variants on the basic boosting approach that have been tested included the use of deuterium gas only and the use of lithium deuteride/tritide, but it isn't known whether any of these approaches have been used in operational weapons.

A later U.S. design was called the Staged Radiation Implosion Weapon (SRIW), or the Teller-Ulam design. When a neutron is absorbed by a molecule of lithium deuteride ($^6Li^2H$ or lithium-6 deuteride), the molecule breaks up into an alpha particle, ^2H (deuterium) and ^3H (tritium). The deuterium then reacts with the tritium in a fusion process. This releases enormous amounts of energy, much greater than possible in a fission reaction. The end products include a free neutron and a helium atom. Schematically:

$$^6Li + n \longrightarrow {}^4He + {}^3H + 4.7\ MeV$$

then

$$^2H + {}^3H \longrightarrow {}^4He + n + 17.6\ MeV$$

The first test of a thermonuclear device was code named IVY MIKE and used an inner fission bomb, followed by a fusion secondary containing liquid deuterium and a little tritium, and further enclosed by U-238. When it was detonated in 1952, the yield was about 10.4 megatons (MT) with some 80% contributed by the fission bomb. This was a device, not a weapon, since it was very large, required sophisticated cryonics to keep deuterium in the liquid state, and weighed some 82 tons. But it proved the theory, and future developments in strategic weapons have continued using fusion devices.

U.S. Army engineers even developed a small nuclear weapon for demolition of bridges, dams, and buildings, to be fired from the gun of a tank-like vehicle. The Army realized that this was not a practical device, and the atomic demolition munition (ADM) was removed from the inventory.

Likewise, the "atomic cannon" of the 1950s, designed to fire an atomic artillery round, experienced only a short life. The atomic cannon was designed to throw a nuclear artillery shell some 32 kilometers (20 miles). It was test fired only once in 1953 in Nevada, firing a shell with a projectile yield of 15 kilotons. Only 20 of the cannon were ever made; a few remain as military museum relics.

As the public became more aware of the dangers of radiation resulting from nuclear weapons, the military responded by making their device more stand-off and less tactical. In fact, only strategic uses of nuclear weapons remain in most nations.

The use of nuclear devices in construction was even proposed in the 1950s and 1960s. There was a serious proposal that a canal parallel to the Panama Canal be built using nuclear charges to blast out the enormous amounts of earth. Test blasts were conducted at the National Testing Site (NTS) in Nevada, and the engineering proved workable; however, the plan was abandoned when modeling calculations showed that thousands of people, including indigenous tribes, would be in the path of fallout debris and resulting radiation.

There was much talk about a so-called "cobalt bomb" in the late 1950s and early 1960s. Rumors were heard that this new bomb was many times more powerful than the hydrogen bomb. In fact, the best fissile material for making a nuclear weapon remains plutonium because of its neutron cross section, availability, machinability, and density; its major drawback is its short half-life. The utilization of cobalt bomb is not due to the fission characteristics, but rather to the generation of intense gamma radiation. Each cobalt bomb is an ordinary atomic bomb encased in a jacket of cobalt. When a cobalt bomb explodes, it spreads a huge amount of radiation. The idea of a cobalt bomb was first proposed by Leo Szilard in February 1950, not as a serious proposal for a weapon, but to generate sufficient fear that nuclear weapons research might cease. To design such a weapon, a radioactive nuclide is needed that can be dispersed worldwide before it decays. The design would be similar to a fission-fusion-fission weapon. A thick cobalt metal blanket captures neutrons to maximize the fallout hazard. Instead of generating additional explosive force from fast fissioning of U-238, the cobalt is transmuted into Co-60 which produces very energetic and penetrating gamma rays. It is not believed that any nation ever built such a weapon; a test device may have been produced by the U.S. or the U.S.S.R.

Another type of device that received a lot of press coverage about the same time was the "neutron bomb," technically, the "enhanced radiation weapon." This bomb was said to produce intense neutron radiation with reduced blast and thermal effects with the result that living creatures would be killed but structures and facilities remain intact for use by the winner. It was suggested that such a weapon could better kill the soldiers inside a tank than existing weapons. President Jimmy Carter made a public announcement that the U.S. did not and would not produce such a device. In fact, there is neutron radiation from the detonation of any nuclear weapon, but this largely depends upon the composition of the bomb and is related to ratios of the blast and thermal effects. It is believed that the U.S. produced some 200 enhanced radiation weapons. While it is possible to make a bomb with enhanced neutron radiation, there would still be blast and thermal damage far beyond that of conventional explosives.

The U.S. and the former U.S.S.R. engaged in what has been termed the "Cold War" from the middle 1950s until the fall of Communism. Both countries produced large numbers of nuclear and thermonuclear weapons, and designed a number of delivery systems ranging from long-range bomber aircraft to missiles to artillery. The U.S. built a large complex near Aiken, South Carolina for the production of tritium to add to the other sites at Oak Ridge and Hanford. Smaller sites were also developed, near Mount Ohio in Idaho, and at (or near) a number of U.S. Air Force bases, where weapons maintenance and repair could be easily undertaken. The Los Alamos site was not practical for large-scale weapons production, so the Atomic Energy Commission (AEC), later the Department of Energy (DOE), which had taken over responsibility for nuclear weapons from the military, built a large weapons construction facility near Amarillo, Texas. This plant, Pantex (from the fact it is in the PANhandle of TEXas), became the sole U.S. site for building thermonuclear weapons. Operations at most of the plants were transferred from the government to contractors. In the U.S.S.R., the military maintained control of all nuclear weapons production, testing, and storage.

MODERN WEAPONS

From the early nuclear tests, it was discovered that there is an intense electromagnetic pulse associated with a nuclear detonation. A high-altitude nuclear detonation produces an immediate flux of gamma rays from the nuclear reactions. These high-energy photons then produce high energy free electrons by Compton scattering. These electrons are trapped by the Earth's magnetic field, giving rise to an electric current. This current gives rise to a strong electromagnetic field called an electromagnetic pulse (EMP).

The pulse can easily span entire continents and affect systems on land, sea, and air. The first recorded EMP incident accompanied a high-altitude U.S. nuclear test over the South Pacific, which resulted in power system failures as far as Hawaii. A large nuclear device detonated at 400 – 500 km over the central U.S. would affect all of the continental U.S., damaging computer chips, magnetic storage devices, and, possibly, electricity distribution systems. The signal from such an event extends to the visual horizon as seen from the burst point. Protection from EMP is extremely difficult in a world using computer chips in everything from coffee makers to automobiles. The primary means of protection would be shielding, or hardening, of the object to be protected. In the 1980s, the U.S. obtained a new model Soviet MIG fighter. Upon examining it, the engineers first thought that the Soviets must be years behind in electronic technology since the computers, radio, and other equipment was largely vacuum tube-based. As they went deeper into the aircraft, the more modern the electronics became. They realized that the Soviets had designed their aircraft to be better hardened against EMP by using electron vacuum tubes which are less affected by these pulses.

One of the problems involved in nuclear weapons is how to deliver them. From the 1940s until the 1960s, due to the large size and weight of these weapons, the only practical delivery method was using manned bomber aircraft. As bomb and warhead design improved and much smaller weapons were possible, most nuclear nations changed their military strategy to the use of missiles. The U.S. and U.S.S.R. abandoned the use of artillery systems for delivery of nuclear weapons for a number of reasons, chiefly out of concern for the troops firing them and those having to later reoccupy the target location. As devices became even smaller and missiles became capable of heavier loads, multiple warheads were placed on one missile.

As nuclear weapons became smaller, the need for large aircraft lessened. Most nations turned to missiles to carry their nuclear weapons. Multiple-impact re-entry vehicles (MIRVs) are thermonuclear devices packed some three to ten in a single missile nosecone. Upon reaching the desired altitude, the nosecone is blown away and the independent warheads adjust their path toward specific targets on the ground or at sea. In this way, one missile can carry a number of warheads and attack multiple locations. Variable yield thermonuclear warheads have been developed with the intent that the blast could be tailored to a specific target.

Conventional militaries would, most likely, use missile warheads in delivery of thermonuclear weapons. Most of the current U.S. arsenal of nuclear weapons are designed to be carried on ballistic missiles launched from underground silos, or cruise missiles either launched from a truck-pulled platform, from a naval ship, or from an aircraft.

Terrorists, on the other hand, lacking such a sophisticated (and expensive) delivery system, would likely sail a weapon into a harbor on a commercial freighter or as cargo on an aircraft. With a device weighing only 60 pounds or so, it is possible that a terrorist could carry one in a suitcase. Due to the very low energy x-rays and alpha emissions from Pu-239 and weak alpha radiation from U-235, detection is very difficult.

ARMS LIMITATIONS

By 1978, the United States and the Union of Soviet Republics had stockpiled large arsenals of nuclear weapons. The U.S. strategic forces had some 2,142 missiles or bombers capable of delivering over 11,000 warheads against the Soviet Union. Soviet forces had 2,550 missiles or bombers with a total of 4,500 warheads against the U.S. It was estimated that the U.S.S.R. would have over 7,500 weapons by the early 1980s. Both sides possessed sufficient nuclear weapons to destroy the other many times over.

The Strategic Arms Limitation Talks (SALT) between the former U.S.S.R. and the U.S. resulted in limits on the numbers and yields of nuclear weapons in both countries. In 1979, Presidents Jimmy Carter of the U.S. and Leonid Brezhev of the U.S.S.R. signed a treaty, SALT II, placing limits on the nuclear arsenals of both nations. The treaty placed a total limit of 2,400 on the number of land-based intercontinental ballistic missiles (ICBMs), submarine-launched ballistic missiles (SLBMs), heavy bombers, and air-to-surface ballistic missiles (ASBMs) until the end of 1981, at which time the limit would drop to 2250. The number of ICBMs with multiple warheads (MIRVs) would be capped at 820; the number of ICBMs and SLBMs with MIRVs would be held at 1200; and the number of ICBMs and SLBMs plus bombers with cruise missiles capped at 1320 for each side. Verification was required by both sides, using satellite imaging as well as on-site observations.

The SALT II treaty did not place a maximum limit of the number of nuclear weapons nor their power. It did, however, limit the number of warheads on MIRVs to 10 for land-based and 14 for submarine-launched. Heavy bombers were limited to no more than 28 cruise missiles each. Both nations also agreed not to test or deploy any new types of ICBMs. Recent amendments to the SALT II have seen the U.S. and Russia agree to further reduce their arsenals to 2,200 warheads each by 2012.

Tactical weapons are not covered by the treaty. Russia is believed to have some 8,000 tactical nuclear weapons, while the U.S. has some 1,000. It is believed that both sides will continue to maintain stockpiles of these smaller devices.

Most nuclear weapons, by both sides, fall generally in the following sizes: 50 kilotons of TNT (kT), 200 kT, 1 megatons of TNT (MT), and 10 MT. For comparison, the bombs dropped at Hiroshima and Nagasaki were both less than 20 kT, equivalent to the force of detonating 20,000 tons or 40,000,000 pounds of TNT. In general, the strategic nuclear weapons of the former Soviet Union are of larger yield than those of the U.S.

TEST BAN TREATIES

In 1963, Hans Bethe was the principal scientific advisor to the U.S. negotiators on the Partial Test Ban Treaty (PTBT). A number of nations met to see if there might be a way to eliminate the hazards of fallout from nuclear tests. The chief outcome of this treaty is that all aboveground, underwater, or outer space testing of nuclear weapons was prohibited. Unfortunately, neither China nor France, both possessing nuclear weapons, signed the treaty, formally known as "The Treaty Banning Nuclear Weapons Tests in the Atmosphere, in Outer Space, and Under Water."

By 1991, it was clear that the PTBT was ineffective, so signatory states met to amend the treaty to make it stronger. With strong support from the United Nations (U.N.) General Assembly, negotiations began in 1993. Intensive efforts were made over the next three years in writing the text and annexes of a new, better treaty. The Comprehensive Test Ban Treaty (CTBT) was adopted on September 10, 1996 by the U.N. General Assembly. The CTBT prohibits all nuclear test explosions in all environments. The Treaty was signed by 71 nations, including the then five nuclear weapon nations.

NUCLEAR NONPROLIFERATION TREATY

The Treaty on the Non-Proliferation of Nuclear Weapons, signed by most of the nations of the world, has its intent to limit the number of nations possessing nuclear weapons. It went into force on March 5, 1970. The treaty limits possession to those nations currently having such devices. The signatory nations, other than those allowed to possess them, agree that they will not design or build nuclear weapons. The states allowed to have the weapons, called the Nuclear Weapons States (NWS), agreed to make progress on nuclear disarmament. Free access to nuclear technology for all non-military uses was agreed to and encouraged. While 187 nations are signatory to this Treaty, a number of nations, notably North Korea, Iran, and Iraq, have either withdrawn from it or never signed it.

WEAPON ACCIDENTS AND MISSING WEAPONS

While security of nuclear and thermonuclear weapons is of primary importance to all the nations possessing them, accidents can occur. Most weapons have been lost as the result of aircraft accidents, almost all in the deep oceans where recovery is impossible or highly unlikely. A few weapons have been destroyed in aircraft crashes like those onboard a U.S. Air Force bomber which crashed at Thule, Greenland. Others have been lost at sea or deliberately dropped into sea like the one off Savannah, Georgia following damage to the B-47 bomber carrying it. A few have been destroyed as a result of burning or explosion of the conventional explosive like two of the four bombs dropped at Palamaris, Spain in 1966 following the collision of a USAF tanker aircraft and a B-52 bomber during refueling (the other two fell into the sea and were recovered). Some nuclear weapons have been lost from error, like two events in 1962 when Thor booster rockets failed during launch, or when a technician dropped a wrench, breaking a fuel tank of a Titan II missile in Arkansas in 1980, and caused an explosion which sent the 740-ton door hundreds of feet into the air and the re-entry vehicle with a 9-megaton warhead almost a mile. Some have been lost as a result of sinking of nuclear missile-equipped submarines, like the Soviet Golf-II class submarine off the Hawaiian Islands in 1968. In no cases have there been any nuclear yield; that is, none of the material in the bombs resulted in any fission or fusion greater than one pound of TNT equivalent.

There is fear that some of the states of the former U.S.S.R. may be experiencing less than efficient control over their weapons, but there has been no proof made available to the public. However, a nuclear institute in the northeastern city of Kharkiv, Ukraine, where Iraq opened an official mission in 2000, is particularly troubling. The Physics and Technology Institute (IPTK) reportedly stores some 75 kilograms (165 pounds) of enriched uranium, enough to build three or more nuclear bombs. The U.S. was eager to buy that uranium off IPTK to make sure it did not end up in the wrong hands, but IPTK's director said that they would never sell it to anyone. The Ukraine is in critical need of cash, which terrorists or Iraq or Iran could supply. Russian authorities have reported as many as 23 attempts to steal fissile material since the dissolution of the Soviet Union. The U.S., in 1994, purchased some 1,300 pounds of highly enriched uranium from Kazakhstan to prevent it being diverted to terrorists.

In early 2003, the Japanese government admitted that it may have lost some 206 kilograms of plutonium extracted during reprocessing of spent commercial power reactor fuel; 6,890 kilograms were apparently produced, but some 3% could not be accounted for.

TODAY

The nations which have admitted to possessing nuclear and thermonuclear weapons include the United States, the United Kingdom, France, Russia, China, Pakistan, and India, but it is believed that other nations possess such weapons. Some of the countries suspected of having nuclear weapons include North Korea and Israel (which may have a hundred or more). Other nations, which may have developed such weapons or have been actively pursuing them, are Iran, Iraq, and Libya. Other countries had nuclear weapons programs, but abandoned them (South Korea, South Africa, and Egypt). Some of the states of the former Soviet Union, like the Ukraine, Kazahkstan, and Belarus, voluntarily gave up those weapons that had been located on their territory. It is suspected that a number of terrorist organizations, such as *Al Qaeda*, are actively seeking nuclear weapons, possibly from states of the former Soviet Union.

TERRORISM

The use of radioactive materials produced in a nuclear reactor to contaminate an area was first proposed in 1941 in a meeting of Arthur Compton's National Academy of Sciences (NAS). The next year, some of the scientists working on the atomic bomb project expressed a fear that the Germans might have an operating reactor to produce highly radioactive materials that could be used to make a "dirty bomb" to attack the Manhattan Project laboratories. In 1943, Enrico Fermi, working on the Manhattan Project, suggested to Robert Oppenheimer that if the atomic bomb couldn't be built, the mixed fission products from nuclear reactors could be used to poison the German food supply. These are probably the first mentions of using radioactivity as a weapon of terrorism (although nations at war might not be considered as conducting terrorism).

It is not possible to detonate an atomic bomb by simply pushing two pieces of uranium together. While this would lead to fission and a lethal dose of radiation, an explosion isn't possible due to the slow speeds involved and the absence of a source of energetic neutrons. Due to the careful control exercised by most nations over their nuclear arsenals and because of the large costs associated with development and construction of these weapons, evidence has been presented that some terrorist organizations have considered the use of a "dirty bomb." Such a bomb would not be a nuclear or thermonuclear device, but contain radioactive materials and conventional explosives. There are large amounts of highly radioactive material in commercial and medical applications. A terrorist could steal or otherwise obtain this material, and, by simply exploding it, spread contamination over an area. Even more likely, a terrorist could grind up the material into small dust-like particles and spread them over a wide area using an aerosol generator similar to those used to spray insecticide from trucks.

The greatest effect such an operation would have would be to spread panic and fear rather than producing serious health effects. Another concern would be the economic impact involving clean-up costs. Most of the reactor waste mixed fission products have very short half-lives so it does not remain radioactive for long. In most other cases, the actual amount of radioactivity would be small due to the size of the original device. The exposure or dose received by members of the public would, most likely, be quite small, and the number of people affected only in a local area. Even the potential for long-term cancer induction is small. Yet, the goal of a terrorist is to generate fear, even if irrational, and cause economic harm, so such a device could be very effective for them to employ.

If a terrorist organization were able to obtain a true nuclear weapon, it would not necessarily have to explode it in a population or industrial center. An electromagnetic pulse is produced by detonation of a nuclear weapon and is capable of causing widespread damage to communications systems, electric power grids, and all forms of computer data. The resulting economic impact could be enormous. Consider the loss of all digital data from the banks in the country. There would be no tax, ownership, or bank balance records. A relatively small nuclear weapon could be taken up in a commercial jet airplane and detonated at an altitude of some 20,000 or 30,000 feet with resulting pulse damage over hundreds or even thousands of square miles. Even proving the identity of those responsible would be difficult.

U.S. power reactor facilities have greatly increased security since the attacks of September 11, 2001, so the likelihood of these being the source of radioactive materials is diminished. Security at airports and airfields has greatly increased to prevent the importation or movement of weapons. Commercial radio and television stations, as well as electric power generation and distribution systems, have been asked to improve their protection against electromagnetic pulse radiation. Most medical and commercial facilities have likewise increased their control over radioactive sources. There is an international movement toward a registry of all radioactive sources, using the assets of the International Atomic Energy Agency (IAEA).

The most likely terrorist scenario would be a friendly state supplying radioactive materials to the terrorists from their power or research reactors in order to make a "dirty bomb." Various national and international agencies are actively investigating this possibility. There are Department of Energy (DOE) nuclear emergency rapid response teams located around the country to deal with a radiological terrorist threat or activity.

WORLD WAR II DIRTY BOMB OPERATION

A little-known incident occurred at the very end of World War II. A German submarine, U-531, was en route to Japan carrying radioactive material and two Japanese scientists, to deliver their cargo to the Japanese military. The operation was to load the radioactive material into conventional bombs to be carried by submarine-launched fighter-bombers for detonation over San Francisco. The date of this "dirty bomb" attack was to have been August 15, 1945. Fortunately, the German U-boat commander surrendered his ship to the U.S. Navy (after killing the two Japanese). Japan surrendered the week before the attack and it never occurred.

4

TYPES OF NUCLEAR WEAPONS

EARLY NUCLEAR WEAPONS DESIGNS

In the Manhattan Project, there were competing designs for the mechanism of the bomb. It was recognized that there would be a certain amount of fissile material needed—the so-called *critical mass*. Only when there are enough atoms present will the chain reaction continue. The concept of critical mass derives from the fact that, for a sphere, the surface area increases as the square of the radius while the volume increases as the cube of the radius. At some point, depending upon the density of the material and its cross section, more neutrons should hit nuclei than escape from the surface. To design the bomb, *fissile material* in an amount greater than the critical mass would be divided into parts and separated. At the moment of detonation, the parts would be forced together very quickly and a chain reaction initiated. When pieces of fission material are pushed together, forming a critical or supercritical mass, the bomb explodes.

Design of the bomb itself was conducted at Los Alamos, New Mexico. There, experiments were conducted in development of shape charges, lens design (focusing of explosive forces), criticality, and yield measurements. One early and simple idea was to push two hemispheres of uranium together using springs, but such a mechanical method might be too slow. Another design was having the two hemispheres pushed together perpendicular to their equators by an explosive charge, but there was concern that the explosive shock wave might shatter or distort the hemispheres.

The design that was settled upon used a bullet made of uranium-235 fired, by cordite, down a gun tube into three rings of uranium-235, producing a supercritical mass. This design became known as the "gun tube" design. This was shown in Figure 3.1 previously. "Little Boy" was the code name for the bomb made using this design; it used 51.55 kilograms (114 pounds) of uranium-235 with average enrichment of 80.4%; 50 kilograms (110 pounds) of 89% enriched uranium combined with 14.1 kilograms (31 pounds) of 50% enriched uranium. Little Boy was the first bomb dropped on Hiroshima, Japan since it was of simpler design and the military did not want to risk a failure with a more complicated system. Its yield was estimated at about 12.5 kilotons. Based upon this yield, it has been estimated that only about one kilogram of the uranium-235 fissioned with the remaining 50 kilograms of the uranium-235 released to the environment, much as fallout. The explosive force of a nuclear weapon is usually given in the equivalent weight of trinitrotoluene (TNT), a conventional high explosive. The force is often called the weapon's yield.

Uranium-235 could be used in the gun tube bomb, but plutonium-239 could not be because it spontaneously fissions at a rate several times faster than uranium-235. In order for plutonium-239 to be used in such a design, it would have to be fired down the gun tube at a speed much greater than any conventional explosives could deliver. Another, more ambitious, design involved forming either an empty spherical shell or subcritical hemispherical shells of fissile material, either uranium or plutonium, around which would be placed high explosives. When the explosives detonated, the uranium or plutonium would be forced into a much smaller volume of more than critical mass. This design was known as the "implosion" design. This design was shown in Figure 3.2 previously. Since plutonium-239 was a much better fissile material, it was decided to use this radionuclide in the bomb. "Fat Man" was the code name for the bomb that used this design because it required a spherical shell of high explosives around the fissile material, and was almost spherical. Each of the high explosive lenses had to be carefully fitted together in a completely static-free location. The bomb contained 6.2 kilograms (13.7 pounds) of plutonium-239 as the nuclear material and used a polonium/beryllium mixture as the initiator. Polonium, an intense alpha emitter, causes the beryllium to release neutrons to begin the nuclear fission reaction within the plutonium. The core of the Fat Man implosion bomb was a sphere of plutonium-239, actually made of two hemispheres separated by a thin gold foil, with the initiator at the center. Around the plutonium was an outer shell, or tamper, of uranium to serve to contain the explosion. The tamper smoothes out the high-explosive shock wave and holds the fuel together by inertia a few microseconds to allow for additional chain reaction generations to take place, increasing the overall efficiency of the explosion. Finally, there was a layer of boron to absorb thermal neutrons and thus prevent pre-detonation. Around this were the 32 shaped charges of high explosive, each separated by a thin piece of felt.

The shaped charges had an outer fast detonating explosive (Composition B) and an small inner slower explosive (Baratol), followed by another block of Composition B, in order to develop a completely spherical shock wave. Composition B was composed of a mixture of wax, trinitrotoluene (TNT), and RDX, a non-melting crystalline powder some 40 times as powerful as TNT alone. Composition B was formed as a heated molten mixture and poured into shaped molds. Baratol was made of a mixture of barium nitrate, aluminum powder, TNT, stearoxyacetic acid, and nitrocellulose. Baratol was made as a slurry which was poured into molds and allowed to solidify.

Fat Man's design was difficult to implement because the high explosive charges would have to detonate at exactly the same time and propagate at exactly the same speed in order to form the small, dense ball of nuclear material. The placement of the detonators had to be very precise, and the composition of the explosive identical throughout and without bubbles or cracks. Fat Man was the bomb dropped on Nagasaki with an estimated yield of 22 kilotons.

Plutonium metal is also a very unusual material; it has a number of metallic phases, some of which can contract upon heating and become either brittle or plastic. It is self-heating from radioactive decay. Machining plutonium into an exact size was very difficult even after learning that a small amount of gallium alloyed with it makes it behave somewhat better. Plutonium also spontaneously generates heat, so it has to be cooled in construction, storage, and flight. The high explosive charges on the implosion bomb look like a soccer ball, with 20 hexagonal lenses and 12 pentagonal lenses. As an aside, the crime for which the infamous Julius and Ethel Rosenberg were convicted and executed in 1953 was the details and placement of the detonators on the high explosive blocks, the so-called "lenses."

Plutonium production facilities were established at a site on the Colorado River in Washington; the site called Hanford Reservation. A number of reactors were built, and plutonium was being produced at a rate capable of fueling some eighteen 5-kilogram bombs by 1945, once the problem of xenon fission product poisoning had been corrected for by enlarging the reactors by about 30% (xenon has a large cross section for neutrons and will dampen a nuclear reaction). Chemical separation of the plutonium from the uranium and mixed fission products was conducted at Hanford, and the plutonium shipped to Los Alamos.

There was so much confidence in the gun tube design that the decision was made not to test it. The implosion design, on the other hand, was much more complex, and it was decided that a test was needed. The test, code named "Trinity" by project leader Robert Oppenheimer, was conducted at Los Alamos on July 16, 1945 at 15 seconds before 5:30 AM. It yielded about 18.6 kilotons of explosive power and was the first nuclear detonation in history.

There is another fissionable isotope of uranium, uranium-233, which was investigated as bomb material. It is even rarer than uranium-235, but can by formed from thorium, a component of monazite sands abundant in Brazil, India, and the U.S. in North and South Carolina. Uranium-233 could be separated, much like plutonium, from the reactor material easily by chemical separation methods. The U.S. tested a few uranium-233 bombs, but the presence of uranium-232 in the uranium-233 was a problem; the uranium-232 is a copious alpha emitter and tended to "poison" the uranium-233 bomb by knocking stray neutrons from impurities in the bomb material, leading to possible pre-detonation. Separation of the uranium-232 from the uranium-233 proved to be very difficult and not practical. The uranium-233 bomb was never deployed since plutonium-239 was becoming plentiful.

Although Fat Man had contained a sphere of solid plutonium metal, research at the very end of World War Two had indicated a better design using a thin outer sphere of uranium with an internal void of air and a small central core of plutonium. This design is called the composite, or levitated, core. The process involved has been described as hitting a nail (the central core) with a hammer (the outer uranium) thrown through an arc rather than just pushing the nail in with head of the hammer (like the high explosive lens in Fat Man). It was known that the implosion design required exacting spherical shapes and forces; how to hold the central sphere in place while being surrounded by an outer sphere of uranium was difficult to design. The wires holding the central sphere had to be light and very strong to resist vibration during transport. But, with the levitated design, much smaller and lighter weapons were possible. Where Fat Man had used 6.2 kilograms of plutonium-239, as little as 3.2 kilograms (7 pounds) of plutonium could now be used (together with some 6.5 kilograms, or 14 pounds, of uranium-235).

SECOND GENERATION NUCLEAR WEAPONS

The U.S., possessing the only functional weapons designs in the late 1940s, continued upgrading and testing weapons, first in New Mexico and Nevada, but later in the South Pacific at Bikini Island and Eniwetok Atoll. Some of the fission bombs tested there were code named Sandstone, Greenhouse, Yoke, X-Ray, Zebra, Dog, Easy, and George, ranging from 18 kT to 81 kT in yield. It was soon proven that there were limits to the yield of a pure fission weapon, so emphasis was placed on development of the thermonuclear bomb for strategic uses. Fission weapons, however, were ideal for tactical devices. Development of nuclear weapons in the U.S. divided into the larger strategic fusion bombs and the smaller tactical fission bombs.

The "atomic cannon," designed to fire an atomic artillery round, experienced only a short life made up mostly of testing in Nevada. The atomic cannon was designed by U.S. Army ordnance engineers to throw a nuclear artillery shell 32 kilometers (20 miles), although in testing, the range was only seven miles. Nicknamed "Atomic Tom," it was test fired only once in Nevada, firing a shell with a projectile yield of 15 kilotons in 1953. The M65 cannon itself was 12.8 meters (42 feet) long and weighed 19,275 kilograms (42,500 pounds); it had a bore of 280 millimeters (11 inches). 20 were made by the Watervliet Arsenal in New York. A few were deployed to Germany, but were never used to fire a nuclear round. An apocryphal story is told of soldiers going into storage igloos and hugging the nuclear artillery shells in the cold German winter because the plutonium in the shells made them warm. Several atomic cannons now rest quietly at the Aberdeen Proving Ground Ordnance Museum, at Yuma Proving Ground, at Fort Sill, in Junction City, Kansas, and a few other locations.

Serious consideration was given to a Fat Man design to yield about 15 kT but with a hardened bombshell so it could be dropped onto and penetrate underground bunkers and submarine pens. It is possible that some of these were produced. By

1952, development of a fusion device, not a weapon, was ready for testing. Just in case it didn't work, a much larger fission bomb, using all U-235 with an expected yield of up to 600 kT, was also being developed. It was code named IVY KING. Even after the fusion bomb had been tested and worked, and done so magnificently, IVY KING was detonated shortly after the fusion bomb, and yielded the expected 500 kT. It was the largest pure fission bomb ever tested. Developments in fission bombs have continued in making them lighter and smaller for use both to detonate fusion weapons as well as in tactical weapons.

THE HYDROGEN BOMB

Work on the hydrogen bomb began in 1942, even before there was an atomic bomb. Because an atomic bomb destroys the fissile material when it detonates, a theoretical upper limit to the yield of a fission bomb is about one megaton; more explosive power would require a different design. Edward Teller realized that deuterium nuclei could be fused to form heavier elements while releasing binding energy. But the deuterium had to be very energetic to overcome the electric repulsion of two positively-charged objects. The temperature would have to be more than 400,000,000 °C, but such temperatures are produced during a nuclear explosion. Teller first proposed two possible reactions. In one, two deuterium nuclei collide and fuse to form helium-3 with the ejection of a neutron and some 3.2 MeV of energy. The second results in the formation of tritium, hydrogen-3, with ejection of a proton and about 4.0 MeV of energy. Another important reaction is deuterium fusing with tritium yielding a neutron and an alpha particle plus 17.6 MeV of energy; later weapons used this system. Deuterium is much easier and cheaper to separate from heavy water than U-235 is from uranium, but tritium was difficult to obtain until later reactors were built to produce tritium. A one megaton bomb could be made using only about 25 pounds of liquid deuterium. Later designs used much less mass of lithium deuteride mixed with lithium tritide in a "dry bomb" which did not need cryogenic cooling. However, a nuclear fission bomb is still needed to initiate the thermonuclear reactions.

The primary reactions for a thermonuclear bomb are:

$$\mathrm{D} + \mathrm{D} \ddagger {}^{3}\mathrm{He} + \mathrm{n} + 3.27\ \mathrm{MeV}$$
$$\mathrm{D} + \mathrm{D} \ddagger \mathrm{T} + \mathrm{p} + 4.03\ \mathrm{MeV}$$
$$\mathrm{D} + \mathrm{T} \ddagger {}^{4}\mathrm{He} + \mathrm{n} + 17.59\ \mathrm{MeV}$$
$$\mathrm{T} + \mathrm{T} \ddagger {}^{4}\mathrm{He} + 2\mathrm{n} + 11.27\ \mathrm{MeV}$$
$${}^{6}\mathrm{Li} + \mathrm{n} \ddagger {}^{4}\mathrm{He} + \mathrm{T} + 4.78\ \mathrm{MeV}$$
$${}^{3}\mathrm{He} + \mathrm{D} \ddagger {}^{4}\mathrm{He} + \mathrm{p} + 18.35\ \mathrm{MeV}$$

where D is deuterium, T is tritium, n is a neutron, p is a proton, He is helium, and Li is lithium. Recall that deuterium is hydrogen-2 and tritium is hydrogen-3.

The energy of each reaction is given in million electron volts (MeV). Each of these reactions contributes to the force of a hydrogen bomb explosion. Yields from the tritium reactions are very high, so tritium became an important component of thermonuclear weapons. Gamma and x-radiation are emitted on a large scale; temperatures of millions of degrees are reached; and atoms of every element are formed, at least briefly.

Radiation implosion is used to compress the hydrogen-containing fuel. The design required a cylindrical casing to hold an atomic bomb and hydrogen fuel at opposite ends. The flash of the exploding bomb would hit the case, causing it to glow and flood the interior of the casing with radiation pressure sufficient to compress and ignite the hydrogen fuel. A representation of a thermonuclear weapon is shown in Figure 4.1.

The first test of a thermonuclear device was code named IVY MIKE and used an inner fission bomb, followed by a fusion secondary containing liquid deuterium and a little tritium, and further enclosed by uranium-238. When it was detonated in 1952, the yield was about 10.4 MT with some 80% contributed by the fission bombs. This was a device, not a weapon, since it was very large, required sophisticated cryonics to keep deuterium in the liquid state, and weighed some 82 tons. But, it proved the theory, and future developments in strategic weapons have continued using fusion devices.

Test CASTLE NECTAR was conducted on May 13, 1954 at 6:20 AM at Bikini Atoll from a barge; it yielded 1.69 Mt. The so-called ZOMBIE device tested in Operation CASTLE NECTAR was a prototype of the later TX-15 lightweight thermonuclear bomb, a deployable weapon. The bomb was a transitional design between fission bombs and the fusion bomb. The ZOMBIE was originally envisioned as a radiation-imploded fission bomb with a yield in the range of hundreds of kilotons, similar to Stanislaw Ulam's original concept of using one fission bomb to compress another. The idea of making the outer case of the secondary out of fissile material of

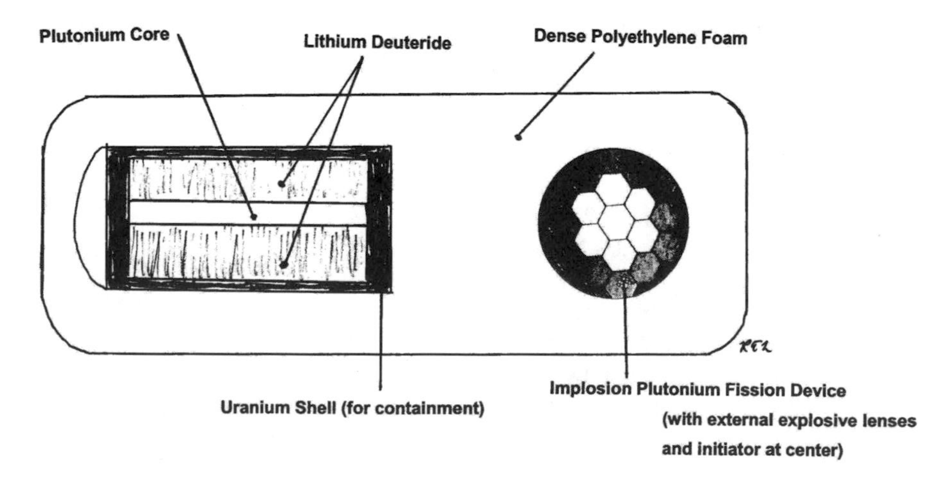

Figure 4.1 Representation of a Simple Thermonuclear Bomb Design

enriched uranium rather than natural uranium was retained, but the design evolved to include fusion fuel of lithium deuteride and gaseous tritium to boost its yield. The final result was basically a radiation-imploded fusion boosted fission bomb. The difference between this approach and the other two-stage thermonuclear systems tested in Operation CASTLE was that the requirements for compression were relaxed. The other thermonuclear systems needed to achieve desired compression of the low density fusion fuel to achieve extremely high densities required for a rapid equilibrium burn when the fission igniter detonated. Although most of total energy release would be due to fission reactions, it was fast fission of ordinary uranium driven by the fusion produced neutrons. The ZOMBIE, by using a secondary case made of highly enriched uranium, would become supercritical through a much less demanding compression process. The rapid release of energy through the fission chain reaction would drive the whole system to thermonuclear temperatures. Due to the intense heating, the fusion fuel inside would burn without nearly as much compression. The ZOMBIE was not as inexpensive as the other thermonuclear systems tested in Operation CASTLE due its use of enriched uranium, enriched lithium, and tritium; but it could be made smaller and lighter than any of them. It was also a relatively low risk design, the fission explosion of the secondary being a very robust process. The yield of ZOMBIE was the closest to original predictions of any of the other CASTLE devices, differing only by a negligible 6%. NECTAR obtained 80% of its energy release from fission (1.35 Mt), the highest percentage of any of the CASTLE devices. The ZOMBIE device was 34.5 inches in diameter, 110 inches long, and weighed 6,520 pounds. It used a fission primary, code named COBRA. The mushroom cloud top reached 71,000 feet.

While the hydrogen-containing material adds much to the total energy at detonation and allows for almost unlimited yields, it must be remembered that a fission bomb is needed to initiate the thermonuclear reaction. A hydrogen bomb is an atomic bomb with an added explosive of low molecular weight materials. The yields from nuclear and thermonuclear reactions can be summarized as:

Fission of U-233: 17.8 kT/kg
Fission of U-235: 17.6 kT/kg
Fission of Pu-239: 17.3 kT/kg
Fusion of pure deuterium: 82.2 kT/kg
Fusion of tritium and deuterium (50/50): 80.4 kT/kg
Fusion of lithium-6 deuteride: 64.0 kT/kg
Fusion of lithium-7 deuteride
Total conversion of matter to energy ($E = mc^2$): 21.47 Mt/kg

MODERN NUCLEAR WEAPONS

There are a number of ways in which the requisite critical mass of fissile material can be brought together. Today, however, almost all devices use the levitated implosion design. The components of the material must be kept separated until explosion is desired. In order to focus the maximum number of neutrons inward toward the fissile material, generally there is an outer sphere of neutron-damping material, often boron.

The concept of critical mass was mentioned earlier. It is necessary to discuss this in a little more detail to understand how nuclear weapons are designed and function. When the idea of an atomic bomb was first proposed, calculations showed that such a weapon would weigh a tremendous amount due to the weight of fissile material plus shielding plus explosive shell. Most scientists in the 1930s used natural uranium in their calculations, resulting in required amounts of uranium on the order of tons to sustain a chain reaction. It was clearly impractical to make a bomb using natural uranium.

However, two facts make nuclear bombs possible. First, one uranium isotope, uranium-235, has a large cross section for neutron absorption, so only a few pounds are sufficient to maintain a chain reaction. Second, the element plutonium possesses an even larger cross section than uranium-235 so even less fissile material was needed. Specifically, the isotope of plutonium most commonly used is plutonium-239. Of course, chemical purity of the fissile material was important in how much would be needed. Since separation of plutonium is from uranium, much of the development of nuclear weapons relates as much to chemistry as to physics or engineering. The purer the uranium-235 or plutonium-239, the less that is needed. The critical mass of ultra-pure plutonium is around 3 kilograms (7 pounds) or so. An example of the size needed for a modern weapon was the atomic artillery shell, weighing some 65 pounds and having a diameter of a little over 6 inches.

The term cross section, usually given in units of square centimeters or "*barns*," refers to the probability of a neutron impacting a fissile nucleus. To understand this concept, imagine you throw a dart at a one-square foot balloon ten times. If the dart bounces off the balloon nine times and bursts it once, the cross section for bursting is 0.1 square foot (1 out of 10 times over 1 square foot), and the cross section for reflection is 0.9 square foot (9 out of 10 times over 1 square foot). Most nuclear processes involve an extremely small target area, the size of the atomic nucleus, on the order of 0.000000000000000000000001 (1.0×10^{-24}) square centimeter. This unit is called a barn because of a joke by the early atomic scientists in that they couldn't hit the side of a barn. The larger the cross section for absorption, the more likely the neutron is to hit a target nucleus. Thus, plutonium is a better fissile material than uranium, and is the material of choice in modern nuclear weapons, both atomic and hydrogen.

However, for nuclear power reactors and for rudimentary weapons in third world nations, like Iran and Iraq, it is possible that uranium separation is an important component of their program. Following the Persian Gulf War, it was found that Iraq had built and was using calutrons for uranium isotope separation. In the following years, it has become obvious that Iran, and possibly Iraq, have sought gas centrifugation equipment; it is known that Iraq attempted to purchase centrifuges from Germany, but was prevented from doing so in the late 1990s. Gas centrifuges were built and used in the Manhattan Project, but proved to be impractical compared to the gaseous diffusion method and plutonium generation. In the 1980s, the U.S. Department of Energy (DOE) developed a gas centrifuge program, including construction of a test cascade. More than 1,300 gas centrifuges were installed and some seven hundred operated using uranium hexafluoride at the Gas Centrifuge Enrichment Plant in Piketon, Ohio. About a hundred machines operated for nine months. However, the idea was abandoned in 1986 in favor of the Advanced Vapor Laser Isotope Separation (AVLIS) process. Research on the AVLIS process was terminated in 2000.

The U.S. began to reconsider gas centrifuge technology as a practical advanced enrichment technology for replacing gaseous diffusion, which is more expensive and requires more energy. By this time, the gas centrifuge process had already been commercially developed in Russia and in the United Kingdom, Germany, and The Netherlands. In the early 1990s, a private U.S. company, Urenco, teamed up with several U.S. utilities to form the LES partnership. In January 1991, the NRC received an application from LES to construct and operate the nation's first privately-owned gas centrifuge enrichment facility. The 1.5 million Separative Work Unit (SWU) plant was to be built in Homer, Louisiana, but because of an extended licensing hearing process, LES decided to withdraw its application in 1998. However, within the next few years, there was renewed interest.

The gas centrifuge uranium enrichment process uses a large number of rotating cylinders in series to enrich uranium in its U-235 isotope. These series of centrifuge machines, called trains, are interconnected to form cascades. Currently, both Louisiana Energy Services (LES) and U.S. Enrichment Corporation (USEC) have plans to submit license applications to construct and operate gas centrifuge uranium enrichment facilities in the U.S. Purchases of centrifuge components by Iran and Iraq indicate that they are considering this technology; if for weapons or not is not known. Details of gas centrifuge isotope separation are either classified or very closely held from the public.

The Louisiana Energy Services (LES) partnership is made up of partners currently consisting of Urenco, Exelon, Duke Power, Louisiana Power and Light, and Fluor Daniel Engineering. On October 25, 2001, Exelon and Duke Energy sent a letter to President George W. Bush indicating that a U.S. consortium was planning to send in an application to the U.S. Nuclear Regulatory Commission (NRC) in 2002 for a license to build a new gas centrifuge uranium enrichment plant. On July 22, 2002, LES announced that it would negotiate with Westinghouse Corporation and Cameco to also become partners. The partnership intended to use Urenco's sixth-generation gas centrifuge technology that was being used in Europe. Currently, Urenco has a capacity of about 15% of the world's enrichment market. Full-scale capacity of 3 million SWUs per year is projected to be in 2010 or 2011, depending on market demand. On September 9, 2002, LES announced its final site selection to be the former TVA reactor site in Hartsville, TN.

On February 12, 2003, U.S. Enrichment Corporation (USEC) submitted a license application to the NRC for a gas centrifuge uranium enrichment test facility or "lead cascade" in late 2002. The lead cascade is based on U.S. Department of Energy (DOE) advanced gas centrifuge technology. The USEC objective is to replicate the existing technology and reduce costs using advances in carbon fiber and other material and manufacturing technologies. The lead cascade site was to be located in Piketon, OH. The lead cascade, consisting of 240 centrifuges, would recycle the enriched and depleted uranium it produces. The only uranium withdrawals from the cascade will be in the form of samples. In a later commercial deployment phase, the commercial plant would have a capacity of 3.5 million SWUs per year, with up to 10% enrichment. USEC plans to submit a license application to operate the lead cascade and the commercial-scale facilities in 2004. These facilities, if built, will be carefully monitored and controlled to prevent the improper use of the enriched uranium.

DELIVERY SYSTEMS

Since the era of the Manhattan Project, delivery of the nuclear weapon has been one of the difficulties in causing destruction or terror. Many of the first devices were very large, like the 164,000-pound Mike thermonuclear device. Even though modern weapons are much smaller, they are not miniature. While terrorists might be willing to settle for a small blast of a few kilotons or of even less yield for primarily psychological impact, modern warfare weapons require sophisticated delivery systems.

The 1950's atomic cannon was replaced by nuclear weapons capable of being fired from conventional artillery pieces, typically in the U.S. by the now-obsolete 8-inch cannon and howitzer or the 155-millimeter cannon and howitzer. The use of nuclear weapons on the battlefield, however, using artillery has been determined not be practical. First, the range is such that the detonation will take place within some 20 miles of the firing. Second, the ground contaminated by the device is likely an area desired for occupation by the successful army. Third, the resulting fallout and residual radioactivity presents problems with refugees and return of any occupants. Fourth, transport and security of the weapon on a modern high-speed battlefield is very problematic. For these reasons, the use of artillery-fired nuclear weapons is remote.

During the Cold War between the Western Allies, led by the U.S., and the Warsaw Pact, led by the U.S.S.R., the primary delivery system was the long-range manned bomber. Because of the weight and size of high-yield weapons, only very large aircraft could transport and deliver these weapons. In the U.S., the B-36 bomber was designed as an atomic bomb delivery system. It was the largest aircraft produced in its day. The B-36 was followed by the B-47, and, later, by the B-52. While the modern B-1 bomber was first envisioned as a nuclear weapon delivery system, its mission was changed to conventional roles as a result of weapons conventions and improvements in ballistic missile accuracy. The U.S.S.R., on its part, has similar bomb delivery systems, most notably the Bear and Backflash bombers (these are NATO code names, they begin with the letter "b" for "bomber"). It was recognized by both sides in that tense period that manned aircraft required a fairly long time to reach their target with a definite likelihood of being brought down by ground-to-air missiles or fighter aircraft. An advantage to these systems is the fact that they could be recalled up to moments before releasing their weapons. However, the costs associated with upkeep of these manned bombers made both sides decide to reach agreements of the numbers the other could have. Manned bombers gave way to ballistic missiles.

The early nuclear weapons were far too heavy to be launched using missiles. But, as weapons design brought about smaller and smaller devices, missiles began to become the delivery system of choice. The possibility of placing nuclear warheads on missiles being carried by submarines was very pleasing to military planners. The submarine platform could move in all the oceans of the world, remain largely undetected, and stay on post for long periods of time. The first missiles carried only a single warhead, but later designs allowed for up to 10 separate nuclear weapons to be carried on one missile. Since it would not be beneficial to drop all 10 on a single target, systems were devised to allow for each to hit a different target (although the targets cannot be extremely separated). In order to protect ground-based missiles from destruction by an enemy, underground silos were built primarily by the U.S. (since the U.S.S.R. seemed to prefer mobile launchers). Major cities and certain borders were protected by batteries of these missiles. Again, the costs associated with construction and upkeep of these systems led to agreements by both sides to limit their numbers and locations. The largest disadvantage to the use of ballistic missiles is that they cannot be recalled once launched since they follow the laws of physics on a ballistic trajectory. Although the nuclear weapon can be deactivated under certain situations and at certain times during flight, the impact would detonate the high explosive with subsequent radioactive contamination of an area around the site.

The current U.S. inventory includes a few silo-launched and a number of submarine-launched ballistic missiles, but chiefly contains warheads for cruise missiles. Cruise missiles are similar to drone aircraft in that they fly from their launch point to the target. Using the U.S. satellite Ground Positioning System (GPS), the cruise missile can follow an indirect route to the target while either flying at altitude or close to the ground. The U.S. inventory includes submarine-launched, surface ship-launched, ground-based mobile platforms, and aircraft-launched cruise missiles. These missiles can be sent to simply crash in an isolated location or retargeted, as the situation changes. For the foreseeable future, the U.S. will rely on the cruise missile as the system of choice for nuclear weapon delivery.

Russia continues largely to rely upon ballistic missiles as their nuclear delivery system, although they have designed and placed into service a limited number of cruise missiles both for ground launch and from ships, including submarines.

WEAPON DETONATOR

In order to initiate the nuclear chain reaction, it is necessary to bring together the critical mass of fissile material very quickly and uniformly. This is done by having a conventional high explosive force the fissile material into a very small volume. A number of explosive mixtures have been used in nuclear weapons. The high explosive must be detonated by a primer. This assembly is often called the detonator. However, just pushing fissile material together will not result in a nuclear explosion. Two factors are involved: containment of the material long enough to have sufficient fissions take place, and a burst of neutrons at just the right time to start the chain reaction. The first is accomplished by bomb design and enclosing materials. The latter is accomplished by use of an initiator.

FISSION INITIATOR

The design of any of the nuclear weapon initiators has never been declassified. A little is known about the urchin device (possibly because it looked like a sea urchin with its wedge-like grooves and spines), but this is an old design and not relevant today. While the details are unknown, it seems possible that the shape of the initiator might be dimpled like a golf ball or spiked like the ball on the end of an medieval ball and chain.

It is known that an initiator must quickly and uniformly mix beryllium and polonium-210, which is the source of the initial alpha particles, while keeping them apart until just before detonation. There were a number of designs from the early urchin (or screwball) design to more modern, but likely also conical shaped charges, designs.

NUCLEAR POWER REACTOR

While a nuclear weapon is an uncontrolled release of energy suddenly, nuclear energy can also be used in a controlled and slow manner. When Enrico Fermi started the first chain reaction under the stands at Stagg Field, there was a small amount of energy produced, maybe one-half watt, but the potential for using nuclear energy to produce energy is great. Today, even a small nuclear power plant can produce 5,000,000 watts; a large one up to 50,000,000 watts.

A nuclear reactor generates heat from the fissions of billions of atoms, this heat can be used to produce steam, which can spin turbines to generate electrical power. Worldwide, some 15% of all electricity is produced by nuclear reactors. In the U.S., about 20% of all electrical power comes from nuclear sources; in France, the number is closer to 90%.

The design of the reactor is such that a detonation is not possible. As was seen at the accident at Chernobyl, The Ukraine, some designs are less stable than others. The reactor there, used to produce electricity, used graphite as a moderator similar to the early Stagg Field and Oak Ridge reactors. Its design was such that it became more unstable as the power level is reduced (the Wagner void concept). As workers were in the process of restarting it, they made a decision to drop the power; this caused the core to become unstable, producing more heat, and eventually leading to an explosion and fire when the graphite was exposed to air. Fission products, uranium, and radioactive graphite were spread over much of the northern hemisphere as high-altitude winds carried the fallout.

Better designs, usually of the boiling water type or the heated water type, are safer. A nuclear explosion is not possible with any kind of reactor, but accidents are possible. At Three Mile Island in Pennsylvania, the failure of an indicator device led operators to think that cooling water was flowing when it was not. The resulting high heat severely damaged the reactor core, and a small amount of radioactive water and gases were released from the plant. While the radiological consequences were not great, the fear and panic caused among the population in the area of Harrisburg, PA were. Because of a fear of radioactive and radiation, possibly since the first the public knew of them was in a bomb detonation, panic is possible whenever radiation is concerned. This would be the greatest effect of a terrorist "dirty bomb." The actual damage would, most likely, be quite small and definitely localized, but the psychological terror could be a very effective weapon.

NUCLEAR REACTOR WASTE

The reactor grade uranium that goes into a commercial power reactor is not very radioactive. However, as fission processes occur, a large number of highly radioactive materials are produced. In addition to producing plutonium, fission fragments and neutron activated materials are produced. As they are produced, they tend to "poison" the reactor by interfering with neutron processes, and the power output of the reactor drops with time. From time to time, the fuel must be replaced. The mixed materials—fission products, fission fragments, and actinides—are highly radioactive; these are the reactor waste.

"DIRTY" BOMB

Some terrorist organizations have considered the use of a "dirty bomb." Detailed plans were discovered in Afghanistan in 2002 for such a device along with possible locations of employment. Technically, this system is called a radiological dispersal device, but the public's attention has been drawn to the term "dirty bomb." This bomb would not be a nuclear or thermonuclear device, but would contain radioactive materials and conventional explosives. There are large amounts of highly radioactive material in commercial and medical applications. Some medical radiotherapy sources contain cobalt or cesium; nuclear power reactors produce a variety of radioactive materials and store them on-site; and many occupations use radioactive sources for x-raying metals, verifying soil moisture, conducting subsurface exploration, etc. The most likely materials for use in a "dirty bomb" would be cesium-137, cobalt-60, iridium-192, strontium-90, and americium-241, due to their availability in commercial applications. Of these, the most likely would be cesium-137 in cesium chloride (CsCl) because of its easy availability and the fact that it, unlike some of the other materials, is a light crystalline material (it looks like blue table salt) that can be made into very tiny particles which can remain in an aerosol for extended periods. These particles could be carried by wind over large distances. A terrorist could steal or otherwise obtain this material, and, by simply exploding it, spread contamination over an area.

The mixed fission products from a commercial nuclear reactor in a friendly nation could be supplied to terrorists; these are intensely radioactive and consist of internal and external radiation hazards (alpha, beta, and gamma radiation as well as x-rays). According to the Ukraine's Academy of Sciences, the trafficking in radioactive metals from Chernobyl's central reactor, which was closed in 2000 following the explosion in 1986, is a serious problem. Even small-scale university research reactor waste could be used as a psychological dirty bomb, striking fear into a population from an existing fear of radiation.

A dirty bomb would cause only local effects and contamination, but the fear component could be great. The greatest medical effect would be the possible increase in cancer rates some ten to thirty years after exposure to the radioactive materials rather than radiation sickness due to the small amount of radioactivity actually present.

Another major consideration would be the clean-up and decontamination of the area where the bomb was exploded. If exploded in a major city, some buildings might have to be demolished or the area barricaded with entry forbidden for many years. Imagine if this were to happen in downtown New York City near Wall Street, a symbol of likely terrorist attack. The stock exchange, banks, offices, and other offices might have to be razed to the ground and all the material buried as low-level radioactive waste. It has been estimated that a single CsCl source commonly used in industrial plants as tank gauges and flow indicators could be used to contaminate an area over 200 meters (65 feet) in diameter with enough radiation to double the normal background level. With a typical wind, detectable contamination could be spread as far as several kilometers in the direction of the wind.

In 2002, the U.N.'s International Atomic Energy Agency (IAEA) discovered a high-level radiation source in the former Soviet Union. It caused two deaths of woodsmen who had huddled near it to keep warm during a snowfall. Upon investigation, it was discovered that the Soviet Union had built thousands of these sources, using very pure strontium-90 (Sr-90), for the purpose of killing weed seeds for agriculture uses. The grain irradiators were never carefully tracked; it is not known how many persist and where they might be. The Sr-90 in a single device could cause 70 times the normal background radiation over an area of some 200 meters (65 feet) in diameter.

Another possible target for the maximum spread of radioactivity would be the explosion of a "dirty bomb" in a large city's subway system or underground automobile tunnel. The passage of vehicles and the concentration of air currents would allow for wide-spread contamination before anyone might detect it. New York City and a few other cities have installed sensitive radiation detectors at vehicular tunnels;

in fact, it was reported that a cancer patient who had received injections of radioactive material actually set off one of the sensors in the Holland Tunnel. Washington, DC has installed similar detectors in its MetroRail subway system.

The use of an actual terrorist nuclear weapon, while very remote, is not an impossibility. It is very unlikely that any state would knowingly transfer a nuclear weapon to a terrorist group because of the potential for retaliation as well as universal condemnation. Yet, if a state thought it could remain secret while transferring the device, there are several nations which might consider selling a weapon.

DULL SWORD

In the U.S., the code for minor damage to a nuclear weapon is "Dull Sword." In these cases, there is no danger of detonation or release of radiation. Events as minor as dropping a wrench onto the housing of a nuclear weapon while it is being installed in an aircraft or missile qualify as Dull Sword incidents and all details must be covered in a detailed investigation.

BROKEN ARROW

In the U.S., the code name for a serious nuclear weapons incident is "Broken Arrow." It is U.S. policy neither to confirm nor deny the existence of nuclear weapons at any specific location; this also applies to Broken Arrow incidents. Only the people who need to know will be informed that a nuclear weapon has been involved in an accident.

5

EFFECTS OF NUCLEAR WEAPONS

INTRODUCTION TO NUCLEAR WEAPONS' EFFECTS

The effects resulting from the detonation of a nuclear weapon are many, ranging from wounds and blast overpressures to future cancers. A nuclear detonation creates a severe environment including blast, thermal pulse, neutrons, x-rays and gamma-rays, particulate radiation, electromagnetic pulse (EMP), and ionization of the upper atmosphere. Depending upon the environment in which the nuclear device is detonated, blast effects are manifested as ground shock, water shock, cratering, and large amounts of dust and radioactive fallout.

The explosion of a nuclear weapon produces tremendous heat, strong blast, and radiation. While conventional explosives produce blast and heat, the amounts are minute compared to even the smallest nuclear weapon. The primary effects on people and structures include blast, heat, and radiation. The energy of a nuclear explosion is transferred to the surrounding medium in three distinct forms: blast; thermal radiation; and nuclear radiation. The relative distribution of energy among these three forms will depend on the yield of the weapon, the location of the burst, and the characteristics of the environment. For a low altitude atmospheric detonation of a moderate sized weapon of 20 kilotons, the energy is distributed roughly as follows:

- 50% as blast (tremendous winds and overpressure)
- 35% as thermal radiation; made up of a wide range of the electromagnetic spectrum, including infrared (IR), visible, and ultraviolet (UV) light as well as some soft x-rays emitted at the time of the explosion

- 15% as nuclear radiation; including 5% as initial ionizing radiation consisting chiefly of neutrons and gamma rays emitted within the first minute after detonation, and 10% as residual nuclear radiation. Residual nuclear radiation is the hazard in fallout.

Considerable variation from this distribution will occur with changes in yield or location of the detonation. Each of these effects will be discussed in detail.

BLAST AND THERMAL EFFECTS

The term "overpressure" refers to an increase in atmospheric pressure as a result of a detonation. When any explosive is detonated, there is a shock wave generated by rapidly expanding air heated by the chemical reaction. Blast overpressures are measured in units of pressure, usually pounds per square inch (psi) in the English system or pascals (Pa) in the metric (or S.I.) system. This shock wave, hitting a solid object, will either move it in the direction of the wave propagation or shatter it, depending upon the strength of the shock wave and the characteristics of the object. Essentially, no building can withstand 10 to 12 psi overpressure. Industrial facilities might survive 5 psi, but almost all houses will be destroyed. Even 2 psi will seriously damage frame and composite houses. Upon hitting a human being, the shock wave will push the person in the direction of propagation and compress the body as the wave passes, resulting in damage ranging from burst eardrums to destruction and liquefaction of the internal organs, depending upon the amount of overpressure.

Because of the tremendous amount of energy liberated in a nuclear detonation, temperatures of several tens of million degrees Celsius are developed in the immediate area of the detonation. This can be compared to the few thousand degrees of a conventional explosion. At these very high temperatures, the non-fissioned parts of the weapon are vaporized. The atoms do not release the energy as kinetic energy but release it in the form of large amounts of electromagnetic radiation. In an atmospheric detonation, this electromagnetic radiation, consisting chiefly of soft x-rays, is absorbed within a few meters of the point of detonation by the surrounding atmosphere, heating it to extremely high temperatures and forming an extremely hot sphere of air and gaseous weapon residues, the so-called *fireball*. Immediately upon formation, the fireball begins to grow rapidly and rise like a hot air balloon due to lower density. Within a millisecond after detonation, the diameter of the fireball from a one megaton (MT) air burst is 150 meters (m) or about 450 feet. This increases to a maximum of 2,200 m (over a mile) within 10 seconds, at which time the fireball is also rising at the rate of 100 meters per second (m/sec). The initial rapid expansion of the fireball severely compresses the surrounding atmosphere, producing a powerful blast wave.

As the fireball expands toward maximum diameter, it cools, and after about a minute, the temperature decreases so that it no longer emits significant amounts of thermal radiation. The combination of the upward movement and the cooling of the fireball gives rise to the formation of the characteristic mushroom-shaped cloud. As the fireball cools, the vaporized materials in it condense to form a cloud of solid and liquid particles. Following an air burst, condensed droplets of water give it a typical white cloud-like appearance. In the case of a surface burst, this cloud will also contain large quantities of dirt and other debris which are vaporized when the fireball touches the earth's surface or are sucked up by the strong updrafts afterwards, giving the cloud a dirty brown appearance. The dirt and other debris become contaminated with the radioisotopes generated by the explosion or activated by neutron radiation and fall to earth as fallout. It is estimated that the radioactivity from detonation of a fission-yield weapon results from some 300 different radionuclides representing some 36 different elements. Figure 5.1 illustrates the stages of a typical nuclear air burst.

Figure 5.1a Detonation of the Bomb

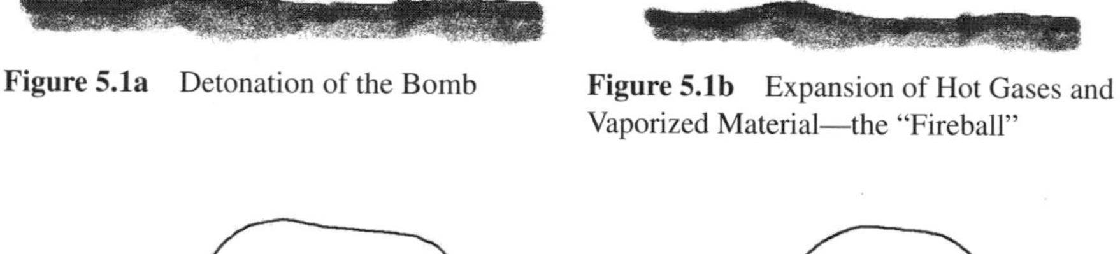

Figure 5.1b Expansion of Hot Gases and Vaporized Material—the "Fireball"

Figure 5.1c Fireball Reaches the Ground

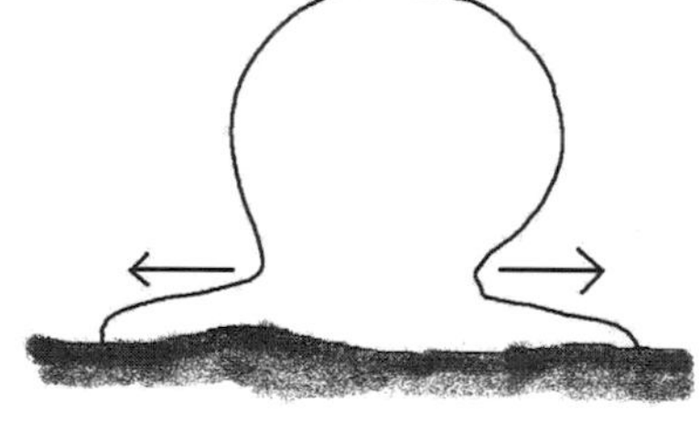

Figure 5.1d Rush of Hot Gases Outward

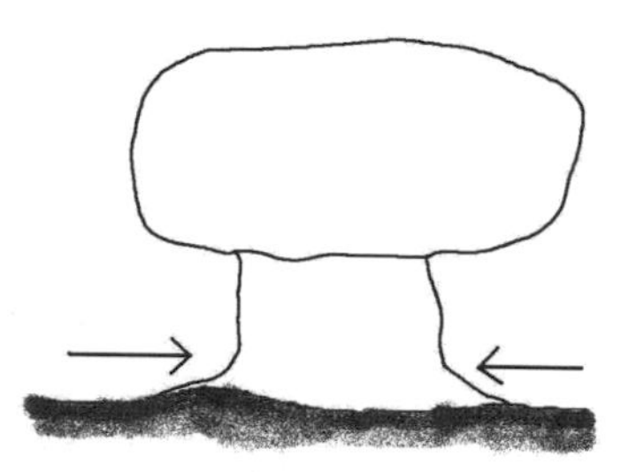

Figure 5.1e As Gases Cool, Rush of Air Inward

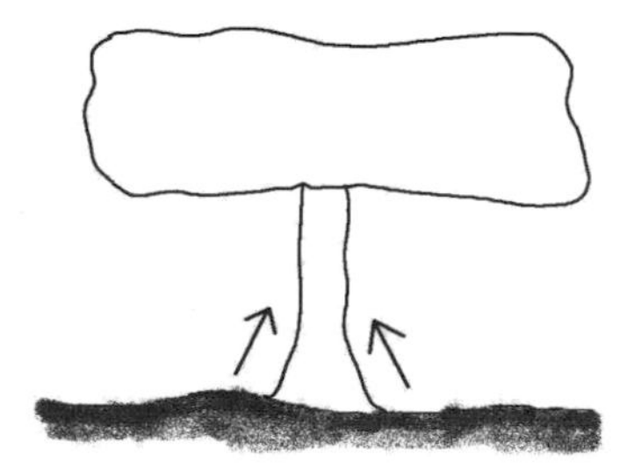

Figure 5.1f Updraft of Vaporized and Fragmented Materials to Form "Mushroom"

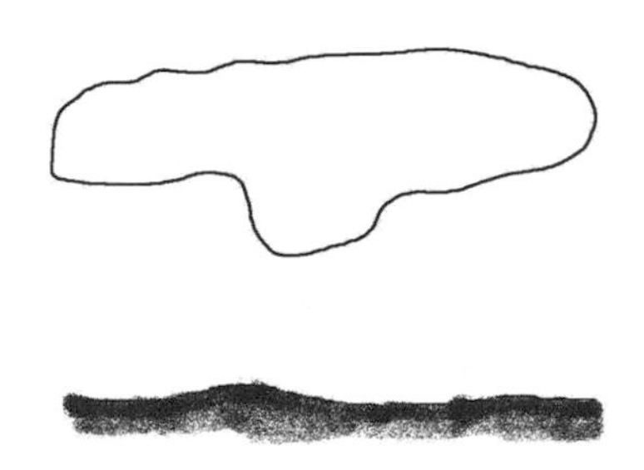

Figure 5.1g Cloud Rises with Thermal Gradients and Adiabatic Lapse rate

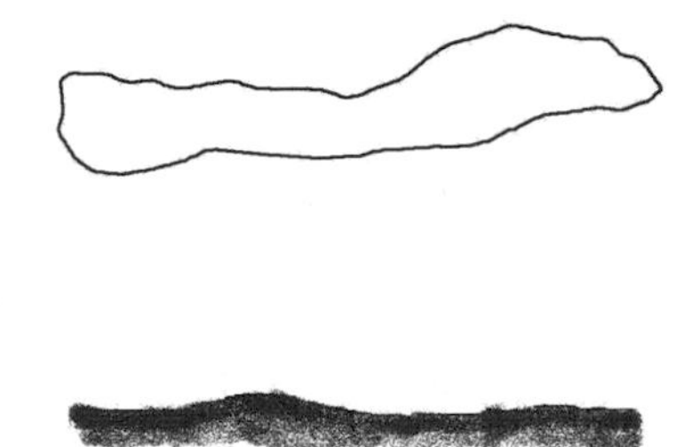

Figure 5.1h Cloud Disperses with Wind, Dropping Larger Particles

The temperature of a conventional explosive at the moment of detonation is about 5,000° C, that of a nuclear weapon thousands of times hotter. Some 35% of the energy of a nuclear weapon is given off as heat, reaching temperatures at the moment of detonation of 100 million degrees Celsius (some 180,000,000° F). This temperature is as hot as the sun, and is so hot that about 1/3 of the energy is converted into x-rays that then heat the surrounding air and form a fireball. The surface temperature of the fireball is about 8,000° C (14,400° F). This tremendous heat incinerates everything within a large distance. At the point of detonation, structures and people are vaporized. Farther away, flammable materials are ignited, humans included.

FALLOUT

Depending upon the height at which the bomb is detonated, there is a possibility that earth and other materials (such as remains of structures, vehicles, and organisms) will be taken up into the fireball and resulting mushroom cloud. This material will be subjected to intense radiation and become radioactive itself. Fallout consists of this material with induced radioactivity as well as unreacted bomb components such as uranium and plutonium as well as beryllium and fission products. Generally, the closer the bomb detonation is to the ground, the greater the fallout. High-altitude bursts produce little or no fallout. Surface bursts produce the most. Subsurface or sub-sea bursts will vaporize soil or water and take it up into the cloud, but may actually cause less fallout due to reduction in blast effects. The amount of fallout depends upon the type of bomb material, altitude, and weather conditions. Fallout in the form of dust can cause beta skin burns. Inhalation or ingestion of fallout can cause a variety of radiological effects, depending upon the type of radiation, the chemical and physical properties of the material, and the length of time exposed.

For nearby fallout, at distances from 0 to 30 kilometers, models are often based on solutions to diffusion equations involving vertical and horizontal spreading of a contaminant plume, including the commonly-used Gaussian plume dispersion model (GPDM) mentioned in Chapter One. Superior simulation methods, requiring considerably more computer power than the GPDM, utilize finite differences rather than Gaussian differential solutions, which assume average values for dispersion. These models are usually set up to handle "puff" releases (single events) and "continuous" releases. Models begin with a simplified picture of the local atmosphere, apply the *source term*, which is the quantity and nature of the original contaminant, and deduce the downwind concentrations of fallout. If the solution is carried out for many different locations and altitudes, a three-dimensional picture can be generated showing the shape and location of the fallout cloud. Models such as these are routinely used to determine the local transport of pollutants from point sources, such as a smokestack; line sources, such as a highway; and area sources, such as a large industrial area; and apply equally well to radiation in fallout. The accuracy of Gaussian plume dispersion models and other simplified solutions to the diffusion equation suffer from the inherently simplified picture of the atmosphere that they begin with.

For each kiloton of nuclear force produced, about 3.0×10^{23} fission product atoms are formed (this is from about 60 grams or 2 ounces of uranium or plutonium mass loss). At one minute after the bomb's detonation, about 1.0×10^{21} nuclear disintegrations per second, or 3.0×10^{10} curies, is the radioactivity per kiloton. There is a "rule of thumb" used to describe the decease in dose rate from fallout. The rule is that for every increase in time after the explosion by a factor of 7, the dose rate will decrease by a factor of 10. For example, after 7 hours the dose rate is $1/10^{th}$ of that one hour after the detonation. After 49 hours (7×7), the dose rate is $1/1000^{th}$, and at 343 hours ($7 \times 7 \times 7$), the dose rate is $1/1000^{th}$ of the dose rate at one hour. This estimation is fairly good up to about 2 weeks following the explosion. It may be seen that fallout radiation decreases fairly quickly.

RADIATION ISOPLETHS

The process of weather map analysis for fallout entails the organization of the plotted information into a logical and, usually, geographical portrayal of the data. Typically, a major part of the analysis phase involves drawing of isopleths, a generic term referring to lines of equal values. The word *isopleth* is derived from the Greek "iso" meaning "equal" and "plethes" meaning "quantity." In this case, they are lines connecting locations having radiation levels of the same magnitude.

At first glance, the array of data plotted on the display map may appear unorganized and overwhelming. However, large scale organized weather systems as well as radiation fallout can be discerned through map analysis. In meteorology, the term weather analysis usually refers to the sequence of operations leading to interpretation of a graphical portrayal of a weather map displaying the distribution of one or more weather elements to increase the visual communication value of the chart. For example, once isobars, or lines of equal barometric pressure, have been drawn upon a surface chart, it is possible to immediately identify regions of high and low atmospheric pressure. Likewise, given information about the radiation levels at different locations, a map can be drawn showing the radiation isopleths.

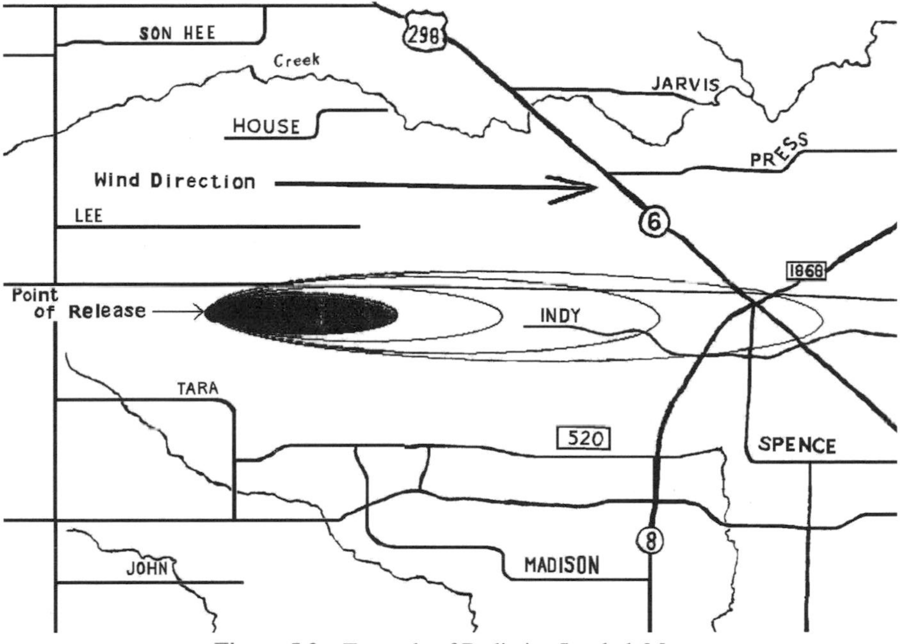

Figure 5.2 Example of Radiation Isopleth Map

Drawing isopleths is a relatively straightforward procedure. The objective is to connect all points reporting the same value of the particular weather element, whether temperature, humidity, pressure, or radiation level. However, if the locations reporting data do not contain the exact value of the desired isopleth, it may be necessary to interpolate between the locations. In other words, it will have to be assumed that a desired isopleth value will lie someplace between a point having a slightly smaller observed value and one with a slightly higher value. Because the field is continuous, this spatial interpolation procedure used to produce uninterrupted isopleths is possible. Additionally, plotted data must exist on both sides of the isopleth. When the isopleth is completed, the value of the element will be greater than the selected isopleth value on one side of an isopleth and less on the other side for the entire length of the isopleth.

A hypothetical map of radiation isopleths is shown as Figure 5.2.

RADIATION EFFECTS

Conventional weapons possess blast effects and thermal effects; a unique characteristic of nuclear weapons is the radiation component. The radiation from a nuclear detonation ranges from very low frequency electromagnetic waves up to and including gamma radiation as well as particulate radiation (alpha and beta).

Some of lower-energy electromagnetic radiation produces an electromagnetic pulse capable of disrupting radio communications, destroying computer chips and transistors, and erasing magnetic tapes and computer memory. A nuclear detonation in the atmosphere, even miles above the earth, could result in loss of all computer data over a wide area.

The detonation of a nuclear device produces electromagnetic radiation including x-rays, x-rays, ultraviolet (UV), visible light, and infrared (IR). The intense visible light is sufficient to cause blindness from corneal and retinal damage. The UV can cause lens and corneal opacity. But, by far, the greatest radiation effects are from radioactivity and nuclear radiation.

ELECTROMAGNETIC PULSE (EMP)

There are a number of electromagnetic effects from a nuclear detonation which vary according to the altitude of the blast as well as the yield and design of the weapon. In case of a high-altitude nuclear detonation, an immense flux of gamma rays from the nuclear reactions occurring within the device occurs. These gamma photons produce high-energy free electrons by Compton scattering, mentioned in Chapter Three, typically at altitudes between 20 and 40 kilometers (km) above the surface (12 to 24 miles). These electrons become trapped in the Earth's geomagnetic field, generating an asymmetric oscillating electric current, which causes a rapidly rising electromagnetic (EM) field, called an *electromagnetic pulse* (EMP). The electrons are trapped within a very short span of time, and results in a very large electromagnetic source that radiates coherently. The pulse can span entire continents and can affect electrical and electronic systems on land, sea, and air. This pulse can erase computer chip memory, magnetic storage devices, and even cause arcing in simple electronic equipment. The first recorded EMP incident accompanied a U.S. high-altitude nuclear test over the South Pacific, and resulted in power system failures as far away as Hawaii. A large nuclear device detonated at some 400–500 kilometers over the central U.S. would affect all the continental U.S. The pulse from such an event extends to the visual horizon as seen from the burst point, so the altitude is important.

The EMP produced by the Compton electrons typically lasts for only about one microsecond (ms). In addition to this prompt EMP, scattered gamma rays and inelastic gamma-rays produced by the weapon's neutrons produce a signal out to about one second. The intensely energetic radioactive debris that enters the ionosphere produces ionization and heating of the E-region of the ionosphere. In turn, this causes the geomagnetic field to increase quickly, producing a magnetohydrodynamic (MHD) EMP, called a heave signal. The size of the signal, in millivolts per meter (mV/m), from this process is not large, but systems connected to long lines (*e.g.*, power lines, telephone wires, railroad rails, tracking wire antennas, some large radio antennas) are at risk because of the large size of induced current due to the length of the conductor. The additive effects of the MHD-EMP can cause damage to unprotected civilian and military systems that depend on or use long-line cables. Small, isolated systems tend to be unaffected by MHD-EMP.

In principle, even a new or small nuclear state could execute such a nuclear strike. But, in reality, it seems unlikely that such a state would use one of its scarce weapons to inflict damage which would be secondary to the primary effects of blast, shock, and thermal pulse. Moreover, an EMP attack must use a relatively large warhead, on the order of a megaton or more, to be effective, and new states or terrorists are unlikely to be able to construct such a device, much less make it small and light enough to be lifted to high altitude by a missile they might possess. Finally, in a tactical situation, an attack by a nation against opposing forces would have been also an attack against its own communications, radar, missile, and power systems since EMP cannot be confined to only one side of the burst.

As opposed to high altitude detonation EMP and MHD-EMP, *Source Region Electromagnetic Pulse* (SREMP) is produced by a low-altitude nuclear burst. An electron current is formed by the unequal deposition of electrons in the atmosphere and the ground, and this current emits a pulse of electromagnetic radiation in a direction perpendicular to the current flow. The asymmetry from a low-altitude explosion occurs because some electrons emitted downward are trapped in the upper portions of the Earth's surface while others, moving upward and outward, can travel long distances in the atmosphere, producing ionization and charge separation. In the region near the detonation, peak electric fields greater than 10^5 volts per meter (V/m) and peak magnetic fields greater than 4,000 amperes per meter (A/m) can exist. These are larger than those from initial EMP, and can pose a serious threat to all electronic and electrical equipment. During the Cold War, SREMP was conceived as a serious threat to electronic and electrical systems even within hardened targets such as missile launch and command and control facilities. SREMP effects are only important if the targeted systems are expected to survive the primary damage-causing mechanisms of blast, shock, and thermal pulse. Components sensitive to magnetic fields have to be specially hardened since SREMP effects are unique nuclear weapons effects. Because SREMP is uniquely associated with nuclear strikes, the research and technology associated with SREMP protection have no commercial application. However, technologies associated with SREMP measurement and mitigation are commercially interesting for lightning protection and electromagnetic compatibility applications.

PROMPT RADIATION

At the moment of detonation, there is an intense amount of radiation, chiefly neutron, but also containing alpha, beta, hard and soft x-ray, and gamma radiation. This *prompt radiation*, or initial radiation, only lasts for a few minutes, but is lethal to all life within a distance of a few thousand meters. However, at these distances, blast and thermal effects are also lethal. Since the particulate radiation cannot travel far, only the electromagnetic radiation affects objects and people at a distance. The electromagnetic radiation (chiefly, gamma rays) interacts with the molecules in the air (and debris) so the intensity is reduced. Also, such radiation follows the *inverse square law* in that intensity from a small source decreases in relation to the square of the distance. For example, if the radiation at 10 meters is 500 sieverts (Sv), at twice that distance or 20 meters, it is 125 Sv ($^1/_4$ of 500). At 30 meters (about 100 feet), it is only about 56 Sv ($^1/_9$ of 500), and so on. Prompt radiation will kill fewer people than the heat and blast at the same distance.

INDUCED RADIATION

Many atoms can be changed into radioactive isotopes by neutron activation. An ordinary non-radioactive atom can absorb a slow neutron and become a radioactive isotope of one additional mass unit. The new atom can emit radiation as alpha, beta, x-ray, or gamma rays. Ordinary soil can be activated and become radioactive. Fortunately, most atoms formed by neutron activation have short half-lives and decay fairly rapidly. The intensity of induced radioactivity diminishes rapidly.

Burns, Trauma, and Wounds

The burns, traumatic injuries, and other wounds are similar to any other high explosive blast, and the medical treatments are the same. These will be discussed in greater detail in the chapter on medical treatment for nuclear injuries.

Radiation Effects

A potential long-term radiation effect to the victim is cancer. From detailed studies of the Japanese atomic bomb victims, increases in several kinds of cancer have been seen. The first to appear was leukemia. Several other cancers appear to have greater incidence in the survivors of the atomic bomb detonations, chiefly skin cancers, soft tissue cancers, and thyroid cancers. It is well known that radiation can cause, or at least induce, cancer formation.

Another long-term effect is genetic damage whereby birth defects are exhibited in the offspring of the victims. If the DNA code has been altered in the sex cells of either parent, resulting in gene defects, the progeny may exhibit any number of birth defects. However, it is important to remember that radiation may increase the incidence of birth defects, but it does not cause any unique forms of birth defects. That is, the deformities or other effects are seen in other children; there are no radiation birth defects not found in nature.

If the victim is pregnant at the time of the exposure to radiation, high doses to the fetus will probably cause fetal death and miscarriage. Lower dose effects will depend upon the amount of radiation, the type of radiation, and the stage of development of the fetus; the more developed the fetus, the less likely are adverse effects. Even low levels of radiation may lead to reduced mental capacities in the child.

It has been suggested that potassium iodide (KI) pills might be protective in a nuclear event. However, the protection of KI is to block radioactive iodine from being concentrated in the thyroid, especially of children. The release of radioactive iodine is possible from an accident at a nuclear power plant, but is highly unlikely from a nuclear weapon detonation or from a "dirty bomb." The use of KI pills following a nuclear weapon detonation is not likely to have any effect at all; in fact, potassium iodide can be toxic, especially to children, since it can cause potassium toxicosis with resulting neurological effects, even death. More on medical effects of radiation are covered in the chapter on medical treatments.

Hematological Effects

Red blood cells, essential for carrying oxygen to the cells of the body, are made in the marrow of the long bones of the body. Platelets are important in the formation of blood clots. White blood cells, such as lymphocytes and granulocytes, are important in the body fighting off infection. All of these can be damaged by radiation. Radiation first causes a reduction in the body's ability to make more red blood cells in the bone marrow of the arms and legs. These are collectively called *hematopoetic effects*. The next cells to be affected by radiation are the lymphocytes, formed in lymphoid tissues, critical in the body's immune response to pathogens. This reduces the body's ability to respond to an infection. The next cells affected by radiation are the platelets, which results in reduced ability to control bleeding, leading to hemorrhaging and even death. Finally, granulocytes, formed in the bone marrow, are reduced in number; these white blood cells are also important in fighting infection. Much of the clinical information on the chapter on medical treatments is related to hematological effects of radiation.

Radiation-Induced Neurological Effects

A number of studies have indicated that large doses of radiation, especially over a short period of time, can cause mental defects. Coordination, reasoning, and concentration have been affected. It seems that most humans will be affected so that they can do tasks routinely familiar to them, but lose the ability to think clearly.

Other Effects

Indirect effects of a nuclear detonation include psychological conditions such as fear, anger, and guilt. These are highly personal and difficult to estimate or even evaluate.

SUMMARY OF NUCLEAR BOMB EFFECTS

This scenario is based upon a report to the U.S. Congress from the Office of Technology Assessment (OTA) in 1979 and a paper by H.F. York in 1976. Even though it is an old paper, the results are still used today. Assuming a 1-MT bomb, a size typical of those in the Russian stockpile, detonated over a major city at an altitude of 1,000 meters (about 3,100 feet), which gives the greatest blast and thermal effects but with somewhat reduced fallout over a surface detonation, the energy released would be about 4.6×10^{15} joules, equal to that of 2,000,000,000 pounds of TNT. However, the comparison with the effects of TNT is not exact. The heat of the TNT explosion is far less than that of a nuclear explosion. Temperatures in the center of the fireball reach 100 million degrees compared to some 5000° C (9000° F) for TNT. This temperature is as hot as the sun and is so hot that about one third of the energy is converted into x-rays that heat the surrounding air and form a fireball. The surface temperature of the fireball is about 8000° C (about 14,000° F), radiating electromagnetic radiation in the UV, visible, and IR regions of the spectrum.

The fireball increases rapidly in size and rises at over 200 miles per hour. Within 10 seconds, the fireball is a mile and a third in diameter. After one minute, it is 4½ miles high, but stops emitting visible radiation. Water vapor begins to condense and form a cloud. The cloud penetrates the tropopause and enters the lower stratosphere after about 10 minutes. It then stops gaining in altitude, but continues to grow in width to produce the mushroom shape characteristic of a nuclear explosion. The cloud will last about an hour before being dispersed by winds, and then merges with natural clouds in the sky. On the ground, wooden homes and occupants are incinerated as far as 9 kilometers (6 miles) from the center of the explosion. Nuclear fallout extends as far as 100 km (62 miles) in concentrations sufficient to kill most people in the open, and as far as 25 km (15 miles) to kill people in basements.

If the 1-MT bomb is detonated at the ground surface rather than at altitude, there is greater cratering and fallout, but slightly reduced prompt radiation and blast. This explosion leaves a crater about 300 meters (1,000 feet) in diameter and 70 meters (200 feet) deep, surrounded by an embankment of highly radioactive soil blown out of the crater. No buildings stand within 2 kilometers (1½ miles) of the center due to some 12 pounds per square inch (psi) overpressure. Between 50,000 and 200,000 people are dead immediately. From that distance out to about 4 kilometers (2½ miles), private homes are destroyed, multi-story buildings are heavily damaged, but some industrial plants might remain functional, depending upon their reliance on computers since the electromagnetic pulse would wipe out all computer chips and memory devices far beyond that distance. Farther out, to about 8½ kilometers (km) (5 miles), there are 50% casualties, but most will be injured, not dead. The percentage drops to about 25% with few deaths and only light damage to buildings between 8½ and 12½ km (5 and 7_ miles), but there are many fires. Many of those people who survive will have been blinded if they looked at or toward the fireball in its initial minutes. The detonation has killed about 250,000 people, injured 500,000, and destroyed some 70 square miles. Many of those killed, especially in the inner city where hospitals are usually concentrated, will have been doctors, nurses, and other medical personnel, so there are few trained people available to help the injured. Likewise, most police and fire fighters are killed or injured so there is little or no order or emergency response. Some survivors will be suffering from fatal doses of radiation or reduced mental reasoning ability due to radiation effects. These are the immediate effects. There is also fallout.

Radioactive fallout pattern depends upon meteorological conditions. The radioactive particles are blown by the wind, slowly falling to earth. Those closest to the detonation will receive the greatest amount of fallout. There will be a plume, roughly elliptical in shape, centered at the point of detonation and extending outward in the direction of the wind. Persons in basements and out of the open will receive some radiation from the fallout as it settles on the buildings and ground. As time passes, the fallout cloud becomes larger, but less concentrated. Much of the induced radiation diminishes quickly since most of these radioisotopes have short half-lives, but there will still be some unreacted uranium or plutonium as well as strontium and other fission products (especially cesium-137, strontium-90, and cobalt-60). The greatest threats at a distance from the explosion will be particles of fallout landing on the skin causing beta burns and from inhalation of the radioactive particles and resulting alpha damage. Immediate washing of the skin will remove much of that fallout. The wear of a respirator would reduce the risk from inhalation, but most people would not have access to these devices; however, even the use of a cloth over the nose and mouth would reduce the exposure. Fallout, depending upon the size and density of the particles, can cover extensive areas. Some very fine particles, taken up into the stratosphere by the fireball, can remain airborne for many years.

People will be killed by the blast overpressure, flying debris, heat, and radiation; some will suffer multiple injuries. Those surviving will be suffering from burns, electrolyte imbalance, blood conditions, wounds, and mental problems. Some survivors will later develop cancer. Some pre-natal infants, depending upon the time of exposure during development, will be born with birth defects, reduced mental abilities, or cancers.

CONSEQUENCES OF A DIRTY BOMB EXPLOSION

The effects from the detonation of a nuclear weapon above are much greater than from the explosion of a so-called "dirty bomb." The blast and heat would result from the conventional explosive and be much less than described above. The blast which destroyed the Murrow Building in Oklahoma City, OK has been estimated to have been a few hundred pounds of ANFO (a mixture of ammonium nitrate and fuel oil) and equivalent to one to two hundred pounds of TNT. While the building was destroyed and several others in the vicinity, including a church across the street, were damaged, the effects were localized. A conventional explosive was used in the terrorist attempt to blow up the World Trade Center in New York City in 1991; estimated at a thousand pounds of TNT equivalent, it failed to destroy the building, but did cause extensive damage to the parking garage and subfloors.

Adding radioactive materials to the bomb will spread them over an area, but the effects will be largely localized. The range of initial contamination is the range of debris from the explosion. If people do not carry the radioactive materials away, as on their shoes and clothing, the area contaminated will not be large. The greatest effect of a dirty bomb will be psychological, spreading fear and panic.

6

NUCLEAR WEAPONS DETECTION, PROTECTION, AND DECONTAMINATION

RADIAC INSTRUMENTATION

Since nuclear radiation, aside from the effects of a detonation, cannot be sensed (smelled, tasted, felt, *etc.*), instruments are needed to learn of its presence and magnitude. Instruments used to detect and measure radiation are, as a class, known as radiac instruments.

The earliest means of detecting radiation was by use of a *gold leaf electrometer*, consisting of thin metal foils capable of picking up an electrostatic charge. If charged, the foils move away from each other, but, being attached at one end, they move apart like the wings of a butterfly. When brought into the vicinity of radiation, the charge would be dissipated and the foils would moved back together. The greater the radiation, the faster the foils would move together. There are obviously drawbacks to this method; the foils have to be continually charged, it is not possible to quantitatively measure the amount of radiation, and different forms of radiation cause the foils to act differently.

Very similar to the electroscope is the *fiber dosimeter*. In this device, a very fine fiber has a static electric charge placed on it, usually by a battery-powered charger. When ionizing radiation passes through the device, some of the charge is dissipated and the fiber moves. A lens system allows for reading the amount of movement, which indicates the amount of radiation received. One advantage to this device is its small size and weight, and the fact that it is a direct reading without development or other process. Its major disadvantage is that it is shows the total amount of radiation received over time; that is, the average dose. Another disadvantage is that the type of radiation cannot be determined.

There are other radiation indicators which change color when exposed to ionizing radiation. This type of device is a quick and simple method of seeing if radiation is present, but they, too, only show the total energy received and not the type or instantaneous levels present.

All forms of radiation, of sufficient energy, are capable of causing ionization. This is actually the reason that the foil method and fiber dosimeter work. Ionization chambers are extremely sensitive detectors of radiation. As a particle or ray passes, ionization of air molecules takes place; since ions carry electrical charges, conduction of electricity takes place. This can be measured using a sensitive electronic meter.

Thermoluminescent detectors (TLDs) are solid state crystals which have their internal crystal structure affected by ionizing radiation. The crystal lattice defects are retained within the crystal until they are heated. Upon heating in an oven, the stored energy is released as light which can be detected and measured by sensitive photocells. One disadvantage to TLDs is the fact that they only provide information on integrated radiation exposure; that is, the energy retained is the total of all received, so the amount at any given time cannot be determined. This is the same drawback to the predecessor system to TLDs, film badges. *Film badges* contain small pieces of photographic film behind different shields. The type of radiation received can be determined from the relative exposures of the film behind the filters of different materials. As with TLDs, the film only indicates the total amount of radiation received, not when or at what rate it was received. Electronic instruments are needed to measure dose rates. The following sections will describe a number of the common electronic radiac instruments.

Ionization Chambers

Some of the oldest and most widely used types of radiation detectors are based on the effects produced when a charged particle passes through a gas. Along the particle track, there is ionization or excitation of the gas molecules as they absorb energy from the particle. The term *ionization chamber* is used for the type of radiation detector in which ion pairs are collected from gases. As a radiation particle passes through the glass wall of the chamber, gas molecules inside are ionized as energy is transferred from the particle to the gas atoms. If the energy is greater than the ionization energy of the atom, ionization results. The resulting positive ion and free electron is called an ion pair. Ions can be formed either by direct interaction with the particle or through a secondary process in which some of the particle's energy is first transferred to an energetic electron (a delta ray). In either case, an ion pair results. The type of gas within the chamber will have a specific ionization energy, but most gases absorb between 30 and 35 electron volts for each ion pair formed. Therefore, an incident 1 MeV particle, when fully stopped within the chamber, will form some

30,000 ion pairs. These ion pairs will allow for conduction of electricity between two metal plates in the chamber which have an applied voltage; this flow of current can be measured on a meter. This is the basic function of the direct current (DC) ion chamber.

Proportional Counters

The *proportional counter* is a type of gas-filled detector first used in the late 1940s. Rather than simply measuring the average current as the ion chamber does, the proportional counter is able to detect pulses using gas multiplication to amplify the charge on the original ion pairs created within the tube. One important application of proportional counters is the detection and measurement of low energy X-rays. Gas multiplication results from increasing the electric field within the gas to a high value. In effect, the formation of one ion pair becomes many. The reason the name proportional counter is used is that the number of electrons and positive ions formed is a multiple of those produced by the radiation particle or rays itself. The actual energy of the x-ray or particle can be determined from the amount of voltage needed to cause the multiplication effect.

Geiger-Mueller Detectors

Ionization can take place inside an electron tube. Most beta-gamma radiation detection and measuring instruments use a Geiger-Mueller tube. The detector consists of a gas-filled tube containing electrodes between which there is an electrical voltage, but no current flowing. When ionizing radiation passes through the tube, a short, intense pulse of current passes from the negative electrode to the positive electrode and is measured or counted. The number of pulses per second measures the intensity of the radiation field. It was named for Hans Geiger and Wolfgang Mueller, who jointly invented it in 1928. It is sometimes called simply a Geiger counter or a G-M counter, and is the most commonly used portable radiation instrument.

G-M counters employ gas multiplication to greatly increase the charge on the original ion pairs formed along the radiation track, but in a different manner than the proportional counter. The electric field within the gas is much greater for the G-M tube. At just the right voltage, one ion pair can cause an avalanche of many ion pairs which can then cause a second avalanche at a different position within the tube, in a manner similar to a chain reaction. All pulses from a G-M tube have the same amplitude regardless of the number of original ion pairs that initiated the process. The G-M detector can only function as a simple counter of radiation because all information about the initial energy is lost, but it is very sensitive.

Scintillation Detectors

When certain materials absorb radiation energy, they re-emit bursts of light. These flashes can be amplified and measured. These materials are of two types: organic scintillators and inorganic scintillators.

Organic scintillators include pure organic crystals of anthracene or stilbene; liquid organic scintillators (solutions of the previous chemicals); plastic scintillators (usually a styrene containing the previous chemicals); thin film scintillators (like the plastic scintillator); and loaded organic scintillators (having added higher molecular weight materials so that gamma rays can better be detected). Liquid scintillators are commonly used to measure low-energy alpha and beta particles because of their short paths in solids.

Inorganic scintillators are similar to transistors in that they are solid state semiconductors, but function as light emitters rather than amplifiers. Some of the common inorganic scintillators include sodium iodide (NaI), cesium iodide (CsI), lithium iodide (LiI), zinc sulfide (ZnS), calcium fluoride (CaF_2), cesium fluoride (CsF), and lithium glass. Most these are doped, or have an added substance like thallium (Tl) or europium (Eu) or silver (Ag), in order to better function as a semiconductor. When incoming radiation passes through the material, energy is absorbed and a small amount of emitted light released.

In both cases, for the organic and inorganic scintillators, the amount of light is extremely small and could not produce a measurable electrical signal. However, a device known as a photomultiplier (PM) tube can receive a small amount of light striking a photoemitter and then multiply it millions of times. This can be measured.

X-rays and gamma rays are uncharged so do not directly cause ionization. Therefore, the scintillator must act as a conversion medium in which electrons are produced and as a detector for these secondary electrons. When these roles are successfully performed, the scintillator can function. An advantage of the scintillator is that the actual energy of the incoming radiation can be detected and measured; this is called spectroscopy. A drawback to scintillators is the low energy resolution; that is, the resulting spectrum is generally broad so nearby peaks might merge into one.

Semiconductor Diode Detectors

The use of a solid detection medium is advantageous for the measure of high-energy electrons or gamma rays because the density of a solid over a gas is thousands of times greater, allowing for many more interactions. In addition, the energy resolution is much greater than for scintillators. Technically, this relates to increase in statistical energy resolution requires an increase in the number of information carriers. In the diode devices, the information carriers are electron-hole pairs created along the path of the charged particle (alpha, beta, or delta) through the detector. Advantages to diode detectors include compact size, fast response times, and thicknesses that can be varied to match the specific application. Drawbacks to diodes include limits on how small they can be and susceptibility to radiation-induced damage.

These diodes, being semiconductors, have a valence band of electrons or holes separated from a conduction band of electrons or holes by a band gap. When radiation impacts the diode, charge can be transferred across the band gap in only one direction. This small charge can be amplified electronically and read on a meter.

Typical materials include silicon (Si) and germanium (Ge), usually doped with lithium (Li), sodium (Na), potassium (K), or barium (Ba). Silicon detectors operated at room temperature are very good detectors of alpha radiation and light ions. Germanium detectors are very good general purpose radiation detectors. The output spectrum from a diode detector appears as a series of peaks like mountains in a range. From the energy peaks, the identity and the magnitude of the radiation source can be determined.

NEUTRON DETECTION

The detection of neutrons, since they are uncharged, is different than for gamma and x-rays or charged particulate radiation. The measurement of neutron kinetic energies is very difficult and beyond the scope of this book. Only the detection of neutrons will be mentioned. All the common reactions used to detect neutrons result in the formation of heavy charged particles, such as recoil nucleus, proton, alpha particle, or fission fragments. One such reaction involves the neutron impacting a boron-10 atom forming lithium-7 and an alpha particle (of either 2.792 MeV or 2.310 MeV). It is the alpha particle that is then detected. Another reaction involves a neutron hitting lithium-6 atom forming tritium and an alpha particle of 4.78 MeV energy. This reaction is much less sensitive than the boron-10 reaction. An additional reaction includes helium-3 to produce a 0.765 MeV proton, but the low-energy proton is difficult to detect.

A widely-used detector for slow neutrons is the boron trifluoride (BF_3) proportional tube. The gaseous BF_3 is both the target for the neutron conversion as well as a proportional gas, as in the proportional counter discussed above. The detection of fast neutrons requires the use of a moderator like in the nuclear reactor (heavy water, or, more likely, graphite or a solid-state material).

TYPES OF FIELD RADIAC INSTRUMENTS

The detection of nuclear radiation away from the laboratory is of vital importance to people since the field is where potential exposures are most likely. Serious injury or death can result from exposure to sufficient quantities of these invisible rays and particles. In considering the effects on personnel exposed to radiation, two kinds of information are needed: (1) the intensity of the radiation field, and (2) the total dose or quantity of radiation received per exposure or time interval (dose rate).

Intensity may be defined as the strength of the radiation. It is expressed as a quantity of radiation per unit of time. The quantity units formerly used (and still commonly seen) were the *roentgen* (R) and the *rad*, and the time unit was usually the hour. Therefore, exposure rates were often expressed as roentgens per hour (R/hr) or as rads per hour (rad/hr). The roentgen relates to the quantity of ionization produced by x-rays or gamma radiation, and is called *exposure*. It is defined as 2.58×10^{-4} coulombs per kilogram of dry air. It is not applicable to particulate radiation like alpha, beta, or neutrons. The rad is defined as 100 ergs absorbed per gram of any substance, and called the *absorbed dose*. Today, the preferred unit for absorbed dose is the *gray* (Gy). The gray is defined as one joule of radiation absorbed per kilogram of matter. Dosage was expressed in two values: the exposure dose, which is measured in R, and the absorbed dose, which is measured in rad. An added factor in the use of rad is that it expresses the dose from any type of radiation, whereas the R relates only to gamma radiation or x-rays. A large number of detection devices scaled in R are still in use. Therefore, R will be considered in this book along with rad and Gy. The rad and R may be assumed to be equivalent for the purpose of this book.

There is another factor to be considered; that is, not all forms of radiation produce the same biological effects per unit of dose. Alpha particles and neutrons produce much more damage along a shorter track than do beta particles. To correct for these differences, the unit formerly used was the *rem* (from "roentgen equivalent man"). Aside from being sexist, it was very useful in comparing the biological effects of different kinds of radiation. The rem is defined as the absorbed dose (in rads) corrected by a Quality Factor (QF), which varies from one type of radiation to another. X-rays, gamma rays, and beta particles are considered to have a QF = 1 so the number of rems equals the number of rads, but alpha particles have QF values which vary from 1 to 20 so the biological effect can be twenty times greater (20 rems from a dose of 1 rad). Neutrons have QF values generally between 2 and 11. Today, the preferred unit is the *sievert* (Sv), which is equal to the absorbed dose in grays times the Quality Factor, which is the same as that used for the rem.

Information on intensity and dosage is essential in measuring the extent and degree of radiological contamination. It permits the calculation of safe entry time and stay time for people in radiologically-contaminated areas. Also, it provides an objective means for withdrawing essential personnel who may be nearing a critical point of radiation exposure. Finally, it is useful in anticipating the severity of radiation sickness. Data needed for these and other calculations can be gathered by various radiac instruments.

No portable radiac currently available will measure both radiation intensity and dose. Therefore, the different kinds of measurements must be made by separate instruments. The device that measures radiation intensity is called a dose rate, or survey, meter. It provides information needed to calculate any radiological hazards of occupying a contaminated area or handling contaminated equipment. It also provides the information necessary to calculate the approximate length of time that a person can safely remain in a radiologically-contaminated area. The device that measures the total radiation received by an individual is called a *dosimeter*. Medical personnel must have dose information to predict the severity of radiation sickness, to make a prognosis, and to provide appropriate medical treatment.

The two types of radiation detection and measuring instruments may be compared to automobile speedometers and odometers (mileage indicators). The dose rate, or survey, meter measures the intensity of radiation in R/hr, rad/hr, or, today, Gy/hr, and is like the speedometer that indicates the speed of an automobile in miles per hour. The dosimeter measures the total exposure in R, rads, or grays, without regard to time. Therefore, it is like the odometer which records the total distance traveled in miles, without regard to time.

FIELD SURVEY METERS

Following a detonation or suspected release of radioactivity, areas should be monitored for radioactivity by means of survey meters (dose rate meters). Field monitoring will reveal both the geographic extent of the radiological contamination as well as how "hot" various parts of the area are. The term "hot", when used in reference to radiological contamination, means that a certain area has a high contamination reading. Survey meters are also used to monitor food, water, and the skin and clothing for radiological contamination.

There are radiac instruments that can be used to perform alpha, beta, gamma, and neutron surveys. The following sections discuss the survey meters currently used by the U.S. military. No attempt has been made to discuss the specialized instruments used around nuclear power plants or propulsion units nor commercially available units.

Alpha Meters

The alpha particle is a relatively large and heavy nuclear particle that has a short range in matter. Even ordinary clothing or a piece of paper can block out alpha particles. These alpha particles require a sensitive scintillation (light detecting) detector since alpha particles have a range of only about one inch in air. The incoming alpha particle strikes a phosphor with the resultant production light, which is converted to electrical energy in a photomultiplier (PM) tube. Since the PM tube is extremely light sensitive, the phosphor must be protected from ambient light. The most common material in use is black Mylar®, which can be made thin enough to allow the passage of alpha particles while blocking out light. The fragile Mylar window is usually held about a quarter of an inch from the surface being surveyed.

The AN/PDR-56 series are the military's standard alpha survey meters. The various units are used to locate and measure alpha contamination of the skin and clothing, and on the surfaces of equipment. The AN/PDR-56 has a meter readout with four ranges of counts per minute (1/60 c/min) up to 1,000,000. The unit becquerel is equal to 1 disintegration per second (DPS), so is equal to 1/60 c/min (assuming 100% efficiency of the detector, which isn't possible).

Gamma and X-ray Survey Meters

The radiac instruments described in the following paragraphs can detect and measure both gamma and x-rays. The detecting elements of some of these instruments are so constructed that by moving an integral shield (the beta shield, often made of aluminum), an operator can expose the ports to the sensitive volumes of the detectors.

This permits the detection of beta particles along with gamma rays. Judicious use of the beta window permits an operator to determine if a surface is contaminated with a beta or a low energy gamma-ray emitter.

The AN/PDR-27 series of meters consist of portable, watertight, battery-powered, low-range survey meters. Two Geiger-Mueller (GM) tubes are mounted in an extendable probe connected by a coiled cord. A spare Geiger-Mueller tube set is included in the carrying case. The probe is fitted with a beta shield. The meters are powered by six alkaline D-cell batteries. The AN/PDR-27 provides both visual and audible indications of gamma and beta radiation levels. The visual indication is shown on a meter; the audible indication is heard through a headset. The radiation measurement is in milliroentgens per hour (mR/hr). The unit is capable of detecting and measuring gamma radiation when the beta shield is in place. It is also capable of detecting beta and gamma radiations together when the beta shield is removed. There are four linear scales on the AN/PDR-27. The scales are 0 to 0.5 mR/hr, 0 to 5 mR/hr, 0 to 50 mR/hr, and 0 to 500 mR/hr. Beta radiation can be detected on the lower two scales only.

The AN/PDR-43 series radiac instruments are standard military watertight, battery-powered, high-range survey radiac meters. The detector consists of one GM tube which is installed in the forward end of the high-intensity survey meter. It is a portable, alkaline D-cell battery-powered unit. There are two check sources. The detector has a spring-loaded beta window controlled by the function switch. The AN/PDR-43 is capable of detecting and measuring gamma radiation alone, or the unit can also detect gamma and beta radiations combined when the beta shield is not used. The unit has three meter scales of 0 to 5 R/hr, 0 to 50 R/hr, and 0 to 500 R/hr.

The AN/PDR-65 is designed to detect and measure gamma radiation, most commonly aboard U.S. Navy ships. The set consists of two primary elements: the detector unit and the radiac meter, which can be separated by a distance. The detector unit must be located where it has a clear view of the radioactively-contaminated areas. The radiac meter can be installed remotely. One or more auxiliary readouts can also be used. The radiac meter has two types of displays: one for the dose rate and the other for accumulated dose. The main meter displays the dose rate. The small counter registers the accumulated dose in rads by counting the rad pulses from the detector. Each time that a dose of 1 rad is accumulated, the radiac meter sounds a loud beep. The range of the small counter is 0 to 9,999 rad. Gamma intensity is indicated on one of the following four ranges: 0 to 10 rad/hr, 0 to 100 rad/hr, 0 to 1,000 rad/hr, and 0 to 10,000 rad/hr. For normal shipboard operation, the AN/PDR-65 is powered by the 115 VAC ship's power. In an emergency, the radiac instrument can be operated on four internal, rechargeable, nickel-cadmium, C-cell batteries. These batteries can power the radiac for approximately 20 hours.

Dosimeters

A number of types of radiation dose-indicating devices (dosimeters) are in use by the U.S. military. The DT-60/PD is a gamma radiation dosimeter with a usable range of

10 to 600 R. It is a solid-state package in the form of a locket, designed to be worn on a chain around the neck. Inside the black plastic casing is a phosphate glass. When the phosphate glass is exposed to ultraviolet light, it emits an orange light with the intensity of the orange light proportional to the amount of radiation the glass has received. The DT-60/PD stores the dose information indefinitely and is a permanent record of the amount of exposure to radiation.

The CP-95/PD is a radiac computer-indicator used to read the amount of radiation a DT-60/PD has been exposed to. A newer model, the CP-95A/PD, is also available. The cover on the DT-60/PD must be removed before the DT-60/PD is inserted into the radiac computer-indicator. Each of these radiac computer-indicators has two scales: 0 to 200 R and 0 to 600 R. However, 10 R is the minimum detectable exposure. These units have an accuracy rate of ±20%. The radiac computer-indicators operate off of a 115 VAC power source.

The IM-9/PD series consists of a pocket dosimeter of the quartz-fiber type. It indicates the gamma radiation dose in the range of 0 to 200 mR. The IM-9/PD is a self-reading dosimeter. By holding the dosimeter up to the light and looking through the lens, the radiation dose received can be read off the scale. The reading is obtained by observing the position of the quartz fiber on the scale of the built-in optical system. The IM-9/PD is primarily a health-physics device that is particularly useful in areas of low dose rates.

The IM-143/PD is identical to the IM-9/PD except in range. The IM-143/PD indicates gamma radiation dose in the range of 0 to 600 R. It is used by personnel involved with survey, monitoring, and decontaminating details. It keeps track of the dose they have received up to the time they read the dosimeter.

The PP-4276A/PD dosimeter charger is used to reset the self-reading dosimeters to zero. This is accomplished by placing the dosimeter into the charger. The charger provides an adjustable voltage source that is applied between the central wire and the shell. Because the quartz fiber and the fixed central wire of the ion chamber are attached, each will receive the same charge. When the dosimeter is charged, the movable fiber is repelled from the fixed wire. By proper adjustment of the voltage applied by the charger, the fiber can be set exactly on the zero line of the scale. When nuclear radiation enters the chamber, ions are produced and are attracted to the wire and the shell. The charge is thus reduced; the fiber moves across the scale because it is less repelled from the fixed wire as previously. The power source for the PP-4276A/PD is one alkaline D-cell battery.

LIMITATIONS OF RADIAC INSTRUMENTS

Although there is radiac equipment to detect and measure nuclear radiation, these instruments do have some limitations. None of these instruments are capable of detecting and measuring alpha, beta, gamma, and neutron radiation at the same time. Even those that can detect both beta and gamma radiation do not automatically separate these two types of radiation. Instead, the operator can keep beta particles from entering the chamber by manually pulling a beta shield over the thin window in order to get the gamma reading only. A separate instrument must be used to measure alpha radiation. No instruments can simultaneously give the radiation intensity and the energy of the radiation particles or gamma rays.

RADIATION SURVEYS

Intensity and dose measurement surveys are required when a radiation hazard potentially exists. Measurements should be started before the expected arrival of the debris or fallout, and continued regularly until the damage has passed. At early times after the burst during the emergency phase, rapid simple measurements are all that can be performed. At later times during the recovery phase, required measurements may become very detailed. This section is largely concerned with monitoring surveys in the recovery phase; however, some of the information can be used during the emergency phase.

Radiation Monitoring Procedures

The following procedures apply to emergency responder organizations, such as police, fire, and medical services utilizing radiation monitoring. While they are not practical for members of the public, some of the details will give insight into what responders are doing in the event of an emergency, or if a member of the public is called upon to assist officials in their actions.

The hazards of radiation and radioactive contamination are different from most ordinary hazards in that radiation is invisible. However, instruments have been developed to measure radiation, and techniques have been developed to estimate the amount of removable contamination. Therefore, the hazards can be identified, their importance can be estimated, and they can be dealt with in a systematic fashion. This section describes techniques for measuring radiation and contamination. These measures provide the information necessary to identify the radiation hazards and bring them under control.

Radiac instruments provide intensity and exposure information on radiation hazards. The two basic types of radiac instruments discussed above are survey meters and dosimeters.

Radiation measurements can be only as accurate as the radiac instruments used to make the measurements and the techniques used to operate the equipment. Some radiac instruments detect the ionizing properties of radiation, which is an electrical phenomenon. They convert this property, through electronic circuitry, to a visible indication. Other radiac instruments, such as the dosimeters, depend on chemical changes due to ionization rather than on electrical effects.

Most radiac instruments with electronic circuitry are delicate and should be handled with care. Gamma-measuring radiac instruments, due to the nature of the rugged Geiger-Mueller tube, are the sturdiest. Beta-detecting radiac instruments usually have detectors with thin glass walls. This delicate wall is visible when the beta window of the detector probe is open. Alpha-measuring radiac instruments are the most delicate. The detector probe of this type of radiac has a window, usually of Mylar®, which is very thin. This allows the short-range alpha particles to pass through it into the detector. However, this window must be gastight to prevent the escape of the special gases used in the detector. Those who use radiac instruments should be well acquainted with the particular precautions to be taken with each type they are using.

It is important to prevent contamination of the detector probe since a contaminated detector probe will give high radiation readings from the material on it. These readings might be interpreted incorrectly as contamination on the surface or person being monitored. Contamination of a detector probe can be avoided in two ways: by covering the instrument with a removable covering and by not touching contaminated surfaces with the detector probe. Gamma radiac instruments can be covered by any convenient material. Transparent polyethylene bags are generally found to be most satisfactory. The instrument dial can be read, and the control knobs can be operated through the bag. The bag can be changed periodically. The detector of beta radiac instruments can be protected from contamination by using a layer of thin plastic film, such as Mylar. The detector of an alpha radiac meter cannot be covered at all because the cover would reduce the sensitivity of the instrument and readings would be incorrect.

It is important to use a standardized monitoring procedure for each type of radiation survey to ensure the maximum value of the radiation data obtained. A uniform monitoring technique will use no more time or effort than a random procedure, and it may actually save time since repeat measurements are not needed.

General pre-attack preparations for radiation monitoring surveys include these steps: establish monitoring points; maintain radiac instruments in usable status; select an area for analyzing wipe samples; prepare master data forms; and regularly perform monitoring exercises. Establish monitoring points in areas that are vital to whatever activity is performed there. Examples are living quarters, medical clinics, food storage and serving areas, shelter compartments, and a few representative remote areas. Each point should be representative of the location of personnel occupying the compartment or area. Identify the points by a distinctive label that includes the location mark and an identifying number. The number of monitoring points should be the fewest that will give a good sampling of the condition of vital areas. The number should also ensure that a complete monitoring survey can be made with only a few monitoring teams in about ten to thirty minutes. Test and maintain radiac instruments routinely to ensure proper functioning and to stay familiar with their operation.

Select an area for analyzing wipe samples and fallout samples. The area should have a desk or workbench, and be located in a shielded compartment. If wipe samples are to be saved, store them apart from the analytical instruments so that they will not increase the background radiation intensity.

Training exercises should be conducted to establish a pattern for calculating and reporting radiological information. When that has been done, prepare masters of data sheets for each step of monitoring, calculation, and reporting procedures (together with a small supply of copies), and keep them on file.

The minimum monitoring team consists of two personnel: a radiac operator (or a dosimeter reader or a wipe sampler, depending on the type of survey being made) and a recorder. The monitoring-team members should be dressed in protective clothing if they are to go into an area suspected of being contaminated. For rapid monitoring, the team should have specific pre-selected monitoring points to survey. They should travel a previously practiced route, or an agreed-upon route if the previously practiced route cannot be used. The recorder should try to keep his or her hands and the data sheets from becoming contaminated.

Prior to Entering the Radiation Area

Select the type of radiac instrument that will measure the type of radiation known or suspected to be present. Use beta/gamma instruments for fission products, and beta/gamma and alpha instruments for weapon incidents/accidents.

- Select the radiac model that has an indicating range sensitive enough yet broad enough to measure the radiation intensity of the area to be surveyed.
- Test the radiac survey meter for proper operation.
- For each survey point, use the lowest range of scaling factor that will give an on-scale reading.
- Record the exact time of departure and return on the data form, as well as the time of all measurements.

During the Survey

- Protect the radiac equipment from contamination.
- Be alert for hot spots, and note their location while not spending too much time searching for hot spots unless specifically directed to do so.
- Read the radiac meter to the nearest scale division, recalling that these instruments are designed for about ±20% accuracy.
- In detailed monitoring surveys, proceed slowly and carefully; thoroughness is more important than speed.
- For each reading, hold the radiac steady, and allow enough time for the meter to reach maximum deflection. This generally takes no more than eight to ten seconds and usually only three or four seconds.

Gamma Monitoring

This monitoring procedure is a routine action for providing a pre-attack baseline value of the gamma-radiation level in the vicinity of working or living areas. It also provides data to detect the start of passing initial radiation, to detect the arrival of fallout during the very early time period after burst, to determine the end of fallout, and to determine the effectiveness of any decontamination operations.

Reporting procedures should be established prior to an attack. These procedures should include who is to be notified when the post-attack gamma intensity changes by some proportional amount, such as every time the radiation intensity doubles or halves.

Area Gamma-Exposure Monitoring Survey

Locate dosimeters at monitoring points in any shelters and at vital locations. Readings can then be made to determine the total gamma exposure from the initial, passing, and deposited radiations, as well as routine checks. These data serve primarily as a rapid check of the exposure levels at various points to identify gamma-radiation hazard areas for assessing the radiological situation.

These data will not normally be used to estimate specific doses to individuals or small groups of people. However, if an individual had a malfunctioning dosimeter or none at all, a rough estimate of the individual's dose could be obtained from this survey. Pre-attack monitoring can be used to identify faulty dosimeters. Exposure readings should also be checked routinely at certain stations such as living areas to ensure that low-intensity gamma radiation undetected by other surveys is not a long-term hazard.

Two methods can be considered for area exposure monitoring. The preferred method is to have one set of dosimeters and several spares. Read and record each dosimeter periodically. If a dosimeter shows over 50% of full-scale exposure, replace it with a freshly-charged dosimeter. The second choice is to have two sets of dosimeters for each location to be monitored. Place one set on the station, and keep the second set in reserve. Collect the exposed dosimeters periodically and replace them with the freshly-charged reserve set. Take the exposed dosimeters to a central location to read and record the exposures.

All self-reading dosimeters (such as the IM-9/PD and the IM-143/PD) can be used for area exposure monitoring. Monitor selected locations with both a low-reading and a high-reading dosimeter.

Rapid Interior Gamma-Monitoring Survey

Conduct a rapid monitoring survey to determine the gamma-radiation levels in shelters and at vital locations. This survey provides the gamma-intensity readings necessary to make a radiological-situation evaluation and to plan personnel rotation or calculate recommended maximum stay times. When this survey is made before and after a decontamination pass, and the standard intensity results are determined from the survey, the effectiveness of the decontamination operations can be estimated.

This survey should be completed in ten minutes or less. The results should be immediately transmitted to the designated person in charge, if possible. This rapid interior gamma-monitoring survey can be combined with an area-exposure monitoring survey to obtain all the radiation data required for the radiological-situation evaluation. When performing this rapid monitoring survey, observe the general rules for monitoring. In addition, when reading gamma-radiation intensity, be sure that the beta shield covers the detector if the radiac meter is so equipped. Hold the radiac instrument at waist height over the designated monitoring point. Take the reading when the indicating needle steadies. Monitoring teams operating earlier than one hour after burst should record the time of reading within an accuracy of one minute. Monitoring teams operating at times greater than one hour after burst should record the time of reading within an accuracy of five minutes so that accurate evaluations may be performed.

Rapid Exterior Gamma-Monitoring Survey

There are three primary purposes for a rapid exterior gamma survey. The first is to obtain data for evaluating the radiological situation at exposed locations. The second is to locate the general areas most in need of decontamination. The third is to determine the effectiveness of any decontamination. The people on the monitoring teams must be dressed in protective clothing. They should use the designated entrances and exits through personnel decontamination stations, if present. For their safety, the personnel should wear a self-reading dosimeter and read it often while conducting the surveys. Except for the above precautions, the procedure for the exterior survey is the same as that for the interior survey.

Rapid Contamination Survey

The purpose of this survey is to obtain samples required to quickly identify areas having removable contaminants. Wipe samples are useful for detecting the presence of contaminants from a weapon-handling accident, usually alpha radiation, or from fallout, usually beta and gamma radiation. This survey should be made after any event suspected of contamination. It should be repeated after known or possible significant changes in the amount or distribution of contamination. An example would be after any decontamination operation or after a suspected scattering of the contaminant through previously clean areas.

The procedure for taking wipe samples requires at least a two-person survey team. These people should wear protective clothing when they survey the contaminated areas or when they attempt to determine the contamination limits. The equipment needed for sampling includes filter papers or absorbent papers such as paper towels, about 3 inches square, for wipes and for beta-contamination samples. An equal number of envelopes to contain the wipes, a pencil and a clip board to mark the envelopes, and a stapler to seal the envelopes are needed. One team member, designated the sampler, will take all the wipe samples. The other team member, designated the recorder, will seal the samples in an envelope after marking the envelope with the time and location. A sample is taken by wiping the clean piece of absorbent paper lightly over a 16-square-inch area of the surface being surveyed. Generally, in a rapid contamination survey, wipes are taken only of general areas. The recorder holds the envelope open, and the sampler drops the sample into the envelope. The recorder then seals the envelope by stapling, taping, or folding it closed, but never by licking the gummed flap.

To determine the extent of the contaminated area, take samples from several types of areas such as the following: important areas presumed to be clean; areas that might be contaminated; and areas presumed to be contaminated. After fallout, these would include selected locations which may need to be occupied; after a weapon handling accident, these would include areas near the accident scene.

Several survey teams might be dispatched; one for each type of area. If only one survey team is available to take samples, the team should take the samples in the order given above (from clean to suspected to known), and should not mix the samples.

The wipe samples protected by the envelopes should then be taken to a low-radiation level area for radiation-intensity measurements. The radiation intensity of samples of fallout contamination should be measured while they are in the envelope. Measure the radiation intensity with a beta-gamma instrument. Be sure that the open beta window is in contact with the envelope. Always use the same radiac instrument. Before samples are read, take and record an initial background reading. Subtract this background reading from the gross sample reading to get a net sample reading. The background reading is merely a radiation reading with no sample present, and should be taken about every ten minutes when a series of samples is being analyzed.

Alpha wipe samples can be read as they are taken. Place the wipe sample on a flat surface with the wiped side facing up. Place the probe within one-eighth inch of the sample's surface. Allow three to five seconds for the alpha radiac meter to stabilize before recording the reading.

Personal Exposure Survey

The purpose of a personal exposure survey is to make an estimate of the radiation exposures of an individual. This estimate is made by reading and recording the exposures indicated by the dosimeters worn by one or more persons in each team.

Protective actions, such as personnel rotation, are based in part on personal dosimeter readings. For this reason, readings of dosimeters must be made quickly to allow adequate planning and scheduling times. The interval between dosimeter checks will depend on the intensity levels where the individual performs his or her duties, on previous individual exposures, and on the designated planning exposure.

Rapid Alpha-Monitoring Survey

The purpose of a rapid alpha survey is to obtain the data necessary to quickly detect any contaminated areas after a nuclear weapons incident. The alpha-contamination limit can be delineated at the same time. A rapid alpha survey should take place immediately after the incident and at designated intervals thereafter. The following procedure should be used to delineate the contaminated areas. From the nature of the incident, tentatively identify those areas that almost certainly are not contaminated and also those areas that almost certainly are contaminated. Take readings, beginning with the presumed uncontaminated area and working to the presumed contaminated area; space the survey points so that there will be at least four points between the presumed clean area and the presumed contaminated area. Survey first those areas where spread of contamination would be most important. Record the results of the survey as readings are made; include time, location, meter reading, and notes on any unusual observations.

The technique for taking readings is as follows. Place the probe within one-eighth inch of the surface. Hold the probe in one position for three to five seconds to allow for the relatively slow response of the alpha radiac instrument. Use earphones if they are available. Recall that an alpha reading cannot be taken on a wet surface. Even though alpha-monitoring teams may be dressed in protective clothing, they should try to stay out of a highly contaminated area because they might spread alpha contamination into the clean area. Recall that the hazard of alpha radiation is internal radiation resulting from inhalation, ingestion, or puncture wounds.

Detailed Monitoring Survey

The detailed monitoring survey is a slow, careful inspection of all accessible areas, equipment, and systems that have been exposed to contamination. This survey is generally conducted after the application of countermeasures and toward the end of the recovery phase. The purpose of this survey is to verify the radiological clearance level of various locations. These checks determine whether or not the next lower clearance level can be assigned to the area, thereby reducing radiological restrictions for being in the area. After fallout contamination, the survey includes both a beta/gamma radiation survey using radiac instruments and a contamination survey using wipe samples. After a nuclear-weapon incident, the survey includes an alpha radiation survey using radiac instruments and an alpha-contamination survey using wipe samples. When the general radiation level has been reduced by decontamination operations, read the wipe samples where they are taken rather than taking them to a low-background laboratory area.

The detailed survey should include all accessible surfaces, both horizontal and vertical, in all suspect areas. It is especially important to check surfaces and areas that tend to capture and hold dirt, such as rust spots, caulking in wood items, canvas covers, and corners. Give extra attention to any locations that were ventilated from the outside during fallout. The finer fallout particles, which can enter through ventilation systems, tend to settle on horizontal surfaces. They can adhere to wire screens, steel wool, and air filters such as those on electronic equipment. Remove access panels and vents along the ventilation system, and take wipe samples of interior duct surfaces. Take air samples at the outlets of any ventilation ducts suspected of being used during fallout.

This gamma monitoring procedure is slightly different from that described under Rapid Exterior Gamma-Monitoring Survey. Hold the gamma detector approximately two inches from the surface being monitored. By using this procedure, a contaminated area one foot in diameter or less can be evaluated. It is also possible to further localize a contaminated area by slowly moving the detector probe, with the beta shield removed, over the surface at about a two-inch distance. Identify the contaminated area by marking it with a crayon or marker. Record, at a minimum, the following: the location of the contamination, the radiation intensity, the method of measurement, and the date and time.

Air Sampling

Air sampling has two purposes: to detect the presence of radioactive aerosols created in a contaminated area which may possibly spread to a clean uncontrolled area, and to establish the requirement for respiratory protection for personnel working in a contaminated area. An air-particle sampler draws air through a replaceable filter at a nearly constant rate of flow. With some air samplers, the total volume of air can be read on a digital counter. With others, the total volume of air must be calculated by multiplying the known flow rate by the air-sampling time. The filter is removed from the air sampler. Then, the radioactivity is analyzed, using instrumentation suited to the type of contaminant suspected.

If at all possible, use the type of filter paper recommended by the manufacturer of the air samplers. If that type of filter paper is unavailable, substitute other types of filter paper. However, note that air samplers might give erroneous readings if substitute filter papers are used since coarse filter papers may allow the finest particles to pass through. This gives low collection efficiencies. In addition, coarse papers may absorb some of the alpha radiation from contaminated particles that have been collected.

Air sampling can be done in two ways, depending on the size of the operation. If the operation being monitored is small, take a sample during the operation, with all workers wearing respiratory protection suitable for the aerosol expected. After the sample is analyzed, modify respiratory protection according to the findings. If the operation is large, make a pilot run under conditions as realistic as possible. Analyze air samples from the pilot run, and project the results to the full-scale operation. Continue air sampling through the full-scale operation, monitoring filter papers periodically as a check on the projection from the pilot run.

When monitoring the spread of contamination, set up the air sampler at a height of about five feet, directly downwind from the operation to be monitored, but not further than the line between the contaminated and clean areas. To monitor the inhalation hazard to personnel, set up the air sampler in the highest suspected airborne contamination in which a person may go.

Fallout Samples

Fallout samples can be taken to check the decay rate of fallout. They can also be used to predict decay better than the assumptions made when measurements are not available. Samples collected before and during the recovery phase will be most useful in planning radiological-recovery operations.

Take a fallout sample in the same manner as for a standard wipe sample except that a larger sample is needed to obtain a high-radiation reading. It is advisable to obtain a fresh sample each time there is decontamination. Decontamination can selectively remove certain radioisotopes, and thereby change the decay rate of the remaining mixture of radioisotopes.

These methods should be used by trained health physicists only. Analyze samples by using laboratory counting techniques. If a laboratory counter is not available, use the procedure for analyzing wipe samples, but cover the detector with the beta shield. Plot the net readings on log paper. Compare the plotted points with the average decay rate. If the slope of the plotted data is greater than average decay, then the predicted intensities and exposures calculated by methods based on average decay will be greater than the actual future intensities and exposures. On the other hand, if the slope of the plotted data is less than the average decay, then the predicted intensities and exposures calculated by methods based on average decay will be lower than the actual future intensities and exposures. These differences may be significant for planning purposes.

RADIATION-MONITORING AND DECONTAMINATION TEAMS

A radiation-monitoring team will generally precede or accompany an emergency response team entering a gamma radiation field. These monitors make sure that appropriate precautions are taken to minimize exposure to gamma radiation and advise as necessary.

A monitoring team should consist of around four persons: a monitor, a recorder, a marker, and a messenger. The monitoring team should be equipped with a gamma-intensity radiac survey meter, a self-reading dosimeter, and materials for tabulating data. The team may be separate from the response team, or it maybe made up of emergency responder personnel. They will use the radiac instruments periodically in the course of their work. One monitoring team can probably serve several response teams working close together. A radiation monitoring team can accompany decontamination teams for both technical and radiation measuring support.

As a part of a radiation monitoring team, the members can help a decontamination team operate more efficiently by taking intensity readings of the area being decontaminated, both before and after decontamination passes. Be especially careful to take the readings consistently, and to inform the decontamination team of the effectiveness of its work. Gamma-intensity measurements are usually sufficient to locate the areas that need the most decontamination. Beta-radiation measurements may be required at hand-operated controls. In this case, use a beta-sensitive radiac meter. A decontamination team, supported by a monitoring team, can concentrate on the locations having the highest radiation intensity.

Vary the monitoring technique in support of operations as required to suit the situation. When time is short and gamma-radiation levels are high, the primary concern is determining whether decontamination is reducing the gamma intensity. Measurements should be taken at a number of locations during decontamination, but the only measurements of importance are those taken at the actual locations to be occupied by personnel assigned to the stations. Improvement resulting from decontamination should be noted and communicated to those in charge. Inform the decontamination crews when further decontamination appears impractical or when the acceptable intensity level has been reached.

After initial decontamination and when radiation levels are low enough so detailed monitoring can be performed without exceeding planning exposure levels, hot spots can be located and identified for the decontamination crews. The general radiation levels in the area should be measured first, and then locate hot spots. Locating hot spots with a radiac meter requires a detailed technique. Initially, hold the radiac instrument at waist height, and read the meter while moving slowly and steadily so that body motion does not jiggle or sway the instrument's needle. Move from side to side, and generally forward, until a higher reading appears. The hot spot can be further localize by moving in the indicated direction until the radiac meter is within a few inches of the surface. Hot spots can often be readily suspected by the nature of the material in a given location. For example, rough surfaces, rust, scale, and pools of water are likely to collect contamination.

After decontamination of the assigned area is completed, the monitoring teams should post the general intensity level and the computed maximum safe stay time, based on the planned exposure or the maximum permissible exposure (MPE) as determined by the person in charge. Radioactive decay should be ignored. Post this information in a conspicuous place, together with the time the reading was made. Teams should then report the final monitoring data, including the general intensity level.

When time is available and radiation levels are not high enough to be limiting, a detailed monitoring survey should be conducted. Take readings before and after each decontamination pass. If you find no significant reduction, inform the decontamination team so that adjustments in the decontamination procedure can be made. After decontamination of the assigned area is completed, note the general intensity level, recheck any hot spots, take and check wipe samples as necessary, and report the final monitoring data. At this time, the monitoring teams should post the general intensity level in a conspicuous place, together with the time at which the reading was made.

DECONTAMINATION

Radioactive contamination can result from neutron-activation of atoms in soils and building materials, and from fallout debris. Most of the radiation from activation processes are short-lived, but a few nuclides remain radioactive longer. Fallout consists of unexploded nuclear fuel, activated soil, and fission fragments. In general, the greatest hazards from fallout are beta-emitters, which can cause skin burns as well as damage from inhalation. Radioactivity cannot be altered by chemical processes, so all methods of decontamination rely upon removal and isolation of the radioactivity. Ordinary soap and water and general surfactants (soaps and detergents) can be used to remove radioactivity. It is important to remember that the radiation is only being moved from one place to another or diluted, not destroyed. Rinse water and runoff must be collected and isolated, or removed. Some of the metallic materials can be removed by chelating agents such as ethylenediamine tetraacetate (EDTA) and other surfactants; again, the radioactivity is still present, just washed off and contained. Protection from radiation depends upon time, distance, and shielding; rinse water should be placed in shielded containers for subsequent removal and disposal away from people. There is one very remote concern about collecting radioactive contamination, that of concentrating fissile material. If fissile material is present, there should be no mixing of container contents or increasing the amount of material in a container. Since fallout and other radioactive debris rarely, if ever, contain sufficient fissile material to cause criticality, this should not be a major concern. Remove surface contamination by washing with soap and water, collect the runoff for isolation and disposal, and monitor to ensure that radiation levels are being reduced. This applies to objects as well as people. Wash people thoroughly, but carefully, without abrading the skin.

7

MEDICAL TREATMENT OF WOUNDS FROM NUCLEAR WEAPONS

GENERAL MEDICAL ASPECTS OF RADIATION

This chapter is primarily intended for medical personnel, both at the scene of the emergency as well as at a medical treatment facility. However, even members of the public can learn how to handle limited medical conditions related to nuclear explosions. With small yield tactical nuclear weapons, there will be comparatively large numbers of casualties from initial or prompt radiation, possibly combined with the blast effects. Burn injuries will be more common as the weapon yield increases. The types of injuries associated with nuclear warfare include flash injuries, blast injuries, thermal injuries, and radiation injuries.

Flash Injury

The intense light of a nuclear fireball can cause flash blindness. The duration of blindness depends upon the length of exposure and the light conditions. However, even at night, it is unlikely that flash blindness will last more than a few minutes unless the victim looked directly at the fireball. Most individuals can continue functioning after the short recovery period. Severe cases may have retinal and optic nerve injuries that lead to permanent blindness; these cases will require intense medical treatment. People should be told to try to avoid looking in the direction of any intense light; however, this is difficult since it is a natural response.

Blast Injury

Blast injuries consist of two types:

- Primary injuries due to overpressures, such as ruptured eardrums and lungs.
- Secondary injuries, such as lacerations and puncture wounds, as well as translational injuries from the severe winds.

Thermal Injury

Thermal injuries are generated by:

- Direct thermal radiation, such as flash burns and eye injuries.
- Indirect (flame) effects.

Radiation Injury

Casualties produced by ionizing radiation alone or with other injuries will be common. Radiation complicates treatment by its synergistic action in that burns together with radiation taking longer to heal than burns alone. The unknown nature of either warfare or terrorist radiation limits the ability to determine the patient's total radiation exposure. Additionally, total exposure may not be received at one time, but as the result of multiple exposures in contaminated areas.

MANAGEMENT OF NUCLEAR PATIENTS

Management of people injured from the immediate effects of nuclear weapons or the explosive effects of a "dirty bomb" (flash, blast, thermal) are the same as for conventional injuries, although the injury severity may be increased. First aid (self-aid or buddy aid) for lacerations, broken bones, and burns should be performed.

MASS CASUALTIES

A mass casualty situation is developed by a nuclear attack; that is, the number of patients requiring care exceeds the capabilities of treatment personnel and equipment. Thus, correct triage and evacuation procedures are essential. Triage classifications for nuclear patients differ from conventional injured patients. Nuclear patient triage classifications are as follows:

- Immediate treatment group (T1): Those requiring immediate lifesaving surgery. Procedures should not be time-consuming and should concern only those with a high chance of survival, such as respiratory obstruction and accessible hemorrhage.

- Delayed treatment group (T2): Those needing surgery, but whose conditions permit delay without unduly endangering safety. Life-sustaining treatment such as intravenous fluids, antibiotics, splinting, catheterization, and relief of pain may be required in this group. Examples are fractured limbs, spinal injuries, and uncomplicated burns.
- Minimal treatment group (T3): Those with relatively minor injuries who can be helped by untrained personnel, or who can look after themselves, such as minor fractures or lacerations. Buddy care is particularly important in this situation.
- Expectant treatment group (T4): Those with serious or multiple injuries requiring intensive treatment, or with a poor chance of survival. These patients receive appropriate supportive treatment compatible with resources, which will include large doses of analgesics as applicable. Examples are severe head and spinal injuries, widespread burns, or high doses of radiation; this is a temporary category.

HANDLING AND MANAGING RADIOACTIVELY CONTAMINATED PATIENTS

Patients from fallout areas may have dust-like fallout on their skin and clothing. Although the patient will not be radioactive, he or she may suffer radiation injury from the contamination. Removal of the contamination should be accomplished as soon as possible; definitely before admission into a clean treatment area. The distinction must be made between a radiation-injured patient and one who is only radiologically contaminated. Although patients may have received substantial radiation exposure, this exposure alone does not result in the individual being contaminated. Normally, contaminated patients do not pose a short-term hazard to the medical staff; rather, the contamination is a hazard to the patient's health. However, without patient decontamination, medical personnel may receive sufficient exposure to create beta burns, especially with extended exposure.

To properly handle radiologically contaminated patients, medical personnel must first detect the contamination. Two military radiac instruments discussed earlier, the AN/PDR27 and the AN/VDR2, can be used to monitor patients for contamination. Generally, a reading on the meter twice the current background reading indicates that the patient is contaminated. Monitoring is conducted when potentially-contaminated patients arrive at the medical facility. This monitoring is conducted at the receiving point before admitting the patient. Contaminated patients must be decontaminated before admission.

Removal of surface radioactive contamination is relatively easy. Removing all outer clothing and a brief washing or brushing of exposed skin will reduce 99% of contamination; vigorous bathing or showering is unnecessary. Do not let radiological contamination interfere with immediate lifesaving treatment or the best possible medical care.

RADIATION INJURIES

The extent of radiation injuries depends upon a number of factors, such as the type of radiation, the duration of the exposure, and any concurrent injuries (*e.g.*, burns, broken bones, traumatic injuries). The amount of radiation received will cause increasing levels of injury, but in general include hematopoetic (blood-forming), mental, and blood loss effects. Smaller amounts of radiation, received acutely, will first affect the lymphocytes then the blood platelets and granulocytes. These effects can lead to infection since the body will likely be less able to fight off pathogens. At higher doses, the reasoning ability of the victims (neurological effects) may be diminished. At very high acute doses, severe ulceration of the digestive tract occurs, with blood loss, fluid imbalance, and death. Neglecting other concurrent injuries, the following ranges of radiation exposures should be expected to produce the indicated symptoms and prognoses.

It may not be possible to know the radiation dose received by the patient, but clinical signs and symptoms may be a clue. In fact, from the clinical signs and blood work, it is usually possible to estimate the radiation dose even in the absence of any other information.

For doses between 0 and 70 centigray (cGy), initial symptoms will be none to slight incidence of transient headache and nausea with up to 5% of the victims vomiting, especially at the high end of the range. The centigray is the same as the older unit, the rad. These symptoms, when present, will appear in about six hours and begin subsiding in about twelve hours. The only clinical manifestation is a mild depression of lymphocyte counts at the upper range of the dosage. Patients should receive rest and, possibly, electrolytes.

For doses between 70 and 150 cGy, symptoms are slightly more intense with transient mild nausea with vomiting by up to 30% of those exposed. Initial symptoms appear somewhat quicker, at about two hours, and last longer, up to twenty-four hours. There is increased risk of opportunistic infections. Clinical results include a moderate drop in lymphocyte, platelet, and granulocyte counts.

For doses between 150 and 300 cGy, there will present transient mild to moderate nausea with vomiting in up to 70% of the victims. 25% to 60% of those exposed will show mild to moderate fatigue and weakness. A few deaths may occur, especially at the upper range of exposure, ranging from 5% to 10% of the victims. Opportunistic infections, with attendant fever and bleeding, are very possible for the survivors, even as delayed as a month. Symptoms may appear as soon as two hours and last as long as two days. Bed rest and supportive care should be provided. Antibiotics should be administered unless otherwise contraindicated. Clinically, if there are more than 1.7×10^9 lymphocytes per liter at two days after the exposure, it is unlikely that the individual has received a lethal dose.

For doses between 300 and 500 cGy, there will be transient moderate nausea and vomiting in up to 80% of the victims. Moderate fatigue and weakness will be common in up to 90% of those exposed. These symptoms will usually appear rapidly, within two hours. Later symptoms include bleeding, ulcers, loss of appetite, and diarrhea. After about two weeks, there may be hair loss. Opportunistic infections will be likely, even up to five weeks following exposure. Deaths will range from less than 10% at the lower end of the range to as many as 50% at the upper. Clinically, there will be moderate to severe depression of the lymphocyte count with moderate drop in platelet and granulocyte counts.

At doses between 500 and 800 cGy, the victims will present moderate to severe vomiting, fatigue, and weakness in almost all those exposed. These symptoms will appear quickly, within the first hour of exposure. Bed rest, electrolyte replacement, antibiotics, and general supportive care are called for. Deaths will occur in some 50% at the low end of the range within six weeks. At the high end, up to 90% fatalities will occur between three and five weeks. The clinical results will show almost no lymphocytes after two days. There will be a subsequent severe drop in platelet and granulocyte counts a few days later.

For doses greater than 800 cGy, severe nausea, vomiting, fatigue, weakness, dizziness, and disorientation will be present. There will be moderate to severe fluid and electrolyte imbalance with high fever and collapse within the first few minutes of exposure and lasting until death. At about 1000 cGy, there will be 100% fatalities at two to three weeks, even with supportive care. Clinically, the bone marrow will be totally depleted in two days.

It is important to remember that concurrent injuries can reduce the time to infection as well as increase the fatality rate. A radiation dose of a few hundred centigray can halve the survival rate for persons with serious burns. In a terrorist attack, using a "dirty bomb" together with biological or chemical agents, the fatality rates can be increased considerably over those from any one of the agents alone. It is also important to remember that the length of time in which the dose was received is important, especially at the lower dose levels. Since the body has certain repair mechanisms, damage to cellular DNA can be fixed if there is sufficient time for these processes to take place and the radiation dose was low enough to produce fewer double-strand breaks in the DNA helix. The medical effects above are based upon an acute dose received from the sudden detonation of a nuclear weapon or even dirty bomb. Fallout radiation doses, even if as great, may produce fewer symptoms and fatalities if received over days or weeks rather than minutes.

LOSS OF MEDICAL PERSONNEL

In a nuclear attack, there is a likelihood that some of the losses will be doctors, nurses, emergency medical personnel, hospitals and clinics. At Hiroshima, all the hospitals were either destroyed or severely damaged; some 60% or more of medical personnel were killed or incapacitated. The availability of medical support can be seriously diminished, depending upon the situation. Treating a single burn patient can require a dozen trained workers, hundreds of liters of electrolytes, gallons of blood, and around-the-clock observation. A mass casualty attack could deplete available supplies of electrolytes, blood, gauze, antibiotics, and other materials quickly.

In a terrorist attack, even if a hospital were targeted, these should be sufficient medical support available, including transport capabilities and supplies, to receive, decontaminate, and treat the victims. However, depending upon the terrorist scenario, these might be loss of medical personnel as well as their ability to travel to where they are needed.

TRANSIENT PSYCHOLOGICAL INCAPACITATION

There is another interesting medical effect of radiation. At doses beginning at about 100 cGy, depending upon the rate at which the dose is received, a condition known as transient psychological incapacitation may appear. In this condition, higher levels of brain activity (*e.g.*, reasoning, detailed study) may be diminished. Lower-level functions, like breathing or rote activities, are not as affected. This is important in a combat situation in which nuclear weapons are used. Soldiers, but more especially pilots, might find themselves unable to make critical decisions involving intense thought. They might still be able to fly the aircraft, but be unable to calculate the exact time to release a bomb or missile. Studies are continuing into these radiation effects, and much of the data are classified.

RADIOPROTECTANTS

The Walter Reed Army Institute of Research (WRAIR), in the 1980s and 1990s, undertook a series of studies to see if there were any chemicals capable of rendering soldiers less susceptible to radiation effects. A very few chemicals, mostly containing sulfide linkages, seemed to reduce cellular damage by a small percentage. The first of these chemicals was code named WR-1065. Later, amifostine (ethyol) was seen to be an even better radioprotectant. Developed by the Army as the best of some 4,400 compounds tested, amifostine remains the best drug to date to be tested as a radioprotector, and is sometimes used in cancer therapy using radiation. It is unlikely that further drug development will occur. Using an oral formulation, the U.S. Army evaluated amifostine doses ranging from 100 mg to 5 g in 36 adult men. Toxicities included nausea and vomiting, diarrhea, and abdominal cramps. Although the study lacked pharmacokinetic monitoring, the toxicities observed, even though all were acute and transient, suggested that oral dosing would not be feasible for further drug development. Clinical trials performed primarily at the University of Pennsylvania at Philadelphia continued to investigate the protective properties of amifostine administered before chemotherapy and radiotherapy in cancer patients. The effects of the drug are too small to be useful in a situation of deliberate exposure to radiation.

SUMMARY

The medical treatment of radiation casualties is first to treat wounds and burns, followed by possible antibiotic treatments and electrolyte replacement, and, depending upon the available medical support and amount of radiation received, make a determination of the survivability of the patient. On a nuclear battlefield or aftermath of a massive nuclear attack, care may have to be given only to those less exposed to radiation since the others might take up too many assets even though they are going to die. Such difficult decisions might be needed in war. In a terrorist scenario, most likely using a dirty bomb, medical support services should be capable of treating all the victims with means appropriate to their injuries: traumatic, infectious, or radiation.

Part 2

BIOLOGICAL WEAPONS

8

BRIEF HISTORY OF BIOLOGICAL WEAPONS

USE OF BIOLOGICAL WEAPONS IN ANCIENT TIMES

The use of biological agents in war is not new, but prior to the twentieth century, biological warfare generally took three forms: deliberate poisoning of food or water supplies with infectious material, use of microorganisms or toxins in a weapon system, or the use of biologically-contaminated fabrics. Drinking wells were poisoned with rye ergot by the Assyrians and Persians in the 6th and 4th centuries B.C. Ergot, a mycotoxin from a species of fungus, causes a very painful death from loss of blood, following nausea, loss of coordination, and diarrhea. Arrows were dipped in blood mixed with manure by Scythian archers, circa 400 B.C.

The Roman general Hannibal used clay pots filled with poisonous snakes in a battle in 184 B.C. with the navy of King Eumenes of Pergamon. The snake pots were thrown onto the decks of the enemy's ships, where they broke, forcing the Pergamene sailors to fight against both the snakes and Hannibal's forces. As a result, Hannibal was victorious. The Romans also used bees and hornets by catapulting them at their enemies in other battles.

The Romans also poisoned wells in a number of places. After defeating Aristonicus in Asia Minor at Caria, the Romans were faced with guerrilla warfare by remnants of Aristonicus' army. The Romans poisoned the wells that the guerillas and their local supporters used. The result was the end of the war leaving Rome in control of the region.

In 67 B.C., the Roman general Pompey was leading a campaign against Mithridates, the King of Pontus. As his troops passed through the Trebizond region (in modern Turkey), the people of the region, the Heptakometes, who were allied with Mithridates, put out honey as an apparent tribute. The honey was a local product, which we now know contained acetylandromedol, a grayanotoxin formed when the bees took pollen from rhododendron flowers. This toxin produced nausea and hallucinations, and perhaps some deaths, among the troops who consumed it, and three maniples of the Roman army were attacked and destroyed while intoxicated.

Biological warfare was also practiced in the Middle Ages when the catapulting of diseased or putrefying corpses into castles became a standard feature of siege warfare. The earliest incidence of this, which is well reported, comes from the siege of Thyne Levesque (now called Thun l'Eveque) in the year 1340 A.D. in Flanders (now in northwest France) during the Hundred Years' War, where French siege engines were used to throw corpses of horses over the castle walls. While no disease appears to have actually broken out, the defenders did subsequently withdraw, and the Duke of Normandy won.

Dropping dead horses or other animals into city wells was a fairly common method of polluting an enemy's water supply. It was apparently fairly common in the Middle Ages that bodies of disease victims were catapulted over castle walls in order to spread contagion among the occupants. During the Crusades, armies would apprehend leprosy or plague victims (human or animal; living or dead), mount them upon catapults, and propel them over the walls of a city under attack. Often, they would place the diseased individual within containers with other carcasses or still-surviving victims along with all forms of refuse and excrement before being hurled over the walls. The end result would be the spread of disease within the castle walls, resulting in a weakening of the castle garrison without ever firing a shot. The major drawback of this technique was that medieval medicine was very archaic and the treatment of disease almost non-existent. The victorious army would enter the poorly-defended fortification to become victims of the same disease that they had intentionally spread upon their former enemies. In fact, this remains one of the major drawbacks to the use of biological weapons.

The contents of cesspools were often used by medieval artillery as a bioweapon. For example, in 1422, during the war between the Protestant people's army (the Hussites) and the Catholic King Sigmund the Luxemburg, the Hussite army emptied all the cesspools in Prague during the blockade of the royal Castle Karlstein near Prague (today in the Czech Republic). Records do not indicate if contagion broke out as a result. The garrison of the Castle resisted successfully the attack by this unusual ammunition.

Legend has it that the Black Death that possibly killed as much as a quarter of Europe's population in the fourteenth century may have been spread by refugees from the siege of the Genoese settlement of Kaffa in the Crimea (today Feodosija, Ukraine) when the Mongols threw the bodies of plague victims over the walls. Actually, plague was present before that battle, hence the victims. Both sides suffered many cases of the disease, but is not known how many were due to the action of throwing corpses into the city.

Chemicals derived from biological organisms have been used as poisons. Batrachotoxin was obtained from tree frogs in Hawaii, and used by the native Hawaiians in warfare by tipping spears with the toxin. Aconite, the dried tuber of the monkshood plant (*Actonitum napellus*) containing the cardiac and respiratory sedative aconitine ($C_{34}H_{47}NO_{11}$), was used by the Moors in Spain in 1483.

The natives of the New World also used biological weapons. Arrows poisoned with curare were used by the Amazon Indians to kill enemies. In South America, the smoke from burning chili peppers was used by the Inca in local conflicts and against the Portuguese as a sort of tear gas, while in North America, there are several incidents that suggest that natives may also have tried to cause smallpox outbreaks among the European soldiers and settlers.

During the French and Indian War (1754–1763), British soldiers may have given contaminated blankets from smallpox victims to a number of Native American tribes. The best known example was in 1763, during the Pontiac Rebellion, when the British commander of Fort Pitt found himself in a difficult situation. Three forts in the Ohio River watershed had been lost to the French and their Native American allies, and many refugees had come to Fort Pitt seeking safety. The Fort was crowded and uncomfortable, and smallpox broke out among the refugees. When two Delaware Indians appeared at the fort seeking a surrender, the British refused, but gave them as a gift two blankets and a handkerchief which had been used by smallpox patients. The Delaware tribe did suffer severely from an epidemic of smallpox, although the infected materials may not have been the only sources of infection.

In 1775, the British in Boston found themselves facing both the Continental Army and a smallpox epidemic. They began to vaccinate their troops. But, they also began to vaccinate some of the civilians fleeing the city. While there is no proof that the British deliberately intended to spread smallpox, it was well known that people shortly after vaccination become infectious. General Washington, the commander of the Continental Army, wrote that he was convinced that the enemy intended to spread smallpox among his troops. He had been encouraged in this belief by the stories of British deserters, about which he had been skeptical until an outbreak of the disease among the evacuees convinced him. He delayed his attack, fearing the smallpox in Boston. When the city was finally evacuated by the British, Washington advanced very carefully, fearing the exposure of his troops to the disease.

A letter sent to British General Cornwallis in 1781 informed him that some 700 blacks suffering from smallpox were to be sent to plantations to spread the disease among the rebel population. The evidence is clear that the intent was deliberate biological warfare. Poisoning of drinking water wells occurred in the nineteenth century during the American Civil War (1861–1865) and in South Africa during the Boer War (1899–1902).

MODERN HISTORY OF BIOWEAPONS

In World War I, German agents attempted to disrupt the food supply of the Allies by deliberately infecting sheep, cattle, horses, mules, and reindeer with glanders and anthrax, both diseases known to ravage populations of grazing animals in natural epidemics. Their target was livestock, the horses, mules, sheep, and cattle, being shipped from neutral countries to the Allies. By infecting just a few animals, through needle injection and pouring bacteria cultures on animal feed, German operatives hoped to spark devastating epidemics in countries supplying the Allies. Secret agents waged this campaign in Romania and the U.S. from 1915–1916, in Argentina from roughly 1916–1918, and in Spain and Norway, but the exact dates are obscure. While some animals died, there was apparently no real impact on the war effort and no zoonotic transmission of the disease. This attempt at biological warfare was largely unsuccessful.

Between 1932 and 1945, Japan killed some 260,000 people in China with biological weapons, chiefly plague. To learn more about bioweapons, the infamous Imperial Japanese Army Unit 731 studied the effects of diseases, like anthrax, typhoid, cholera, and plague, on as many as 10,000 prisoners in occupied China. They often dissected their victims while still alive, without anesthesia. Having learned much about the use of bioweapons, the Japanese went on to kill as many as 250,000 more in their war in China. In active military campaigns, several hundred thousand people, mostly Chinese civilians, fell victim. In 1940, the Japanese dropped paper bags filled with plague-infested fleas over the cities of Ningbo and Quzhou in Zhejiang province. Other attacks involved contaminating wells and distributing poisoned foods. The Japanese army never succeeded in producing advanced biological munitions, such as pathogen-laced bombs. At the end of the war, the U.S. granted amnesty to the researchers of Unit 731 in exchange for their data.

It is interesting that Japan's ally, Germany, which committed atrocities by the scores in horrible "medical" research, apparently did not have as extensive a bioweapons program as the Japanese during World War II. Perhaps, the German military understood that the relatively small contiguous areas in which they were fighting would possible subject their own troops to any biological weapons. The Japanese were careful to use them only in far-away areas of China and Manchuria. Germany did conduct research, but did not use any biological weapons, so far as it known.

While not used in warfare, British military tests in 1942 contaminated Gruinard Island in Scotland with anthrax spores and rendered it unusable for the following 48 years. A release of anthrax spores in U.S.S.R. killed at least 68 people in 1979 in the city of Sverdlovsk. An accident at the bioweapons facility allowed a plume of anthrax spores to spread over the city and surrounding farmland. In addition to the people who died, large numbers of cattle were also killed by inhalational anthrax, the most virulent form.

Toward the end of the 1970s, reports began coming out of Afghanistan, Kampuchea, and Laos that people, mostly the Miao and Hmong, were dying a horrible death shortly after military aircraft would fly over. They told stories of people bleeding from their ears, eyes, mouth, and skin after getting a yellow-colored dusty material on themselves. They called the material "yellow rain." Communist leaders of Viet Nam, Laos, the Soviet Union, and China denied that there was a toxin involved. A massive disinformation campaign was undertaken to make those who thought the "yellow rain" might be a military attack appear to be crazy. Years later, U.S. Army researchers discovered that the material was a toxin produced by fungi, but weaponized to be spread using bomblets or aerosol dispersion units. Called trichothecenes, these materials are highly toxic and exposure leads to a quick and painful death by hemorrhaging. It seems that there had been natural occurrences in rural areas of Russia with the fungus growing on wheat or rye producing the toxin, similar in nature to ergot from other fungi. The scientists in the Soviet Union were able to concentrate this toxin into a military weapon. A test method has been developed by the U.S. Army to indicate the presence of these toxins.

With the declaration that smallpox has been eradicated on Earth, smallpox became a potential bioweapon since people were no longer vaccinated. Some nations are known to have been working on gene splicing, that is, introducing different DNA fragments into genes, to produce more virulent forms of smallpox resistant to vaccines. The U.S. and the U.K. are considering a return to routine vaccination against smallpox, at least for members of the military. This action might, however, increase the potential for use of the bioweapon in terrorism since the military would be largely protected but the civilian population susceptible.

ADVANTAGES AND DRAWBACKS FOR THE USE OF BIOLOGICAL WEAPONS

Ideal characteristics of biological weapons are low visibility, high potency, accessibility, and easy delivery. Diseases most likely to be considered for use as biological weapons are those which have high lethality if delivered efficiently, and general robustness to make aerosol delivery feasible. The primary difficulty in use of a biological weapon is not the production of the biological agent, but rather the delivery of the agent in a form in which it will infect a large number of people. An attack using anthrax would, for example, require the creation of aerosol particles of precise size. If the particles are too large, they would be filtered out by the nasal portion of the respiratory system. If the particles are too small, the aerosol would be inhaled and promptly exhaled. Delivery of the aerosol in such a way that the organisms would not be destroyed or weakened by weather, and packaging the anthrax so that it would remain active are two of the technological difficulties involved in mounting such a biowarfare attack.

Biological weapons can be manufactured without much difficulty and in a relatively short time, given the proper conditions and media. A biologist with an average education and a minimum of tools and space can manufacture these weapons in a small space. The main advantages of biological weapons are information availability, the restricted number of resources necessary to carry on the project, and the possibility to test the final agent before employment. Diseases likely to be considered for use as biological weapons include anthrax, ebola, pneumonic plague, cholera, tularemia, brucellosis, Q fever, Venezuelan Equine Encephalitis (VEE), Staphylococcal Enterotoxin B (SEB), and smallpox. Naturally occurring toxins that might be used in weapons include ricin, botulism toxin, and mycotoxins.

In general, the difficulties with use of biological weapons lie in packaging and delivery of the agent. The organism or toxin must not be subjected to high heat typical of bombs and artillery shells because they can be destroyed or rendered nonviable. Most, but not all, bioweapons require inhalation; the particles must be within a certain size range, usually around one micron in diameter. To place liquid or solid materials into the air for breathing exposure requires the formation of an aerosol. The generation of an aerosol depends upon weather conditions as well as characteristics of the particles. Most bombs, bomblets, and shells do not produce a good aerosol over a large area. It is much easier to generate an aerosol using spray equipment like that used to dispense insect fogging agents or crop dusting. However, this requires, in time of war, to place equipment behind the lines.

There is another major drawback to the use of bioweapons in war: soldiers of the side using them must be protected from the disease. No vaccine is 100% effective, and some of these agents have no antidote or treatment. Release of a bioweapon could contaminate the very land that an invading army desires to later occupy. There are political risks as well for the nation using bioweapons. Finally, soldiers of the other side would, most likely, have been vaccinated against the agent. Like chemical weapons, the most likely scenario for the use of bioweapons is against civilians, food crops, or livestock.

The most likely biological weapon attack would be from a terrorist group, possibly with state help. The costs are low, the technology simple and well known, the effects are frightening, and the possibility of detection is low. The most likely method of spreading the material would be by use of an aerosol generator like a pesticide spray apparatus or a crop dusting airplane.

Weather can play a vital role use of biological weapons. Dispersion of the aerosol cloud depends upon prevailing winds; hopefully, in the direction away from the using party. Yet, winds can change directions quickly. There is always a danger of self-inflicted casualties due to wind direction and speed changes. The area covered by aerosol can be reduced if the wind speed falls. In general, the total area affected will be an elongated oval with a tip at the point of aerosol release and long axis in the direction of the wind. Temperature can also be important for viable organisms; most can only reproduce within a narrow range of temperatures. Use of bioweapons in extremely cold weather is limited. Rain and high humidity can affect toxins by reducing their concentration, and, thus, effectiveness.

As with any weapon system, whether armor, aircraft, artillery, ships, or satellites, there are optimal times, locations, and situations for the use of biological weapons, should a nation or group decide to employ them. The advantages of bioweapons are: they are cheap; they are easy to produce in small laboratories; they can be highly lethal to unprotected populations; they often have delayed symptoms; they are frightening; and they are able to affect large numbers of victims. Some of the disadvantages of bioweapons are: they are difficult to package and deliver; they are often affected by weather and climate; they can affect populations other than the target; and their use is outlawed and considered inhumane.

U.S. BIOWEAPONS PROGRAMS

The United States' bioweapons program began during World War II, and the government developed and produced a number of biological weapons. The U.S. had an extensive offensive biological weapons program that took place at a number of facilities such as Fort Detrick in Maryland, Dugway Proving Ground in Utah, and Pine Bluff Arsenal in Arkansas. The U.S. developed tularemia as a standardized, that is ready for use by troops, lethal weapon, and Venezuelan equine encephalitis (VEE) as a standardized non-lethal weapon. The U.S. also developed brucellosis weapons and anti-crop fungal weapons. The program culminated in a large series of successful aerosol tests, using monkeys and rodents as test animals, in the Pacific Ocean in 1969. The U.S. investigated a large number of organisms for possible use as a bioweapon: anthrax, smallpox, botulism, fungi, etc. A building at Fort Detrick, in Maryland, became contaminated with anthrax spores. It was cleaned, and is thought to be completely safe, yet no one is allowed inside or even near it. This research came to a halt at the end of 1969, when President Richard Nixon terminated the program. President Nixon, after reviewing all the options, decided not only to terminate the biological weapons program, but to even renounce the option for the U.S. to have any offensive biological weapons program at all. The U.S. bioweapons program was ahead of that of the U.S.S.R. at that time. The U.S. then took the lead in pushing for a comprehensive biological weapons treaty to outlaw such weapons.

In 1972, with the signing of the Biological Weapons Convention, the U.S. agreed to permanently continue the ban that had been in effect since 1969. The U.S. may have terminated all offensive research, but continued limited studies for defensive purposes. Just prior to that, a number of sheep suddenly died near Dugway Proving Ground in Utah. Anthrax spores, being used in a research study, had escaped. Compensation was paid to the ranchers, and they were told not to tell anybody.

Today, the U.S. continues a defensive program to study vaccines and treatments for bioweapons. To do this, however, requires that the actual agents be available to cause the effects being prevented or cured. The remaining defensive work is largely carried on by the U.S. Army Medical Research Institute for Infectious Diseases (USAMRIID) at Fort Detrick, Maryland, which also studies natural diseases and treatment.

U.S.S.R. BIOWEAPONS PROGRAMS

During the Cold War, the U.S.S.R. had as many as 60,000 workers at some 50 locations conducting bioweapons research. Most of the laboratories were in Russia or

Kazakhstan, partially out of distrust of some of the other republics. While it is claimed that the research facilities have been locked up and their stocks destroyed, the knowledge possessed by the scientists remains.

The U.S.S.R. issued a declaration in 1992, required under the Biological Weapons Convention, recounting the history of their program. The Soviets felt that the U.S. was far ahead of them in bioweapons in 1969. They decided to be able to produce biological weapons as soon as a call came from Moscow. They constructed major facilities for making biological warfare agents. According to the Soviets, they were all in standby status under a loophole in the Biological Weapons Convention. The Convention prohibits the development of, production of, and stockpiling of biological weapons, but it doesn't prohibit the construction of facilities to produce such weapons. That was left out and, so that's what they did. In fact, some of these facilities actually produced and stockpiled bioweapons.

It is known that Soviet scientists weaponized inhalational anthrax, smallpox, botulism, and plague. The U.S.S.R. bioweapons program, which continued in spite of that nation being signatory to the Biological Weapons Convention, is suspected of also studying gene splicing of anthrax DNA into *Escherichia coli* bacteria. This bacterium is habituated to live in the human digestive tract; in fact, much of the mass of feces is composed of bacteria. An altered organism could live in the intestines and cause anthrax systemically rather than by inhalation. Other diseases studied by the Soviets were tularemia, cholera, and other animal diseases, probably to attack cattle.

In 1978, the Bulgarian dissident Georgi Markov was assassinated by the Bulgarian secret police, who stabbed him with a modified umbrella tipped with a platinum pellet containing ricin. He died of unknown causes in the 1970s. Upon autopsy, there was nothing out of the ordinary except for a small puncture wound on his ankle. Inside the wound was found a small metal pellet coated with ricin. Ricin is a chemical found in small quantities in the castor bean. It is highly toxic, causes death within a few days, and requires only 12–24 hours to act. Symptoms include nausea, vomiting, fever, bloody diarrhea, cramps, breathing difficulties, kidney failure, and circulatory collapse. The use of ricin as a bioweapon was developed by the Soviet Union. In 2003, ricin was discovered in a London apartment occupied by Middle Easterners.

Bread accidentally made from *Fusarium*-infected wheat killed thousands of Russian civilians shortly after World War II, focusing Soviet attention on the chemistry of mycotoxins called trichothecenes. When the Soviets discovered these mycotoxins, they developed a research program to weaponize them. During the late 1970s, these mycotoxins, chiefly the T-2 Toxin, were used in Afghanistan, Kampuchea, and Laos as a chemical warfare agent by Russian troops. Because of the appearance, the locals called it "yellow rain."

In the 1970s and early 1980s, the U.S.S.R. began developing new biological weapons such as Marburg, Ebola, and Bolivian hemorrhagic fever. The Soviet Union had two directorates responsible for developing and manufacturing biological weapons, but all the weapons were stored at Ministry of Defense military facilities. The Kirov facility was responsible for storing about 20 tons of plague. The Zagorsk facility (now called Sergiev Posad) was responsible for storing about 20 tons of smallpox biological weapons. The Sverdlovsk facility (now called Ekaterinburg) was responsible for continuous manufacturing of hundreds of tons of anthrax biological weapons. News of the immensity of the Soviets' biological weapons program began to reach the West in 1989, when biologist Vladimir Pasechenik defected to Britain. The stories he told of genetically altered "superplague," antibiotic-resistant anthrax, and long-range missiles designed to spread disease that were confirmed by later defectors and others after the fall of the Soviet Union.

The most potent known form of inhalational anthrax, code named Anthrax 836, was manufactured by the Soviet Union at a number of sites including Sverdlovsk, Sergiyev Posad, Zagorsk, and Stepnagorsk in Kazakhstan, beginning in 1987. The spores are a very fine, gray-brown, silk-like powder which can be carried for many miles by air currents. It is too small to be seen with the naked eye. They were grown in enormous fermentation tanks, very similar to those used to make alcoholic drinks like beer and whisky. An accidental release from one plant in Sverdlovsk, Compound 19, caused the death of 68 people in 1979. The Soviet authorities denied that it was a bioweapons accident even to the families of the victims. Later, after news leaked out, they claimed it was a natural outbreak that can occur in rural communities.

It is also known that Russian scientists were successful in gene splicing portions of the DNA of the smallpox virus in order to make it no longer susceptible to vaccines available in the West. A vaccine was developed to counter this man-made strain of smallpox.

If Russia were to desire to restart manufacturing biological weapons, it would take no more than two to three months to start this activity again because there are at least four military facilities that could be used for manufacturing them. In addition to these plants, Russia continues operating several *Biopreparat* facilities that could be used for mobilization capacity. It is very likely that Russia, and other nations like Israel, have extensive defensive biological weapons program.

CHINESE BIOWEAPONS PROGRAM

China is widely reported to have active programs related to the development of biological weapons, although essentially no details of these programs have appeared in the open literature. In 1939 the Imperial Japanese Army established their Unit 731 germ-warfare research center in Harbin, China, where Japanese medical experts experimented on Chinese, Soviet, Korean, British and other prisoners. It is possible that the Chinese used these facilities as a starting point for bioweapons research.

China possesses an advanced biotechnology infrastructure as well as the requisite munitions production capabilities necessary to develop, produce, and weaponize biological agents. Although China has consistently claimed that it has never researched or produced biological weapons, it is nonetheless believed likely that it retains a biological warfare capability begun before acceding to the Biological Weapons Convention. China is commonly considered to have an active biological warfare program, including dedicated research and development activities funded and supported by the government for this purpose. There is essentially no open source data on the subject of Chinese biological warfare activities, and many legitimate research programs use similar, if not identical equipment and facilities.

GENEVA PROTOCOL OF 1925

After the horrors of thousands gassed in World War I, many of the nations of the world met in Geneva, Switzerland to set limits on the use of weapons in war. The resulting Protocol, in addition to banning chemical weapons, also outlawed the use of all forms of biological weapons in warfare.

The Protocol for the Prohibition of the Use in War of Asphyxiating, Poisonous or other Gases, and of Bacteriological Methods of Warfare, usually called the Geneva Protocol, was signed at Geneva on June 17, 1925 and was registered on September 7, 1929. It prohibited the use of biological weapons, but had nothing to say about production, storage, or transfer. The Protocol prohibited all signatory nations from using bioweapons; however, its provisions were largely ignored by all the great powers, including Germany, Britain, the U.S., and the U.S.S.R. In fact, it took fifty years for the U.S. Senate to ratify it, even though the U.S. had signed it. Japan refused to sign at all.

BIOLOGICAL WEAPONS CONVENTION OF 1972

The general failure of the Geneva Protocol led many nations to fear the use of biological weapons in future wars. In 1972, under the auspices of the United Nations, a meeting was held to create a more enforceable prohibition. The Convention on the Prohibition of the Development, Production, and Stockpiling of Bacteriological and Toxin Weapons and on their Destruction, usually called the Biological and Toxin Weapons Convention or just the Biological Weapons Convention (BWC), was signed on April 10, 1972 and entered into force on March 26, 1975. It endorsed the Geneva Protocol of 1925 in prohibiting the use of biological weapons and extended it with prohibitions on development, storage, or transfer, with exceptions for medical and defensive purposes in small quantities. However, it lacked inspection provisions. The Biological Weapons Convention banned the research, production, stockpiling, and use of almost every form of biological agent, including toxins. It did allow for research into defense against bioweapons and did not specifically outlaw the building of production facilities.

The U.S.S.R., in spite of being a signer, violated the Convention from the beginning. There is no evidence that any of the Soviet research laboratories or production facilities were eliminated or scaled back.

The U.S., on the other hand, ceased all research into offensive uses of bioweapons in 1969, destroyed all stocks and culture media, and shut down all facilities capable of producing bioweapons. The Biological Weapons Convention bans all offensive biological weapons, but the goal of the treaty is total elimination of bioweapon systems; however, defensive work is allowed. By mid-1996, 137 countries had signed the treaty.

FUTURE USE OF BIOWEAPONS IN WAR

It is known that Iraq under Saddam Hussein possessed offensive bioweapons, including weaponized anthrax and trichothecenes; the fate of these materials is unknown. Other nations suspected of either having or researching bioweapons include Russia, China, Iran, Libya, the Sudan, Israel, Syria, India, North Korea, and Egypt.

By the time of the Persian Gulf War cease-fire in 1991, Iraq had weaponized anthrax, botulinum toxin, and aflatoxin, and had other lethal agents in development. Inspectors from the U.N. Special Commission (UNSCOM) spent frustrating years chasing down evidence of the program, which Iraq repeatedly denied existed, then admitted in lists of agents which they claimed to have destroyed. The UNSCOM team found that Iraq's stockpile included Scud missiles loaded to deliver disease. Iraq is known to have unleashed chemical weapons in the 1980s, both during the Iran-Iraq war and against rebellious Kurds in northern Iraq, but there is no evidence that the Iraqi state has ever used its biological arsenal.

The military utility of biological weapons is limited because of intrinsic properties of the pathogens and toxins. Some of these are difficulties in effective dispersion, unpredictability of the dispersion, and weather and climatic conditions. Airborne release of biological warfare agents must be so that a relative high proportion of the agent filling survives the separation from the delivery system. Biological agents travel in the atmosphere as particles of microscopic clusters of individual pathogens or droplets of toxin that must be of the right size in order to enter the body and cause disease and in sufficient quantity. The separation of the agent from the delivery system can happen by means of a small explosion to break the casing, but the heat and shock may kill a majority of pathogens or destroy the toxins. Similarly, an aerosol generator also requires advanced design in order to produce sufficient particles of the right size. Such considerations are particularly important if a bioweapons user selects a non-contagious pathogen that requires each victim be exposed individually to a sufficient infective dose.

Because biological warfare agents travel as aerosol particles, they are dependent on environmental variables, such as air flow at different altitudes, air density, humidity, etc. These variables are beyond control of the bioweapons user after the release of the agent. If a bioweapons user selects a highly contagious agent so that the disease can be transmitted from victim to victim, then a risk exists that his own troops or civilian population will also fall victim to the disease, which means that such agents can only be considered for use in the deep rear of an enemy.

Ultraviolet rays in sunlight and oxygen in the atmosphere may kill most of pathogens surviving the separation from the delivery system, and humidity can weaken toxins, so biological weapons can only be effectively released during certain parts of the day or under certain weather conditions.

Diseases require time to develop. The delays may range from hours to days with respect to a single victim and may be even longer before there are militarily significant effects in the target population. While the delay also makes it difficult to identify the bioweapons user and give time to escape detection, it means that the effects of the attack might not be known for a while. In most military operations, timing is critical and, often, immediate.

A drawback to the use of biological weapons in war is that soldiers of the side using them must be protected from the disease or toxin. Vaccines are not absolutely effective, and some agents have no antidote or treatment. Release of a bioweapon could contaminate the very land that an invading army desires to occupy. Enemy soldiers would, most likely, have been vaccinated against the agent, or, at least, have some protection against it. As mentioned earlier, for these reasons, the most likely scenario for the use of bioweapons in war is against civilians, food crops, and/or livestock.

USE OF BIOLOGICAL WEAPONS IN TERRORISM

In 1984, followers of the Indian guru, Bagwan Shree Rajneesh, living on a rural compound in Oregon, sprinkled *Salmonella* on salad bars throughout their county. It was a trial run for a possible later larger attack The Rajneeshees' scheme was to sicken local citizens and thus prevent them from voting in an upcoming election which contained candidates and items not to the cult's liking. The trial attack triggered more than 750 cases of food poisoning, 45 of which required hospitalization. The Centers for Disease Control and Prevention (CDC) launched an investigation, but wrongly concluded that the outbreak was natural. It took a year, and an independent police investigation, to discover the true source of the attack.

It is known that, in 1995, the Japanese apocalyptic religious sect Aum Shinrikyo released Sarin (Agent GB) gas in a Tokyo subway, killing 12 commuters and injuring thousands, but the cult also enlisted scientists to develop bioweapons and to launch biological attacks. Between 1993 and 1995, Aum Shinrikyo tried as many as 10 times to spray botulinum toxin and anthrax in downtown Tokyo, but without success. Just why the attacks failed is not known, but some experts suspect the cult did not sufficiently refine the particle size of its agents and that it was working with an avirulent strain of anthrax.

A week after the terrorist attacks of September 11, 2001, a letter containing weaponized anthrax spores was mailed to Tom Brokaw at NBC News in New York City. Two other letters with identical handwriting, venomous messages, and lethal spores arrived at the offices of the *New York Post* newspaper and Senator Tom Daschle in Washington, DC. In all, 23 people were infected with inhalational anthrax, five died of the disease, and hundreds of millions more were struck by anxiety and fear of the unknown. Many post offices and mailrooms, including the U.S.

Capitol, had to be decontaminated using chlorine dioxide gas to render the spores inactive.

The most likely direct biological weapon attack would be from a terrorist group, possibly with state help. The costs are low, the technology well known, the effects are frightening, and the possibility of detection is low. The most likely method of spreading the material would be by use of an aerosol generator like a pesticide spray apparatus or a crop dusting airplane.

Another possible area of concern is a terrorist group contaminating a water or food supply. The risk of biological contamination of potable water systems in the U.S. at least is very low since modern water treatment plants utilize possesses which would inactivate or destroy the pathogen or toxin, as well as having routine checks on the quality of the water being produced. The only possible way to affect a drinking water system would be post-treatment, but this limits the extent of the infection. Also, in the U.S. and many other nations, there is a requirement for a chlorine residual in drinking water that would oxidize pathogens and many toxins. In Europe, where ultraviolet light is commonly used to disinfect water, there is no residual that could kill pathogens; they would be more at risk.

Likewise, biological contamination of food would have to be done after processing for most foods. The extent of disease would be limited to a local region in most scenarios. The fear factor, however, would offer the terrorists justification for such actions. For example, when a number of people were infected from eating undercooked hamburgers in a national chain restaurant and a young boy died, thousands became frightened. In these cases, the organism was a variant of *Escherichia coli* (*E. coli*) bacteria, a serotype called *E. coli* O157:H7. In 1982, following outbreaks of foodborne illness that involved several cases of bloody diarrhea, *E. coli* O157:H7 was firmly associated with hemorrhagic colitis. As a result of this association, *E. coli* O157:H7 had been designated as enterohemorrhagic *E. coli*, or EHEC. Most cases involved undercooked ground beef, but a few resulted from apple juice, raw vegetables, and other foods. While these outbreaks resulted in only a few deaths, fear and panic was spread in the communities. A terrorist organization, seeking to disrupt a society, could use similar pathogens to cause fear. Use of bioweapons by terrorists is far more likely than their use in war between armies.

9

BIOWEAPON DISEASES AND MEDICAL TREATMENTS

INTRODUCTION TO BIOWEAPONS

There are many organisms capable of causing disease. Most of these are bacteria, rickettsia, fungi, and viruses. Bacteria are single-celled organisms that possess cell walls and cell membranes and belong to the plant family, and have chromosomes composed of DNA. They reproduce under most circumstances via cell division. Most are described by their appearance, that is, spherical are known as cocci, rod-shaped as bacilli, chains as strepto-, and clumps as straphylo-. Some bacteria are commonly found as pairs, like the diplococci.

Viruses are submicroscopic infectious obligate intracellular parasites and cannot be grown outside of an appropriate cell; the only way to reproduce is to use a host-cell system. Viruses contain either RNA or DNA template molecules, normally encased in a protective coat of protein or lipoproteins. The viral nucleic acid is able to organize its own replication only within suitable host cells.

Rickettsiae are infectious bacteria that are obligate intracellular parasites; the only way to reproduce is to use a host-cell system. In general, rickettsiae retain all the intracellular components of bacteria including a cell wall, a plasma membrane, and cytoplasm with ribosomes and DNA. They replicate by cell division. Rickettsia are most often transmitted by the bite of infected ticks or fleas. Some other rickettsiae are transmitted by lice, including *Bartonella (Rochalimaea) quintana*, the agent of trench fever. Trench fever was a significant medical problem during World

War I, and has recently reappeared among AIDS victims and homeless and alcoholic persons.

Fungi are a form of primitive vegetable life. Examples include mushrooms, yeasts, and molds. They are plant-like eukaryotes, that is, having a cellular nucleus, with cell walls. They can reproduce both sexually and asexually. Some cause disease, that is, any abnormal condition, only in certain species; others only in plants or animals. Many act by interfering with biological functions of the host.

There are also toxins, chemicals naturally produced by living organisms, which can cause disease. Several rare diseases can be caused by proteins. These are known as prions.

It is important to remember that the use of any biological organism or toxin is in violation of international accords. However, a rogue state or terrorist group might be willing to risk condemnation because of the availability, ease, and cost of producing biological weapons. An additional fact is that an agent does not necessarily have to be fatal to be effective; tying up medical workers and the health care system might serve a terrorist better than causing deaths. Finally, an agent can also be used against food crops and farm animals as well as against humans. All of the following have either been considered or actually stockpiled as bioweapons.

AFLATOXIN

Description

Aflatoxins (from the Latin name of the fungus *Aspergillus flavus* + toxin) are toxic metabolites produced by certain fungi in and on foods and feeds. The most likely products they are found on are peanuts, other legumes, and some grains. They are probably the best known and most intensively researched mycotoxins in the world. Aflatoxins have been associated with various diseases, such as aflatoxicosis, in livestock, domestic animals and humans throughout the world. The occurrence of aflatoxins is influenced by certain environmental factors; the extent of contamination will vary with geographic location, agricultural and agronomic practices, and the susceptibility of commodities to fungal invasion during preharvest, storage, and processing periods. Aflatoxins have received greater attention than any other mycotoxins because of their demonstrated potent carcinogenic effect in susceptible laboratory animals and their acute toxicological effects in humans. Aflatoxin is a naturally occurring mycotoxin produced primarily by two types of mold: *Aspergillus flavus* and *Aspergillus parasiticus*. *Aspergillus flavus* (*A. flavus*) is common and widespread in nature, and is most often found when certain grains are grown under stressful conditions such as drought. The mold occurs in soil, decaying vegetation, hay, and grains undergoing microbiological deterioration and invades all types of organic substrates whenever and wherever the conditions are favorable for its growth. Favorable conditions include high moisture content and high temperature. Other fungi of the same family that produce aflatoxins include *A. nomius* and *A. niger*. At least 13 different types of aflatoxin are produced in nature with aflatoxin

B1 considered as the most toxic. While the presence of *Aspergillus flavus* does not always indicate harmful levels of aflatoxin, it does mean that the potential for aflatoxin production is present. Chemically, Aflatoxin B1 is 2,3,6a,9a-tetrahydro-4-methoxycyclopenta[c]furo[3',2:4,5-]furo[2,3-h][1][benzopyran-1,11-dione. Aflatoxin B1 is known to cause liver cancer.

Because of the debilitating symptoms and easy availability, aflatoxins are likely terrorist bioweapons. They are less likely to be used as general warfare weapons because they are easily destroyed by oxidizers.

Signs, Symptoms, and Prognosis

Aflatoxicosis is primarily a hepatic (liver) disease. The susceptibility of individual animals to aflatoxins varies considerably depending on species, age, sex, and nutrition. In fact, aflatoxins cause liver damage, decreased milk and egg production, recurrent infection (*e.g.*, salmonellosis) as a result of immunity suppression, in addition to embryo toxicity in animals consuming low dietary concentrations. While the young of a species are most susceptible, all ages are affected but in different degrees for different species. Clinical signs of aflatoxicosis in animals include gastrointestinal dysfunction, reduced reproductivity, reduced feed utilization and efficiency, anemia, and jaundice. Nursing animals may be affected as a result of the conversion of aflatoxin B1 to the metabolite aflatoxin M1 excreted in milk of dairy cattle. The induction of cancer by aflatoxins has been extensively studied. Aflatoxin B1, aflatoxin M1, and aflatoxin G1 have been shown to cause various types of cancer in different animal species. However, only aflatoxin B1 is considered by the International Agency for Research on Cancer (IARC) as having produced sufficient evidence of carcinogenicity in experimental animals to be identified as a carcinogen.

Humans are exposed to aflatoxins by consuming foods contaminated with products of fungal growth, such as peanuts or grains. Such exposure is difficult to avoid because fungal growth in foods is not easy to prevent or observe. Even though heavily contaminated food supplies are not permitted in the market place in developed countries, concern still remains for the possible adverse effects resulting from long-term exposure to low levels of aflatoxins in the food supply. Evidence of acute aflatoxicosis in humans has been reported from many parts of the world, chiefly Third World Countries like Taiwan, Uganda, India, and many others. The syndrome is characterized by vomiting, abdominal pain, pulmonary edema, convulsions, coma, and death with cerebral edema and fatty involvement of the liver, kidneys, and heart. Conditions increasing the likelihood of acute aflatoxicosis in humans include limited availability of food, environmental conditions that favor fungal development in crops and commodities, and lack of regulatory systems for aflatoxin monitoring and control. Because aflatoxins, especially aflatoxin B1, are potent carcinogens in some animals, there is interest in the effects of long-term exposure to low levels of these important mycotoxins on humans. In 1988, the IARC placed aflatoxin B1 on the list of human carcinogens. This is supported by a number of epidemiological studies done in Asia and Africa that have demonstrated a positive association between dietary aflatoxins and Liver Cell Cancer (LCC). Additionally, aflatoxin-related diseases in humans may be influenced by factors such as age, sex, nutritional status, or concurrent exposure to other causative agents such as viral hepatitis (HBV) or parasite infestation.

Treatment

Avoidance of food containing aflatoxins is the primary measure. Once exposed, there is a risk of future cancer; there is no treatment to prevent this. For an acute exposure, amphotericin B (Fungizone®) may be useful. Itraconazole may also be helpful.

AFRICAN HEMORRHAGIC FEVER

Description

African Hemorrhagic Fever (AHF) is caused by a virus. The natural hosts are not known, but monkeys are probably accidental hosts, along with humans. It is transmitted by direct contact with infected blood, secretions, organs, or semen.

Signs, Symptoms, and Prognosis

Typical symptoms are fever, headache, malaise, followed by chest discomfort, vomiting, diarrhea, and hemorrhage from all orifices. The normal incubation period ranges from four to more than sixteen days. The untreated fatality rate can be as high as 90%.

Treatment

There is no specific treatment; supportive care with electrolytes and transfusions may be of help.

ANTHRAX

Description

Anthrax is an acute infectious disease caused by the bacteria *Bacillus anthracis* and is highly lethal in its most virulent form, inhalational anthrax. Anthrax most commonly occurs in wild and domestic herbivores, but can also occur in humans when they are exposed to infected animals, tissue from infected animals, or high concentrations of anthrax spores. The name *anthrax* originates from the Greek word *anthrax* meaning "coal," because victims commonly develop black skin lesions.

The most common hosts are sheep, cattle, and other grazing animals including bison, elk, kudu, and wildebeest; spores can be found in meat, milk, hides, and hair of infected animals. The *Bacillus anthracis* is a rod-shaped gram-positive bacterium of size about 1 by 6 micrometers (mm). It was the first bacterium ever to be shown to cause disease, by Robert Koch in 1877. The bacteria normally rest in spore form in the soil, and can survive for decades in this state. Once taken in by a herbivore, the bacteria start to multiply inside the animal and eventually kills it, then continues to reproduce in the carcass. Once they run out of nutrients there, they revert back to the dormant spore state. Anthrax bacteria left exposed to light and air for over two hours form spores; it is these spores that are used in biological weapons, and are often implicated in natural outbreaks. Spores can be ingested, inhaled, or inoculated into minor skin abrasions. In a biological warfare scenario, weapons would burst to create an aerosol of airborne spores, leading to inhalation anthrax. During inhalation anthrax's initial phase, spores are carried to the lymph nodes causing nonspecific flu-like symptoms, mild fever, malaise, fatigue, cough, and, in some cases, a sense of tightness in the chest. This phase can last for several days, or for as little as twenty-four hours. The second phase develops suddenly and is characterized by severe shortness of breath; hypotension and shock occur. Perspiration is often profuse. The second, acute phase typically lasts less than twenty-four hours and usually ends in death despite any attempt at treatment.

Natural anthrax infection is rare, but not totally uncommon, in herbivores such as cattle, sheep, goats, camels, and antelopes. Anthrax can be found around the world, but it is more common in developing countries or countries without veterinary public health programs. Certain regions of the world, such as Central and South America, Southern and Eastern Europe, Africa, Asia, the Caribbean, and the Middle East, report more anthrax in animals than other areas.

When anthrax naturally affects humans, it is usually due to an occupational exposure to infected animals or their products such as skin or meat. Hunters and workers who are exposed to dead animals and animal products from other countries where anthrax is more common may become infected with *B. anthracis*, and anthrax in wild livestock has occurred in the United States. Although many such workers are routinely exposed to significant levels of anthrax spores, most are not sufficiently exposed to develop symptoms.

Anthrax can enter the human body through the intestines, lungs by inhalation, or skin by cutaneous exposure. Anthrax is non-contagious, and is highly unlikely to be spread from person to person.

Anthrax was tested as a biological warfare agent by Unit 731 of the Japanese Kwantung Army in Manchuria during the 1930s; some of this testing involved intentional infection of prisoners of war, thousands of whom died. Anthrax, designated at the time as Agent N, was also investigated by the Allies in the 1940s. The British army tested experimental anthrax weapons on Gruinard Island, off the northwest coast of Scotland, in 1943. Gruinard was burned over at least once, yet as of the late 1980s it was still too heavily contaminated with spores to allow unprotected human access, indicating the hardiness of anthrax spores. Weaponized anthrax was part of the U.S. stockpile prior to its destruction in 1972.

Anthrax spores can be killed by heat, which prevents a greater incidence of food-borne infections in areas where anthrax is common. An outbreak of anthrax occurred during April, 1979, among people who lived or worked in a narrow zone downwind of a Soviet military microbiology facility in Sverdlovsk (now Ekaterinburg), Russia. In addition, livestock died of anthrax within a larger downwind zone. The facility was suspected by western intelligence of being a biological warfare research facility. Intelligence analysts attributed the outbreak to the accidental airborne release of anthrax spores. The Soviets maintained that the outbreak was due to ingestion of contaminated meat purchased on the black market. Finally, in 1992, President Yeltsin of Russia admitted that the facility had been part of an offensive biological weapons program, and that the disease in animals and people resulted from an accidental release of anthrax spores. The 2002 anthrax attacks on news media and politicians used a weaponized form of the spores. Contamination of postal facilities was extensive.

Signs, Symptoms, and Prognosis

Pulmonary anthrax, also called pneumonic, respiratory, or inhalation anthrax, infection initially presents with cold or flu-like symptoms for several days, followed by severe, and often fatal, respiratory problems. If not treated soon after exposure, inhalation infection is the most deadly, with a nearly 100% mortality rate. A lethal case of anthrax has been reported to result from inhaling as few as 10,000 to 20,000 spores. Pulmonary anthrax is also been known as woolsorters' disease since it was common among people working with wool. Other routes have included the cutting of imported animal horns for the manufacture of buttons and handling natural bristles used for the manufacturing of brushes.

Gastrointestinal or gastroenteric anthrax infection often has symptoms of serious gastrointestinal difficulty, vomiting of blood, and severe diarrhea. Untreated, intestinal infection results in a 25–75% death rate.

Cutaneous anthrax infection symptoms include a large, painless necrotic ulcer which begins as a irritating and itchy skin lesion or blister which is dark in color, often concentrated as a black dot at the site of infection, forming about a week or two after exposure. Unlike bruises or most other lesions, cutaneous anthrax does not cause any pain. Cutaneous infection is the least deadly form of anthrax; without treatment, approximately 20% of all skin infection cases are fatal. Treated cutaneous anthrax is rarely fatal.

Treatment

Treatment for anthrax infections includes large doses of intravenous and oral antibiotics, such as penicillin, ciprofloxacin (Cipro®), erythromycin, vancomycin, and doxycycline. Tetracyclines, erythromycin, or chloramphenicol could also be used. Treatment with antibiotics must begin prior to the onset of symptoms and must include vaccination prior to discontinuing their use. For inhalation cases, antibiotic treatment is not very effective if initiated more than twenty-four hours after infection, that is, after symptoms appear. Antibiotic prophylaxis is crucial in cases of pulmonary anthrax. Unfortunately, some antibiotic-resistant strains are known. A vaccine, produced from one component of the toxin of a non-virulent strand, is also available. The vaccine must be given at least four weeks before exposure to anthrax, and annual booster injections are required to maintain immunity.

The normal anthrax spores can be trapped with a simple HEPA or P100 filter. There are rumors that some forms of weaponized spores are small enough to pass through a filter. Anthrax as an airborne threat can usually be prevented with a full-facemask respirator. Unbroken skin can be decontaminated simply with soap and water.

BOTULISM

Description

Botulism is the disease caused by intoxification with botulinum toxin. Botulinum toxin is the toxic compound produced by the bacterium *Clostridium botulinum* (*C. botulinum*). The bacterium is an anaerobic spore-forming bacillus. The bacteria *per se* are not the biological agent; rather, the isolated toxin is generally considered to be the warfare agent. The toxin is an enzyme that breaks down one of the fusion proteins that allow neurons to release acetylcholine at a neuromuscular junction. By interfering with nerve impulses in this way, it causes paralysis of muscles in botulism. In this way, it acts somewhat like a reverse of some of the chemical nerve agents. The toxin itself is a two-chain polypeptide with a 100-kilodalton (kd) heavy chain joined by a disulfide bond to a 50-kd lighter chain. It is a neurotropic poison which inhibits the formation of acetylcholine and thus blocks neurotransmission, paralyzing the voluntary musculature, quite the opposite of chemical nerve agents. Most nerve agents inhibit acetylcholinesterase, leading to a buildup of too much acetylcholine, whereas the mechanism in botulism is lack of the neurotransmitter in the nerve synapse. Thus, nerve agent antidotes such as atropine are not helpful in botulism, and could make the condition worse.

Botulinum toxin has long been considered an ideal agent for warfare because it oxidizes rapidly on exposure to air so an area attacked with the toxin aerosol would be safe to enter fairly soon after attack. There are no documented cases of the toxin actually being used in warfare, but the use of botulinum toxin as a military weapon was studied by the Imperial Japanese Army in the 1930s. It is thought by some to have been the weapon provided by British MI6 to the Czech partisans who assassinated Reinhard Heyrich in 1942. The toxin was part of the U.S. bioweapons stockpile prior to its destruction in 1972. More recently, Iraq admitted to a United Nations inspection team in August of 1991 that it had done research on the offensive use of botulinum toxin prior to the Gulf War. Further information given to the U.N. in 1995 after the defection of a leading Iraqi official revealed that Iraq had filled and deployed over 100 munitions with botulinum toxin. The toxin's properties did not escape the attention of the Aum Shinrikyo ("Supreme Truth") cult in Japan, which actually set up a plant for bulk production of this agent, although their terrorist and assassination attacks used the nerve agent Sarin instead because it is easier to disperse and is faster acting.

Two related bacteria, *C. butyricum* and *C. baratii*, produce similar neurotoxins and are suspected of causing some reported cases of botulism. There are two natural forms of botulism, foodborne and infant. Foodborne botulism is a severe intoxication resulting from ingestion of the toxin present in contaminated foods. Infant botulism, the most common form in the U.S., results from spore ingestion (sometimes from honey) and subsequent growth of the bacteria and formation of the toxin. It usually affects children under one year of age, but can be seen in adults with altered gastrointestinal anatomy and microflora. The toxin is produced in poorly processed, low-acid or alkaline foods, usually in cans, or in lightly cured foods held without refrigeration, especially in airtight packaging.

Botulinum toxin is used, under a trademarked name such as "Botox®," for producing paralysis of muscles for several months. This was first intended for the relief of uncontrollable muscle spasms, but is increasingly being abused for cosmetic purposes, to paralyze facial muscles as a means of concealing wrinkles. The vials of toxin used therapeutically are considered impractical for use by terrorists because each vial has only an extremely small fraction of the lethal dose for humans.

Signs, Symptoms, and Prognosis

Foodborne botulism is characterized by acute bilateral cranial nerve impairment and descending weakness or paralysis. The incubation period is usually between 12 and 36 hours for ingestion. The first complaints are usually blurred or double vision, dysphagia, and dry mouth, followed by vomiting and constipation. Even though the victim may appear alert, paralysis is common. Fever is absent without a complicating infection being present. The onset of symptoms after inhaling botulinum toxin may vary from twenty-four to thirty-six hours. Initial symptoms include headache, queasiness, increased flow of saliva, vomiting, dizziness, acute pain in the limbs, and involuntary defecation. Effects on the central nervous system result in blurred vision, mouth dryness, paralysis, respiratory paralysis and death. Recovery, when the patient survives, may take many months. The untreated fatality rate is 5–10%. The human LD_{50} is about 0.001 mg/kg for the A toxin.

Infant botulism typically begins with constipation, followed by lethargy, poor feeding or appetite, difficulty swallowing, generalized weakness, and respiratory insufficiency and arrest. There is a wide spectrum of severity, ranging from mild to death. It may be a cause of some cases of sudden infant death syndrome (SIDS). With treatment, the fatality rate is about 1%; without, much higher depending upon the severity of the case.

Treatment

The toxin is destroyed by boiling, so prevention is the best course of action. To destroy the spores requires greater heating than boiling. Hospital treatment for foodborne botulism includes intravenous administration of trivalent botulinum antitoxin and supportive case. The antitoxin should not be used in cases of infant botulism. In neither case are antibiotics of use unless there is a secondary infection.

BOLIVIAN HEMORRHAGIC FEVER (BHF) OR MACHUPO HEMORRHAGIC FEVER

Description

Machupo hemorrhagic fever is also known as Bolivian hemorrhagic fever (BHF). This hemorrhagic fever was first identified in 1959 as a sporadic illness in rural areas of Bolivia. In 1963, the Machupo virus, a member of the *Arenaviridae* family, was isolated from patients. Investigations established that the rodent *Calomys callosus*, which is indigenous to northern Bolivia, is the animal reservoir for the Machupo virus.

Machupo viral infection in *C. callosus* results in asymptomatic infection with shedding of virus in saliva, urine, and feces; 50% of experimentally-infected *C. callosus* are chronically viremic and shed virus in their bodily excretions or secretions. Although the infectious dose of Machupo virus in humans is unknown, exposed persons may become infected by inhaling virus shed in aerosolized secretions or excretions of infected rodents, by eating food contaminated with rodent excreta, or by direct contact of excreta with abraded skin or oropharyngeal mucous membranes. Reports of person-to-person transmission are uncommon; however, hospital contact with a patient resulted in person-to-person spread of Machupo virus to nursing and pathology laboratory staff, and the infection of six members of the same family from one infected individual suggests the potential for person-to-person transmission.

During BHF epidemics in the 1960s, rodent control was recognized as the primary method for the prevention of Machupo virus transmission. Rodent control programs were successful in stopping the epidemics. Since there are survivors of BHF epidemics, it is suspected that some of them were immune to Machupo virus. BHF, like other related fevers, remains a possible biological weapon.

Signs, Symptoms, and Prognosis

Experimental infection of rhesus monkeys with Machupo virus demonstrated an incubation period of seven to fourteen days, which is consistent with clinical observations in human infection. Early clinical manifestations in humans are characterized by nonspecific signs and symptoms including fever, headache, fatigue, myalgia, and arthralgia. Later in the course of disease (usually within seven days of onset), patients may develop hemorrhagic signs, including bleeding from the oral and nasal mucosa and from the bronchopulmonary, gastrointestinal, and genitourinary tracts. Fatality rates range from 15–30% or possibly more.

Treatment

Rodent control is the best preventive measure. Once infected, the course of treatment is supportive care with use of antivirals such as ribavirin. There is an experimental vaccine, but it is unlicensed in the U.S.

BRUCELLOSIS

Description

Brucellosis is a bacterial systemic disease with either acute or insidious onset. It is also known as undulant fever, Mediterranean fever, or Malta fever. The usual causative organism is *Brucella melitenis*, *B. abortus*, *B. suis*, or *B. canis*. The significant strains of *Brucella* include *B. abortus*, occurring naturally in cattle and sheep, this strain has the most serious effects in humans and is the most severe biological warfare threat; *B. melitensis*, occurring naturally in sheep and goats; and *B. suis*, occurring naturally in swine, was part of the U.S. bioweapons stockpile prior to its destruction in 1972. It is usually an occupational disease of persons working with animals or their tissues, especially farm hands, veterinarians, and abattoir workers, although some cases have resulted from consumption of raw milk and milk products such as soft cheese. Rarely, cases have come from contact with dogs. It occurs in cattle, sheep, swine, and goats. The use of brucellosis as a bioweapon is possible, but not very likely unless a more virulent form has been produced. There are few deaths, and most adults could continue to function even with the disease.

Signs, Symptoms, and Prognosis.

The usual natural incubation period is three days up to several weeks. Brucellosis is characterized by continued intermittent fever, headache, general weakness, profuse sweating, chills, arthralgia, depression, weight loss, and generalized aches. Chronic undulant fever is characterized by periods of normal temperature between acute attacks; symptoms may persist for years, either continuously or intermittently. Infection of the liver and spleen may occur. Recovery is usual, but disability is common, chiefly of osteoarticular complications. The incubation period is highly variable, usually five to sixty days after exposure. While it is a zoonotic disease, there is no evidence of person-to-person spread. The case-fatality rate is about 5% for *B. suis*, but is higher for *B. abortus* and *B. melitensis*.

Treatment

A combination of doxycycline with either rifampin or streptomycin for at least six weeks is the treatment of choice. Avoidance of infected animals is preferred.

CHIKUNGUNYA

Description

Chikungunya epidemics, affecting up to 40% of the population, have occurred since the early 1950s in Tanzania and Thailand. Smaller outbreaks have appeared in Zimbabwe, the Democratic Republic of the Congo, India, and Kampuchea. The causative agent is an arbovirus carried by a variety of mosquito species. Non-human primates and perhaps other animals are considered the natural reservoir. The virus grows well in tissue culture. Susceptibility is almost universal in temperate climates. Aerosol delivery and entry would allow large numbers of the virus to enter the lungs with incapacitation within one to two days.

Signs, Symptoms, and Prognosis

The incubation period is between three and twelve days with a sudden onset. Severe joint pains in the limbs and spine incapacitate within hours. The patient doubles up and remains immobile with a high fever, which lasts about six days. Often a second bout of fever appears after remission of a few days, accompanied by a irritating rash on the trunk and limbs. The second stage lasts about a week, although crippling joint pain can last for as long as four months. Fatalities are generally few, less than 5% even when untreated.

Treatment

There is no treatment other than supportive care, and preventive vaccines are not available. Antivirals might be of use, but are unproven.

CHOLERA

Description

Cholera is a bacterial disease caused by *Vibrio cholerae*. The natural host is human, and the disease is transmitted by eating food or drinking water contaminated by the feces or vomitus of someone already infected with cholera. Current research also implicates shellfish in an estuarine environment; this is as yet poorly understood, but may involve the plankton on which the shellfish feed. *V. cholerae* was spread by the Germans in 1915 in Russia and Italy in an attempt at biological warfare.

Signs, Symptoms, and Prognosis

The most common symptom is a severe gastroenteritis. Other symptoms include mild to severe watery diarrhea, sometimes called "rice water" diarrhea, vomiting, and dehydration with fluid loss of up to one liter per hour. The usual incubation period is between two to three days. The untreated fatality rate is around 50%.

Treatment

Due to fluid loss, intravenous or oral replacement of fluids and salts is critical. Broad spectrum antibiotics are effective.

COCCIDIOIDOMYCOSIS

Description

Coccidioidomycosis is a dust-borne disease caused by the fungus *Coccidioides immitis*. The organism occurs in the soil, mainly in arid and semi-arid regions. It is diphasic, with a hyphal or saprophylic form and an arthrosporic form that is very resistant to environmental influences. The fungus is endemic in the southwestern U.S., northern Mexico, and Venezuela, and has been reported in parts of Russia. Infection is usually by inhalation of the infective arthrospores. The organism is very easy to grow in the laboratory, and produce large number of arthrospores. Use as a bioweapon is possible, with general disruption of medical care facilities as well as widespread fear since there is no satisfactory prophylaxis or chemotherapy.

Signs, Symptoms, and Prognosis

After a typical incubation period of ten to twenty days, there is an influenza-like illness with fever, chills, cough, pleural pain, headache, and backache. The primary infection may resolve without any detectable sequelae, or lung lesions may persist with fibrosis and calcification. Rarely, there is damage to the central nervous system (CNS) and bone abscesses. Untreated, the mortality rate is around 50%. Treated, the fatality rate is quite low. There is no effective preventive vaccine.

Treatment

The drug of choice is amphotericin B; however, this is fairly toxic and relapses are possible when treatment is stopped.

CRIMEAN-CONGO HEMORRHAGIC FEVER

Description

Crimean-Congo Hemorrhagic Fever (CCHF) is caused by a virus. The natural host is cattle. It is transmitted primarily by *Hyalomma marginatum marginatum* ticks and direct contact with infected blood. Other tick species such as *Hyalomma*, *Dermacentor*, and *Rhipicephalus* can transmit the virus to humans.

Signs, Symptoms, and Prognosis

The natural incubation period is between three and twelve days. Symptoms include: headache, fever, chills, body pains, diarrhea and vomiting, dizziness, confusion, abnormal behavior, pharyngitis, conjunctivitis, reddish face and neck, hemorrhage, and, ultimately, multi-organ failure. The untreated fatality rate is up to 70%.

Treatment

Antibiotic treatment with ribavirin may be helpful. Blood transfusions and supportive care are the treatments most commonly used.

DENGUE FEVER

Description

Dengue fever is caused by the dengue arbovirus. It is also known as breakbone fever due to a large number of people suffering from broken bones when suffering from the fever. Natural hosts are not known, but it is transmitted by insects, principally *Aedes aegypti* mosquitoes, as well as *A. albopictus*, *A. scutellaris*, and *A. polynesiensis*. Other arboviruses which cause similar diseases to dengue include: West Nile Virus; Rift Valley Fever Virus; Sindbis Virus; Wesselsbron Virus; and Chikungunya Virus. All have fairly similar symptoms.

Signs, Symptoms, and Prognosis

The normal incubation period is five to six days. Common symptoms include sudden onset of fever; painful headaches; eye, joint, and muscle pain; and rash. The rash typically begins on the arms or legs some three to four days after the onset of fever. Although uncommon, in severe cases dengue fever can be hemorrhagic. Fatalities are rare, even if untreated, but the patient is quite ill for a period of time.

Treatment

There is no treatment for dengue fever at this time. Supportive care is the only course of action.

EBOLA HEMORRHAGIC FEVER (EHF)

Description

Ebola hemorrhagic fever is a recently identified, severe, and often fatal infectious disease occurring in humans and other primates, and caused by the Ebola virus. The virus comes from the *filoviridae* family, similar to the Marburg virus. It is named after the Ebola River in Zaire, Africa, near where the first outbreak was noted in 1976. Two other strains were identified in 1976, and were named Ebola–Zaire (EBO-Z) and Ebola–Sudan (EBO-S). In 1990, a similar virus was identified in Reston, Virginia among imported monkeys; this was named Ebola–Reston. It was found to have originated from the Philippines. Further outbreaks have occurred in Zaire in 1995; Gabon in 1994, 1995 and 1996; and Uganda in 2000. A new subtype was identified from a single human case in the Côte d'Ivoire (Ivory Coast) in 1994, and was named EBO-CI. The reservoir which sustains the virus has not been identified. Among humans, the virus is transmitted by direct contact with infected body fluids such as blood. The cause of the index case is not known.

Signs, Symptoms, and Prognosis

The incubation period of Ebola hemorrhagic fever is variable from two days to three weeks. Symptoms are variable, but the onset is usually sudden and characterized by high fever, prostration, myalgia, abdominal pains, and headache. These symptoms progress to vomiting, diarrhea, oropharyngeal lesions, conjunctivitis, organ damage of the kidney and liver by co-localized necrosis, proteinuria, and bleeding both internally and externally, commonly through the gastrointestinal tract. Death or recovery with convalescence occurs within six to ten days.

Of the first 602 identified cases in Zaire, there were 397 deaths, for a fatality rate of 66%. Ebola-Zaire (EBO-Z) and Ebola-Sudan (EBO-S) have shown lower fatality rates of about 50%. Of the total 1,500 identified cases of all strains since 1976, two-thirds have died.

Treatment

No specific treatment has been proved effective or vaccine exists. Supportive care is the only course of action. Protection from transmission to health care workers is essential.

EPIDEMIC TYPHUS

Description

Epidemic typhus is caused by the rickettsial organism *Rickettsia prowazekii*. It is transmitted primarily by lice, and the natural hosts are humans and squirrels. Epidemic typhus, which is caused by *Rickettsia prowazekii*, is a different disease from typhoid fever, which is caused by *Salmonella typhi*, and results from eating or drinking food or water contaminated with bacteria.

Signs, Symptoms, and Prognosis

The typical incubation period is six to fifteen days following the louse bite. Symptoms include fever, generalized aches, headaches, weakness, and pain, stupor and delirium. The untreated fatality rate can be as high as 30%.

Treatment

Doxycycline, tetracycline, and chloramphenicol have been successfully used in treatment of epidemic typhus.

HANTAVIRAL DISEASES

Description

There are a number of diseases, all with similar symptoms, cause by the hantaviruses. Sin Nombre virus is the cause of Hantavirus Pulmonary Syndrome (HPS) in the southwestern U.S. This virus was discovered in 1993 when there was a cluster of cases in one place at one time. It is also known as Four Corners Virus and Muerto Canyon Virus. The natural hosts are rodents. Transmission of the disease may occur when aerosols of infective saliva, urine, or feces are inhaled, or when dried materials contaminated by rodent excreta are disturbed. Hantaviruses have been isolated from many rodents in the U.S., and serological studies have documented human infections with hantaviruses. However, acute disease associated with infection by pathogenic hantaviruses was not reported in the Western Hemisphere until 1993, when the New Mexico Department of Health was notified of two persons who had died within five days of each other. Additional cases of persons who had recently died under similar circumstances were reported by the Indian Health Service and the health services of Arizona, Colorado, and Utah. Blood and tissue specimens were sent to the Centers for Disease Control and Prevention. In June 1993, only a few weeks after the first reported deaths, the CDC was able to prove that a new hantavirus was responsible, a different strain from the previously-known Korean strain. Acute hantavirus infection has been confirmed in 42 persons in the Southwestern U.S.

Epidemic and sporadic hantavirus-associated disease has occurred since the 1930s in Scandinavia and Northeastern Asia. This disease is a relatively low mortality chronic disease. In 1950–1953, thousands of United Nations military personnel were infected with hantaviral diseases during the Korean War; more recently, U.S. military personnel stationed in the Republic of Korea have contracted disease from hantaviruses. Isolation of the first recognized hantavirus (Hantaan virus) was reported from the Republic of Korea in 1978. The Asian hanta strains are not considered to be potential biological warfare agents.

Signs, Symptoms, and Prognosis

The typical incubation period for the natural disease is one to three weeks, but can be shorter. Symptoms include fever, myalgia, headache, and cough, followed by the rapid development of respiratory failure due to fluid accumulation in the lungs. The untreated fatality rate can be as high as 65%.

Treatment

There is no specific treatment. Avoidance of rodents and their droppings is the best preventive method. If the disease is contracted, the treatment of choice is respiratory support.

INFLUENZA

Description

Influenza is an airborne infection, and aerosol dissemination is easily possible. There are a number of strains of virus which cause influenza. Pandemics usually occur when a shift in the composition of the virus results in a new strain. Certain changes in the viral protein coat appear to allow for increased infectivity. In 1918, worldwide, some 20,000,000 people died in the pandemic, mostly young males, but not young females (it is not known why, but may be related to immune system characteristics). Recombinant techniques have been used in biological agent laboratories to produce more virulent strains resistant to current vaccines. For most known strains, there are effective vaccines with effectiveness of between 50–70%. If a new strain were to appear, either naturally or by recombinant techniques, it would take some three months to produce sufficient vaccine to immunize a population.

Signs, Symptoms, and Prognosis

The incubation period can vary from a few days to a week. The symptoms include fever, general malaise, respiratory distress, headache, myalgia, and prostration, usually lasting several days. One to two weeks are usually required for full recovery. Pneumonia can occur as a complication, but can often be controlled by antibiotics. Ear infections are common in the young, including bacterial infections. Influenza most often appears in the very young and the elderly; the 1918 pandemic was an exception. Mortalities from current strains are usually no more than 2%, but could be much higher with a new strain, especially if bioengineered.

Treatment

Preventive vaccines are effective. Post-infection, complications can be prevented by broad-spectrum antibiotics. Supportive care, including analgesics and fluids, is usually called for. Respiratory support may be needed.

MARBURG HEMORRHAGIC FEVER

Description

Marburg hemorrhagic fever is a rare, severe hemorrhagic fever which affects humans and non-human primates. It is caused by a unique RNA virus of the filovirus family; its recognition led to the creation of this virus family. The four species of the Ebola virus are the only other known members of the filovirus family. Marburg virus was first recognized in 1967, when outbreaks of hemorrhagic fever occurred simultaneously in laboratories in Marburg and Frankfurt, Germany and in Belgrade,

Yugoslavia (now Serbia). A total of 37 people became ill; they included laboratory workers as well as several medical personnel and family members who had cared for them. The first people infected had been exposed to African green monkeys or their tissues. In Marburg, the monkeys had been imported for research and to prepare polio vaccine. The disease is also known as African Hemorrhagic Fever, Green Monkey Disease, or Vervet Monkey Disease.

Recorded cases of the disease are rare, and have appeared in only a few locations. While the 1967 outbreak occurred in Europe, the disease agent had arrived with imported monkeys from Uganda. No other case was recorded until 1975, when a traveler exposed in Zimbabwe became ill in South Africa, and passed the virus to two other people. Other cases have been observed in western Kenya, Uganda, and the Democratic Republic of the Congo. The Marburg virus is indigenous to Africa, including Uganda, western Kenya, and perhaps Zimbabwe. As with Ebola virus, the animal reservoir for the Marburg virus remains a mystery; however, person to person contact has been shown for Marburg hemorrhagic fever. Droplets of body fluids, or direct contact with persons, equipment, or other objects contaminated with infectious blood or tissues are possible sources of the disease.

A fuller understanding of Marburg hemorrhagic fever will not be possible until the ecology and identity of the virus reservoir are established. The impact of the disease will remain unknown until the actual incidence of the disease and its endemic areas are determined. For these reasons, Marburg remains a potential biological weapon. There is evidence that the Russians were investigating Marburg as a bioweapon in the 1970s.

Signs, Symptoms, and Prognosis

After an incubation period of 5 to10 days, the onset of the disease is sudden and marked by fever, chills, headache, and myalgia. Around the fifth day after the onset of symptoms, a maculopapular rash, prominent on the trunk of the body (chest, back, and stomach), may occur. Nausea, vomiting, chest pain, sore throat, abdominal pain, and diarrhea may appear. Symptoms become increasingly severe, and may include jaundice, inflammation of the pancreas, severe weight loss, delirium, shock, liver failure, massive hemorrhaging, and multiple organ dysfunction. The case fatality rate for Marburg hemorrhagic fever is around 25%. Because many of the signs and symptoms of Marburg hemorrhagic fever are similar to other infectious diseases, such as malaria or typhoid fever, diagnosis of the disease can be difficult.

Recovery from Marburg hemorrhagic fever may be prolonged and accompanied by orchititis, recurrent hepatitis, or transverse myelitis or uvetis. Other possible complications include inflammation of the testis, spinal cord, eye, parotid gland, or by prolonged hepatitis.

Treatment

Specific treatment is unknown; however, supportive hospital therapy should be utilized. This includes balancing the patient's fluids and electrolytes, maintaining their oxygen status and blood pressure, replacing lost blood and clotting factors and treating them for any complicating infections. Sometimes treatment also has used transfusion of fresh-frozen plasma and other preparations to replace the blood proteins important in clotting. One controversial treatment is the use of heparin to prevent the consumption of clotting factors. Some researchers believe the consumption of clotting factors is part of the disease process.

MURINE TYPHUS

Description

Murine typhus is a disease caused by the rickettsial organism *Rickettsia typhi.* It is also known as flea-borne typhus, endemic typhus, and urban typhus. The natural hosts include rats and mice, and possibly other rodents. Endemic typhus, which is caused by *Rickettsia typhi*, is a different disease from typhoid fever, which is caused by *Salmonella typhi* and results from eating or drinking food or water contaminated with bacteria.

Signs, Symptoms, and Prognosis

The incubation period is usually one to two weeks following the bite of the flea or lice. The is a gradual onset of fever with severe headache, chills, generalized pains and dry cough, sometimes developing to bronchopneumonia, after about two weeks. A macular rash often appears after about five days, first appearing on the trunk and lasting about six days. Central nervous system (CNS) manifestations are possible. Damage is caused to vascular endothelia by invasion of rickettsiae, possibly leading to thrombosis and hemorrhage. If untreated, the fatality rate is only 1–2%.

Treatment

Doxycycline, tetracycline, and chloramphenicol have been successfully used to treat murine typhus.

O'NYONG-NYONG

Description

This is another arbovirus disease, closely related to the Chinkungunya virus. A large epidemic appeared in Uganda in the last 1950s and early 1960s. The attack rate in this outbreak was over 70%. The vectors are *Anopheles* species mosquitoes. The reservoir is not known, but is suspected to be non-human primates.

Signs, Symptoms, and Prognosis

The incubation period is between three and twelve days with sudden onset of symptoms including chills, severe back pain, severe joint pain, headache, rash, and neck lymphadenitis. Unlike Chinkungunya, there is often no secondary fever period. Recovery without sequelae occurs in about a week. Fatality rates are very low, essentially zero.

Treatment

There is no treatment other than supportive care, including analgesics. There are no preventive vaccines.

PLAGUE

Description

Bubonic plague is a contagious, sometimes epidemic or pandemic, disease caused by the bacillus *Yersinia pestis,* and is usually transmitted by the bite of fleas from an infected host, often a rat, ground squirrel, rabbit, hare, or even a domestic cat. The bacteria are transferred from the blood of infected rats to the Rat Flea (*Xenopsylla cheopsis*). The bacillus multiplies in the stomach of the flea until it fills it and blocks passage out. When the flea next bites a mammal, the consumed blood is regurgitated, along with the bacillus, into the bloodstream of the bitten animal. Any serious outbreak of plague is usually started by other disease outbreaks in the rodent population. During these outbreaks, infected fleas that have lost their normal hosts seek other sources of blood, such as humans. This is one reason never to kill a rodent, since the ectoparasites like fleas will leave it when it is no longer alive and warm to seek another host.

Most scientists believe that the Black Death of the Middle Ages was an outbreak of bubonic plague. However, other theories have now been advanced, suggesting that the Black Death may have been an outbreak of some other disease, possibly a hemorrhagic fever similar to Ebola.

Signs, Symptoms, and Prognosis

As a bioweapon, the plague bacterium can be made to be inhalable. If taken into the lungs, it can cause pneumonic plague directly without the formation of the buboes of the lymphatic system. With pneumonic plague the infected lungs also raise the possibility of person-to-person transmission through respiratory droplets. After two to four days of incubation, the initial symptoms of headache, weakness, and coughing with hemoptysis are indistinguishable from other respiratory illnesses. Without diagnosis and treatment, the infection can be fatal in one to seven days, mortality in untreated cases may be as high as 95%. As a bioweapon, aerosolized pneumonic plague is the only effective plague agent. There is a vaccine effective against bubonic, but not pneumonic, plague for a period of several months after administration.

Treatment

Prevention is the best course of action, rat-proofing buildings, use of pesticides and repellants, and wearing of gloves when handling wildlife. Patients with bubonic plague must first have their clothing and body free from fleas by using an insecticide; then implement precautions against spread of secretions and begin antibiotic treatment with streptomycin, gentamycin, tetracyclines, or chloramphenicol. After treatment with antibiotics, any remaining buboes may have to be incised and drained.

Any patients with pneumonic plague must be strictly isolated with precautions against airborne spread of droplets. Antibiotic treatment with streptomycin is preferred, gentamycin may be used, and tetracyclines or chloramphenicol are possibly alternatives.

Q FEVER

Description

Q fever is a zoonotic disease caused by the rickettsial organism *Coxiella Burnetti*, found worldwide. It derives its name from Query fever, and is also known as Balkan fever, Balkan grippe, pneumorickettsiosis, and abattoir fever. *Coxiella burnetti* are typically spread in nature by the Brown Dog Tick, *Rhipicephalus sanguineus*; the Rocky Mountain Wood Tick, *Dermacentor andersoni*; and the Lone Star Tick, *Amblyomma americanum*. Unlike most other rickettsiae, *C. burnetti* is very resistant to environmental degradation. Many human infections are unapparent. Cattle, sheep, and goats are the primary reservoirs of *C. burnetii*, and infection has been noted in a wide variety of other animals, including livestock and domesticated pets. *Coxiella burnetii* does not usually cause clinical disease in these animals, although abortion in goats and sheep has been linked to *C. burnetii* infection. The organisms are excreted in milk, urine, and feces of infected animals. Most importantly, during birthing the organisms are shed in high numbers within the amniotic fluids and the placenta. The organisms are resistant to heat, drying, and many common disinfectants. These features enable the bacteria to survive for long periods in the environment. Infection of humans usually occurs by inhalation of these organisms from air that contains airborne barnyard dust that has been contaminated by dried placental material, birth fluids, and excreta of infected herd animals. Humans are often very susceptible to the disease, and very few organisms may be required to cause infection. Ingestion of contaminated milk, followed by regurgitation and inspiration of the contaminated food, is a less common mode of transmission. Other modes of transmission to humans, including tick bites and human to human transmission, are rare.

In the United States, Q fever outbreaks have resulted mainly from occupational exposure involving veterinarians, meat processing plant workers, sheep and dairy workers, livestock farmers, and researchers at facilities housing sheep. Prevention and control efforts should be directed primarily toward these groups and environments.

Coxiella burnetii is a highly infectious agent that is resistant to heat and drying. It can become airborne and inhaled by humans. A single *C. burnetii* organism may cause disease in a susceptible person. This agent could be developed for use in biological warfare and is considered a potential terrorist threat. *Coxiella burnetti* was part of U.S. bioweapons stockpile prior to its destruction in 1972.

Signs, Symptoms, and Prognosis

Only about one-half of people infected with *C. burnetii* show signs of clinical illness. Most acute cases of Q fever begin with sudden onset of one or more of the following: high fevers (up to 43° C or 105° F), severe headache, general malaise, myalgia, confusion, sore throat, chills, sweats, non-productive cough, nausea, vomiting, diarrhea, abdominal pain, and chest pain. Fever usually lasts for 1 to 2 weeks. Weight loss can occur and persist for some time. 30–50% of patients with a symptomatic infection will develop pneumonia. Additionally, a majority of patients have abnormal results on liver function tests and some will develop hepatitis. In general, most patients will recover to good health within several months without any treatment; only 1–2% of people with acute Q fever die of the disease.

Chronic Q fever, characterized by infection that persists for more than 6 months is uncommon, but is a much more serious disease. Patients who have had acute Q fever may develop the chronic form as soon as 1 year or as long as twenty years after initial infection. A serious complication of chronic Q fever is endocarditis, generally involving the aortic heart valves, less commonly, the mitral valve. Most patients who develop chronic Q fever have pre-existing valvular heart disease or have a history of vascular graft. Transplant recipients, patients with cancer, and those with chronic kidney disease are also at risk of developing chronic Q fever. As many as 65% of persons with chronic Q fever may die of the disease.

The incubation period for Q fever varies depending on the number of organisms that initially infect the patient; infection with greater numbers of organisms will result in shorter incubation periods. Most patients become ill within two to three weeks after exposure. Those who recover fully from infection may possess lifelong immunity against re-infection.

A vaccine for Q fever has been developed and has successfully protected humans in occupational settings in Australia. However, this vaccine is not commercially available in the United States. Persons wishing to be vaccinated should first have a skin test to determine a history of previous exposure. Individuals who have previously been exposed to *C. burnetii* should not receive the vaccine because severe reactions, localized to the area of the injected vaccine, may occur. A vaccine for use in animals has also been developed, but it is also not available in the United States.

Treatment

Doxycycline is the treatment of choice for acute Q fever. Antibiotic treatment is most effective when initiated within the first 3 days of illness. A dose of 100 mg of doxycycline taken orally twice daily for 15–21 days is a frequently prescribed therapy. Quinolone antibiotics have demonstrated good *in vitro* activity against *C. burnetii* and may be considered. Therapy should be started again if the disease relapses. Chronic Q fever endocarditis is much more difficult to treat effectively and often requires the use of multiple drugs. Two different treatment protocols have been evaluated: doxycycline in combination with quinolones for at least four years, and doxycycline in combination with hydroxychloroquine for one and a half to three years. The second therapy leads to fewer relapses, but requires routine eye exams to detect accumulation of chloroquine. Surgery to remove damaged valves may be required for some cases of *C. burnetii* endocarditis.

RICIN INTOXIFICATION

Description

Ricin is a poisonous water-soluble, heat coagulatable globulin protein derived from the castor bean (*Ricinus communis*) of the family *Euphorbiaceae*. The Chemical Abstract System (CAS) Registry Number is 9009-86-3. Although the castor bean plant had long been noted for its toxicity, ricin was first isolated and named in 1888 by H. Stillmark. Modern feed-making techniques break down the ricin in castor beans by heating at 140° C for 20 minutes, although some studies suggest that some residual toxic effects may linger. The most notorious uses of ricin have been as assassination weapons. In 1978, the Bulgarian dissident Georgi Markov was assassinated by Bulgarian secret police who stabbed him with a modified umbrella tipped with a platinum pellet containing ricin. Crude castor bean extracts also contain ricinine (1,2-dihydro-4-methoxy-1-methyl-2-oxo-3-pyridinenitrile), CAS Reg. No. 524-40-3, melting point 201.5° C. Ricinine is sparingly soluble in water, and is also toxic; however, it has somewhat different symptoms from ricin poisoning and is less toxic than the ricin protein. Some sources incorrectly attribute the toxicity of ricin to ricinine. Ricin was once referred to as Agent W.

With an average lethal dose of 0.2 milligrams (1/5,000th of a gram), it is more than twice as deadly as cobra venom. Ricin is poisonous if inhaled, injected, or ingested, acting as a toxin by the inhibition of protein synthesis. There is no known antidote. In small doses, such as the typical dose consumed of castor oil, ricin causes the human digestive tract to convulse. In large doses, ricin causes severe diarrhea, and victims can die of shock.

Ricin consists of two parts: Ricin A, common to many foods and toxic within the cell by stopping RNA replication and thus protein synthesis; and Ricin B, unique to ricin and required for bringing Ricin A into the cell by meshing with a cell surface component. Ricin has a molecular weight of 64,000, and consists of the two peptide chains linked by a disulfide bond; each chain has a molecular weight of approximately 32 kilodaltons (kD). Inducing Ricin B antibodies in humans may be a protective measure for selected military or intelligence operatives. The estimated LD_{50} for humans is on the order of 3 mg/kg; ingestion of two castor beans has been fatal for humans. Ricin is toxic by inhalation, ingestion, and injection.

Despite ricin's extreme toxicity and utility as an agent of warfare, it is difficult to control the production of the toxin since it comes from a commercial crop bean. Under both the 1972 Biological and Toxin Weapons Convention and the 1997 Chemical Weapons Convention, ricin is defined as a "Schedule One" controlled substance, the highest category. Despite this, more than 100 million metric tons of castor beans are processed each year, and approximately 5% of the total is rendered into a waste containing high concentrations of ricin toxin.

In August of 2002, U.S. officials asserted that the Islamic militant group Ansar al-Islam tested ricin, along with other chemical and biological agents, in northern Iraq. British agents arrested a number of Muslim militants in London in 2003 who possessed quantities of ricin with the suspected purpose of poisoning food for military troops.

Signs, Symptoms, and Prognosis

Intoxification with ricin usually shows the first signs fairly quickly, within a few hours. The victim exhibits dizziness and confusion, followed by collapse. The onset of symptoms after inhaling ricin may vary from minutes up to six hours, depending on the dose. Symptoms include a drop in body temperature, drop in blood pressure, edema of the lungs and the stomach and intestines, if ingested, mild degeneration of the intestinal epithelium, necrosis of the liver, and hemorrhage. Death is almost inevitable.

Treatment

There is no treatment available for ricin intoxification. Supportive care is the only course of action, but death usually occurs.

RIFT VALLEY FEVER

Description

A natural disease of central, and eastern and southern Africa, Rift Valley Fever was named after the Rift Valley, a geological formation in east Africa's Afar Region. An epidemic occurred in South Africa in the early 1950s, during which an estimated 20,000 people became infected. The virus can be cultivated in chicken embryos and tissue culture. Infection in humans is usually by inhalation or from handling infected animals. Transmission in nature occurs through infected mosquitoes. The reservoir is thought to be wild animals, possibly rodents.

Signs, Symptoms, and Prognosis

In humans, the disease is severe, but rarely fatal. The usual incubation period is between four to six days, following that period, the onset is sudden with fever, malaise, nausea and vomiting, severe headache, muscle pains, and dizziness. After two or three days of fever, there are one to two days of remission, followed by another two to three days of fever. Recovery, even untreated, is usually uneventful, but malaise and weakness may persist for weeks. Rarely, ocular complications have been seen.

Treatment

There is no specific treatment other than supportive care, including analgesics. There are no preventive vaccines.

ROCKY MOUNTAIN SPOTTED FEVER

Description

Rocky Mountain spotted fever is caused by the rickettsial organism *Rickettsia rickettsii*. The natural hosts are most mammals, including dogs, rodents, and other small animals. The disease is transmitted by ticks that feed on infected animals. The ticks that generally transmit Rocky Mountain spotted fever are large ticks (American Dog Tick, *Dermacentor variabilis*; Rocky Mountain Wood Tick, *D. andersoni*), in contrast to the small ticks (Deer Tick, *Ixodes scapularis*; Pacific Black-Legged Tick, *I. pacificus*) that generally transmit the spiral-shaped bacterium that causes Lyme disease. However, the larger Lone Star Tick, *Amblyomma americanum,* is capable of transmitting both diseases

Signs, Symptoms, and Prognosis

The normal incubation period of the fever is between three and fourteen days. Typical symptoms include sudden onset of fever that can last for two or three weeks, severe headache, fatigue, red swollen eyes, muscle pain, and chills. After two to six more days, a maculopapular rash typically starts on the wrists and ankles, and quickly spreads to the rest of the body and may become hemorrhagic. The untreated fatality rate is about 20–25%.

Treatment

Doxycycline, tetracycline, and chloramphenicol have all been effective in treating Rocky Mountain spotted fever.

SHIGELLOSIS

Description

Shigellosis is a bacterial disease caused by *Shigella dysenteriae*; it is also known as bacillary dysentery. Naturally, it commonly appears among preschool-age children, usually resulting from fecal-oral transmission. The normal host is humans.

Signs, Symptoms, and Prognosis

The symptoms include fever, nausea, vomiting, abdominal cramps, watery diarrhea, and occasionally, traces of blood in the feces, usually appearing after an incubation period of one to three days. The untreated fatality rate can range from 1–10%.

Treatment

Intravenous or oral replacement of fluids and salts along with ampicillin is usually effective as treatment. Other antibiotics may also be used.

STAPHYLOCOCCAL ENTEROTOXIN B (SEB) INTOXIFICATION

Description

Staphylococcal enterotoxin B intoxification is the disease caused by staphylococcal enterotoxin B (SEB), which is one of several exotoxins produced by the bacterium *Staphylococcus aureus*. It may be aerosolized or used to sabotage food supplies. Many of the effects of staphylococcal enterotoxins are mediated by stimulation of T lymphocytes of the host's own immune system. However, it would not be a likely weapon to produce significant mortality rates. Its effectiveness in causing illnesses would be compelling because those who become ill would be incapacitated for up to two weeks. The normal hosts are humans, but also cows, fowl, and dogs have been implicated. An aerosol delivery produces a distinct syndrome. A milder form of disease caused by toxins from *Staphylococcus aureus* is staphylococcal gastroenteritis, or toxic shock syndrome.

The Staphylococcal Enterotoxin B (SEB) has a molecular weight of 28 kilodaltons (kD). SEB is one of at least five distinct proteins which have been identified (SEA, SEB, SEC, SEE, TSS T-1). These toxins are all heat stable. The intoxification can result from either ingestion or inhalation. The human LD_{50} is about 1.7 milligram by inhalation, while the incapacitating dose is only 30 nanograms (a ng is one-billionth of a gram) by inhalation.

SEB has caused many cases of food poisoning in humans after the toxin is produced in improperly handled foodstuffs that are subsequently ingested. Typically, these cases form clusters due to a common source in a setting such as a church picnic or passengers on an airliner eating the same contaminated food.

Staphylococci can be found on the skin and mucosa of most healthy humans. Sneezing and coughing can contaminate food. Animal-contaminated milk and eggs can also be a source of infection. Many outbreaks have been produced by consumption of inadequately refrigerated raw milk or cheeses. In developing countries, where refrigeration after milking is often inadequate, milk and milk products may be an important source of staphylococcal infection. In the U.S., unrefrigerated cream puffs or potato salad at summer picnics are common sources. If temperature and humidity are favorable, one or more strains of *S. aureus* can multiply in the food and produce enterotoxin proteins, including staphylcoccal enterotoxin B. Once made, the toxin is not destroyed even if the food is subjected to boiling while being cooked. Thus, the toxin may be found in the food where no live organisms are present. An important factor in food-borne incidents is holding food at room temperature, which permits multiplication of staphylococci. Other toxins in addition to SEB include SEA, SEC, and SEE, as well as TS T-1, the toxin produced in toxic shock syndrome. *S. aureus* is not considered to be a potential biological warfare agent; it is the isolated toxin that is a potential agent. The toxin was part of the U.S. bioweapons stockpile prior to its destruction in 1972.

Signs, Symptoms, and Prognosis

The incubation period is a very short one to six hours. From three to twelve hours after aerosol exposure, a high fever (103–106° F), chills, headache, myalgia, and nonproductive cough may appear. Some patients may develop shortness of breath and retrosternal chest pain. If the patient develops pulmonary edema or adult respiratory distress syndrome (ARDS), there may be a cough with frothy sputum. The onset of symptoms after inhaling SEB may vary from one to six hours. Symptoms of ingested SEB include vomiting and diarrhea. Initial symptoms of inhaled SEB are headache, fever, and chills. Inhalation of SEB can induce extensive pathophysiological changes including shock. The fever may last two to five days, and cough may persist for up to four weeks. Ingestion of the toxin causes acute salivation, nausea, and vomiting followed by abdominal cramps and diarrhea. Fever and respiratory involvement are not seen in foodborne SEB intoxication. Higher exposure can lead to septic shock and death if left untreated. Physical examination is often unremarkable. Postural hypotension may be present, particularly with ingestion of SEB, due to fluid loss. Conjunctivitis may be seen, and rales may be present. Fatalities are rare for the foodborne infection, but can range from 3–5% for toxic shock syndrome which resulted from use of vaginal tampons.

An epidemic of influenza, adenovirus, parainfluenza, or mycoplasma could cause fever, nonproductive cough, myalgias, and headache occurring in large numbers of people in a short time. Early clinical manifestations of SEB may be similar to those of inhalation anthrax, tularemia, plague, or Q fever, but the rapid progression of respiratory signs and symptoms to a stable state distinguishes SEB intoxication. Chemical agents, such as mustard gas, would show marked vessication of the skin as well as pulmonary injury.

Treatment

Treatment should include supportive care with close attention to oxygenation and hydration, and in severe cases, ventilation with positive end expiratory pressure and diuretics. Acetaminophen and cough suppressants may make the patient more comfortable. The antibiotics methicillin or vancomycin may be effective.

SMALLPOX

Description

Smallpox, also known by the Latin names "variola" or "variola vera," is a disease caused by the *Variola major* virus, which is a very contagious and highly deadly disease in humans. Humans are the normal host. Estimates are that, untreated, it is 20 to 60% fatal, and many survivors were left blind. Skin scars from smallpox were nearly universal. As recently as 1967, smallpox killed two million people in a single year. Smallpox was responsible for an estimated 300–500 million deaths between 1900 and 2000.

Signs, Symptoms, and Prognosis

Before it was eradicated, smallpox would appear in 80–90% of a population with up to one-half of those dying. It was responsible for much of the blindness in natives of tropical regions. Typically, the disease begins with small, red pustules on the skin simultaneous with a fever, nausea, swollen and painful lymph glands, and myalgia. Malaise, fever, rigors, vomiting, headache, and backache appear after an average twelve-day incubation period. Two to three days later, lesions appear which quickly progress from macules to papules, and eventually to pustular vesicles. The pustules may be concentrated on the chest, back, and face, although all parts of the body can be affected; scarring is very likely after the pustules break open and scabs form.

Treatment

There is no treatment once the disease is contracted; supportive care is the mainstay of therapy. Currently, there are no anti-viral drugs of proven efficacy. Although adefovir, dipivoxil, and ribavirin have significant *in vitro* antiviral activity against poxviruses, their efficacy as therapeutic agents for smallpox is currently uncertain. Vistide (cidofovir), an antiviral drug approved by the U.S. Food and Drug Administration (FDA) to treat cytomegalovirus (CMV) retinitis, a sight-threatening infection commonly found in people with late stage human immunodeficiency virus (HIV) disease, may be effective in treating smallpox.

Patients should be placed in a closed-door, negative pressure room with six to twelve air exchanges per hour and high-efficiency particulate (HEPA) filtration of exhausted air. Patients with smallpox should be placed on strict isolation from the onset of eruptive exanthem until all pox scabs have separated (generally in 14–28 days). Healthcare workers and others entering the room should wear appropriate respiratory protection meeting the minimal U.S. National Institutes of Occupational Safety and Health (NIOSH) standard for particulate respirators (N95). Healthcare providers should wear clean gloves and gowns for all patient contact.

In the event of a large-scale smallpox outbreak due to a bioterrorist attack, there may be massive numbers of victims. In this case, there may be a need to cohort patients due to limited availability of respiratory isolation rooms. If this is done, then all patients should receive smallpox vaccine or vaccine immune globulin within three days of exposure, if available, in the event that some of these patients are misdiagnosed with smallpox.

TULAREMIA

Description

Tularemia, also known as rabbit fever, Francis' disease, deer-fly fever, and O'Hara disease, is caused by the hardy gram-negative bacterium, *Francisella tularensis*, a non-sporulating aerobe. It was named after Edward Francis who described the first case in Tulare County, California. It generally is a non-lethal disease that is extremely incapacitating and which shows weight loss, fever, headaches, and often pneumonia. It is typically found in animals, especially rodents, rabbits, beavers, and hares, but occasionally in pet animals, such as cats. It can also be transmitted by biting flies, mosquitoes, and ticks. It appears that it can also be transmitted by contaminated water, plant material, and aerosolized particles. Normally, tularemia is a rural disease. Typically, persons become infected through the bites of arthropods (most commonly, ticks or deerflies), by handling infected animal carcasses, by eating or drinking contaminated food or water, or by inhaling infected aerosols. *Francisella tularensis* can remain alive for weeks in water and soil. Swallowing the bacteria directly from infected meat or untreated water while in an area where animals are commonly infected may cause throat infection, stomach pain, diarrhea, and vomiting. Inhaling the infectious material by breathing dust particles from contaminated soil or from handling contaminated skins can produce fever and a pneumonia-like illness. There are two main strains, F. tularensis biovar palaerctica [Type B] and F. tularensis biovar tularensis [Type A], with the Type A considered the more virulent. Studies have indicated that as few as five to ten organisms are capable of causing the disease.

Signs, Symptoms, and Prognosis

Depending on the route of exposure, the tularemia bacteria may cause skin ulcers, swollen and painful lymph glands, swollen and inflamed eyes, severe sore throat, oral ulcers, or pneumonia. Early symptoms almost always include the abrupt onset of fever, chills, headache, muscle aches, joint pain, dry cough, and progressive weakness. Persons with pneumonia can develop chest pain, difficulty breathing, bloody sputum, and respiratory failure. This is usually accompanied by swelling of the draining lymph nodes. Tularemia eschar (a dry crusty scab) appears at the site where *F. tularensis* penetrated the skin. Symptoms usually appear two to ten days after exposure to the bacteria, but can take as long as 14 days. Tularemia is not known to be spread from person to person, so people who have tularemia do not need to be isolated. People who have been exposed to *F. tularensis* should be treated as soon as possible. The disease can be severe or even fatal if it is not treated promptly. The untreated fatality rate is only about 5% for naturally-acquired tularemia; the pneumonic form of the disease would be higher, perhaps up to 35% for Type A. Much higher rates are possible for weaponized forms. *F. tularensis* was part of the U.S. bioweapons stockpile prior to its destruction in 1972. A vaccine for tularemia has been used to protect laboratory workers, but it is currently under review by the Food and Drug Administration (FDA) for general use.

Treatment

It is cured by a number of broad spectrum antibiotics. Early treatment is recommended with an antibiotic from the tetracycline class, such as doxycycline, or members of the fluoroquinolone class, such as ciprofloxacin, which are taken orally, or the antibiotics streptomycin or gentamicin, which are given intramuscularly or intravenously. Streptomycin is the usual drug of choice.

TRICHOTHECENE INTOXIFICATION

Description

Fungi from the genus *Fusarium*, growing on barley, corn, oats, rye, or wheat, produce dozens of derivatives of tetracyclic sesquiterpenes called trichothecenes. The best known of these mycotoxins are nivalenol, deoxynivalenol, diacetoxyscirpenol, and T-2 Toxin. Bread inadvertently made from *Fusarium*-infected wheat killed thousands of Russian civilians shortly after World War II, focusing attention on the chemistry of the trichothecenes. Other outbreaks have been suspected from natural events. The Russians, when they discovered these mycotoxins, developed a research program to weaponize them. During the late 1970s and early 1980s, there was use in northern Thailand, Kampuchea, and Laos as chemical warfare agents by Russian troops. Because of the appearance, the locals called it "yellow rain." While some scientists, chiefly those of anti-war political movements, have denied that these mycotoxins were "yellow rain," military evidence is fairly clear that these were the agents used. Initial suspicions were raised in the summer of 1975, when there were multiple reports among Hmong and Cambodian refugees of light aircraft dumping a yellow-green powdery "yellow rain," causing vomiting and involuntary defecation. The powder was followed by the dropping of a munition that detonated in the air to produce a dense red haze that drifted down into the green-yellow dust, causing oral hemorrhaging followed by asphyxiation. Mortality rates were described as high. A U.N. commission was formed, but investigators were denied access to the areas where attacks reportedly took place.

There are similar reports dating from the late 1960s in Yemen as well as from Ethiopia and Afghanistan in 1979–1981. It has been estimated that there were more than 6,300 deaths in Laos, 1,000 in Cambodia, and 3,000 in Afghanistan. All attacks were alleged to have occurred in remote areas, which made confirmation of attacks and recovery of the agent extremely difficult. Much controversy has centered about the veracity of eyewitness and victim accounts, as well as on the quality of the forensic evidence.

Fungi apparently use trichothecenes to enhance their infective attack on their plant hosts by breaking down or dissolving cell walls. The chemicals extracted from Fusaria for chemical warfare use all belong to the trichothecene group, the most notorious being "Mycotoxin Trichothecene 2," "T-2 Toxin," or "Fusariotoxin." The Trichothecenes are easily chemically distinguished from the other major group of Fusarium toxins, the Fumonisins, in that they do not possess an amine function so the trichothecenes are less soluble in water. Because of low water solubility, they tend to contaminate the area where used for a longer periods than the Fumonisins and any other more polar Fusaria compounds.

Trichothecenes are useful as a warfare agent because they can enter the body through the skin, by inhalation or by ingestion. Trichothecenes are best deployed in an aerosol for military uses since both dermal exposure and inhalation are possible. Most uses of yellow rain were from low explosive bursting shells. Preparation of these materials is very simple. *Fusarium* fungi growing on cereal grains biosynthesize the trichothecenes; the need is simply to isolate the mycotoxins. These toxins are heat and ultraviolet light stable, easily made, and capable of long-term storage, so they are superb candidates for stockpiling as weapons. In addition, due to the small quantities needed to cause incapacitation or death, a bioreactor to produce these mycotoxins could be relatively small, on the size of a kitchen refrigerator, so it could be easy to hide from weapons inspectors.

Since there are no antidotes, protective clothing and respirators offer the only defense. The United Nations Special Commission inspectors established the presence of trichothecenes in Iraqi facilities following the Persian Gulf War, but, at the time of this writing, evidence has not been found of them following the liberation of Iraq in 2003.

Signs, Symptoms, and Prognosis

Fusaria compounds are used as chemical warfare agents because of the effects they have on all living cells whether human, plant, and other organisms. T-2 Toxin has experimentally killed 50% of female rats at a dose of only 4 mg/kg. The T-2 Toxin is highly irritating to skin and mucous membranes. Direct contact may cause extensive inflammation and tissue necrosis. Topical exposure has lead to systemic toxicity and death in experimental animals. The adult human LD_{50} of trichothecenes, chiefly T-2, is estimated to be only 35 milligrams.

Vomiting and bleeding, effective means of incapacitating troops and civilians, result from even mild mycotoxicosis. Severe poisoning leads to a protracted death. Symptoms are bleeding, even from the eyes, skin, and ears; dizziness; loss of consciousness; vomiting; diarrhea; and severe electrolyte imbalance.

Since these toxins also kill plants, there have been suggestions to use them as mycoherbicides against drug crops in production countries such as Colombia and Afghanistan. A bioweapon such as this could also be used on food crops, especially in countries already having famine conditions.

Treatment

Supportive care, especially maintenance of electrolyte balance, is the only treatment. There are no antidotes, so avoidance of contact is the only preventive measure. Topical dermal antibiotic creams may be of help to prevent dermal pain and secondary infection.

TYPHOID FEVER

Description

Typhoid fever is caused by the bacterium *Salmonella typhi*. The natural host is humans, and the disease is transmitted by eating or drinking food or water contaminated by the stool of infected individuals, such as shellfish taken from sewage-contaminated waters or fruits and vegetables fertilized with human feces. Typhoid fever is caused by *Salmonella typhi*, which is also known a *S. typhosa*; it is a different disease from typhus, which is caused by *Rickettsia typhi* (endemic typhus) and *Rickettsia prowazekii* (epidemic typhus), both of which are spread by lice.

Signs, Symptoms, and Prognosis

The typical incubation period is one to three weeks. Typical symptoms include sustained fever, headaches, constipation more often than diarrhea in adults, fatigue, and fleeting rose-colored spots, especially on the abdomen. The symptoms can be mild to very severe. Untreated, the fatality rate is about 10%.

Treatment

Broad spectrum antibiotics are used, but some strains are resistant to chloramphenicol.

VENEZUELAN EQUINE ENCEPHALITIS (VEE)

Description

Venezuelan equine encephalitis (VEE) is caused by an alphavirus, and causes encephalitis in horses and humans. It is an important veterinary and public health problem in Central and South America. Occasionally, large regional epizootics and epidemics can occur resulting in thousands of equine and human infections. Epizootic strains of VEE virus can infect and be transmitted by a large number of mosquito species. The natural reservoir host for the epizootic strains is not known. A large epizootic that began in South America in 1969 reached Texas in 1971. It was estimated that over 200,000 horses died in that outbreak, which was controlled by a massive equine vaccination program using an experimental live attenuated VEE vaccine. There were several thousand human infections. A more recent VEE epidemic occurred in the fall of 1995 in Venezuela and Colombia with an estimated 90,000 human infections. Infection of humans with VEE virus is less severe than with some of the other encephalitis viruses, and fatalities are rare. Adults usually develop only an influenza-like illness, and overt encephalitis is usually confined to children. Effective VEE virus vaccines are available for equines. The natural host is the horse. The disease is transmitted by mosquitoes, including *Aedes taeniorhynchus*, *A. sollicitans*, *Culex portesi*, and *Psorophora ferox*. The incubation period is typically one to five days. VEE attacks the central nervous system (CNS), in some cases causing swelling of the brain and spinal cord. Symptoms include fever, severe headache, and muscle pains. Cough, sore throat, and vomiting and diarrhea may follow. Some cases progress to encephalitis, which is more frequent in young children and is marked by meningitis, convulsions, and coma. VEE was part of the U.S. bioweapons stockpile prior to 1972.

There are several other encephalitis viruses, including the following:

- Japanese B Encephalitis transmitted by *Culex tritaeniorhynchus*, *C. gelidus*, and *C. vishnui* mosquitoes
- St. Louis Encephalitis transmitted by *Culex tarsalis*, *C. quinquefasciatus*, and *C. nigripalpus* mosquitoes
- Murray Valley Encephalitis transmitted by *Culex annulirostris* mosquitoes
- Eastern Equine Encephalitis (EEE) transmitted by *Aedes ochlerotatus* mosquitoes
- Western Equine Encephalitis (WEE) transmitted by *Culex tarsalis* mosquitoes
- Tick Borne Encephalitis (TBE) transmitted by *Ixodes ricinus* and *I. Persulcatus* ticks

Enzootic strains of VEE virus have a wide geographic distribution in the Americas. These viruses are maintained in cycles involving forest dwelling rodents and mosquito vectors, mainly *Culex* (*Melanoconion*) species. Occasional cases or small outbreaks of human disease are associated with these viruses, the most recent outbreaks were in Venezuela in 1992, Peru in 1994 and Mexico in 1995–1996.

The causitive organism for VEE is similar to a number of other arthropod-borne viruses, *i.e.*, arboviruses. These are viruses that are maintained in nature through biological transmission between susceptible vertebrate hosts by blood feeding arthropods (mosquitoes, psychodids, ceratopogonids, and ticks). Vertebrate infection occurs when the infected arthropod takes a blood meal. The term "arbovirus" has no taxonomic significance in that it does not identify any specific viral family. Arboviruses that cause human encephalitis are members of three virus families: the *Togaviridae* (genus Alphavirus), *Flaviviridae*, and *Bunyaviridae*.

All arboviral encephalitides are zoonotic, being maintained by complex life cycles involving a non-human primary vertebrate host and a primary arthropod vector. These cycles usually remain undetected until humans encroach on a natural focus, or the virus escapes this focus via a secondary vector or vertebrate host as the result of some ecologic change. Humans and domestic animals can develop clinical illness but usually are "dead-end" hosts because they do not produce significant viremia, and do not contribute to the transmission cycle. Many arboviruses that cause encephalitis have a variety of different vertebrate hosts and some are transmitted by more than one vector. Maintenance of the viruses in nature may be facilitated by vertical transmission (*e.g.*, the virus is transmitted from the female through the eggs to the offspring). There are no commercially available human vaccines for these diseases. An equine vaccine is available for VEE. Arboviral encephalitis can be prevented in two major ways: personal protective measures and public health measures to reduce the population of infected mosquitoes. Personal measures include reducing time outdoors particularly in early evening hours, wearing long pants and long sleeved shirts, and applying mosquito repellent to exposed skin areas. Public health measures often require spraying of insecticides to kill juvenile (larvae) and adult mosquitoes.

Signs, Symptoms, and Prognosis

The majority of human infections are asymptomatic or may result in a nonspecific flu-like syndrome. Onset may be insidious or sudden with fever, headache, myalgias, malaise, and occasionally prostration. Infection may, however, lead to encephalitis, with a fatal outcome or permanent neurologic sequelae. Fortunately, only a small proportion of infected persons progress to frank encephalitis. The fatality rate is about 1%.

Experimental studies have shown that invasion of the central nervous system (CNS) generally follows initial virus replication in various peripheral sites and a period of viremia. Viral transfer from the blood to the CNS through the olfactory tract has been suggested. Because the arboviral encephalitides are viral diseases, antibiotics are not effective for treatment and no effective antiviral drugs have yet been discovered.

Treatment

Treatment is supportive, attempting to deal with problems such as swelling of the brain, loss of the automatic breathing activity of the brain, and other treatable complications like bacterial pneumonia.

GENETICALLY MODIFIED ORGANISMS

All of the above organisms can be modified using modern genetic engineering techniques. The modified organisms can be designed to be:

- more virulent
- less susceptible to current treatment regimens and medicines
- more difficult to detect using standard techniques

In addition, it is possible through recombinant DNA technology for a benign organism to be given a gene-encoding toxin or other pathogenic substance produced by the pathogenic organism.

FUNGI AS ANTI-CROP WEAPONS

Prior to 1972, the U.S. maintained a stockpile of several biological agents capable of destroying the food and industrial crops of potential enemy countries. Examples of these follow.

Wheat Stem Rust

Wheat stem rust is a plant disease which causes elongated ragged pustules on the stems, leaf sheathes, blades, or chaff which usually begin to appear in mid-June in natural cases. The causative organism is the fungus *Puccinia graminis tritici*. The pustules rupture tissue, exposing a powdery, brick red mass of summer spores. As wheat nears maturity, black pustules filled with black spores, teliospores, appear. Wheat varieties entirely immune to stem rust do not exist, but varieties with various degrees of resistance do exist. Eradication of common barberry has been ongoing in the U.S. and is nearly complete. Fungicides are available for control of rust when needed.

Puccinia graminis has many specialized varieties. *P. graminis tritici* attacks wheat and barley. *P. graminis avenae* occurs chiefly on oats. *P. graminis secalis* occurs chiefly on rye. An alternate host is common barberry, a tall erect woody shrub with bristle-toothed leaves. Wheat stem rust was part of the U.S. bioweapons stockpile prior to 1972.

Rye Stem Rust

Rye stem rust is a plant disease caused by the fungus *Puccinia graminis secalis*. Symptoms are similar to those for wheat above including elongated ragged pustules on stems, leaf sheathes, blades, or chaff usually beginning to appear in mid-June for natural cases. The pustules rupture tissue, exposing a powdery, brick red mass of summer spores. As wheat nears maturity, black pustules filled with black teliospores appear.

Puccinia graminis has many specialized varieties. *P. graminis tritici* attacks wheat and barley. *P. graminis avenae* occurs chiefly on oats. *P. graminis secalis* occurs chiefly on rye. An alternate host is common barberry, a tall erect woody shrub with bristle-toothed leaves. Rye stem rust was part of the U.S. bioweapons stockpile prior to 1972.

Rice Blast

Rice blast, also called rice rust, is a plant disease caused by the fungus *Pyricularia grisea*. It appears as early off-white to gray-green spots of the leaves of the rice plants, which rapidly enlarge and turn gray or gray-white in the center with a brown to reddish-brown border. About a week after infection, the fungus sporulates within the lesion producing a dark, gray powdery appearance. The blast kills the entire panicle before grain fill is complete.

Resistant varieties are readily available, *e.g*., Katy, Kaybonnet, and Drew. Blast damage can be prevented or greatly reduced by proper irrigation. A flood depth of at least 2" and preferably 4" must be maintained from green ring through heading to be most effective. Rice blast was part of U.S. stockpile prior to 1972.

ZOONOTIC DISEASES

There are also possible bioweapons against both humans and animals. Some of these affect only animals, but a number can be transmitted from animal hosts to humans. Bioweapons used against animals are generally intended to interfere with the food supply. Zoonotic diseases, those which can be transmitted from animals to humans, are usually employed as bioweapons against humans. A few examples will be presented.

GLANDERS

Description

Glanders is a disease caused by the bacterium *Burkholderia mallei*. The natural hosts include horses, donkeys, and mules. It can be transmitted to humans by inhalation of the bacteria in aerosols or dust or contact with infected animals. *Burkholderia pseudomallei*, a saprophyte of soil and water in tropical areas of southeast Asia, causes a glanders-like disease called mellioidosis; this disease may take a rather benign pulmonary form, but may also develop into a rapidly fatal septicemia.

According to the U.S. Department of Agriculture, glanders was eradicated from the U.S. animal population in 1934.

German saboteurs employed *B. mallei* as a biological weapon during World War I. At that time, horses and mules were very important battlefield resources, bringing supplies to the front and moving artillery. Cultures confiscated from the German legation in Bucharest, Romania, were identified as *B. mallei*, and livestock were infected in Mesopotamia, France, and Argentina. Before the U.S. entered the war, German saboteurs are believed to have inoculated horses destined for export with *B. mallei*. It is known that many horses died at sea or were dying when they arrived in Europe.

Signs, Symptoms, and Prognosis

There are two forms: pulmonary-affecting the respiratory tract, and cutaneous-affecting the skin. Signs and symptoms of the pulmonary form include general respiratory conditions like cough, nasal discharge, acute or chronic pneumonia, usually appearing after two to five days incubation. The cutaneous form shows multiple purulent cutaneous eruptions, often following lymphatics, after some three to six days incubation. The untreated fatality rate in humans is as high as 95%.

Treatment

Sulfadiazine may be effective in some cases. Doxycycline, rifampin, trimethoprim-sulfamethoxazole, and ciprofloxacin have been effective in experimental infection in hamsters.

PSITTACOSIS

Description

Psittacosis is a natural disease of parrots, parakeets, and other birds, which is transmitted to humans by handling the blood, tissues, feathers, or discharges from infected birds, or by breathing in dust particles from the dried droppings of infected birds. It is caused by the bacterium *Chlamydia psittaci*.

Signs, Symptoms, and Prognosis

The most common symptoms include fever, headache, loss of appetite, vomiting, neck and back pain, muscle aches, chills, fatigue, and cough. The usual incubation period is between four and fifteen days. In severe cases, extensive pneumonia may develop and be fatal, but this is rare.

Treatment

Antibiotics such as doxycycline or erythromycin are effective. Fluid intake and electrolyte balance should be maintained.

TOXINS

Toxins are poisons produced by living organisms. In addition to those specifically discussed above, there are a number of other toxins capable of bioweapons use. In general, toxins are extremely poisonous; they have a toxicity several orders of magnitude greater than the nerve agents.

Toxins can be classified by their mechanism of toxicity: cytotoxins cause cellular destruction, ricin is an example; enterotoxins affect the digestive tract, Staphlococcal enterotoxin B (SEB) is an example; hemorrhagic toxins cause bleeding, mycotoxin T-2 is an example; hepatotoxins cause liver damage; and nephrotoxins cause kidney damage; other toxins that inflame the skin and mucous membranes; and neurotoxins affect the central nervous system (CNS).

Neurotoxins may be further classified as:

- presynaptic and postsynaptic neurotoxins; botulinum toxin, and saxitoxin are examples
- ion-channel and sodium-ion binding toxins; tetrodotoxin is an example of an ion-channel neurotoxin
- ionophores

Mixed toxins are toxins showing multiple mechanisms from different categories above. A few well-known toxins follow.

SAXITOXIN

Description

Saxitoxin is produced by bacteria that grow in other organisms, including the dinoflagellates *Gonyaulax catenella* and *G. tamarensis*, which are consumed by the Alaskan butter clam, *Saxidomus giganteus*, and the California sea mussel, *Mytilus californianeus*. Saxitoxin is also known as Mytilotoxin. The toxin can be isolated from *S. giganteus* or *M. californianeus*. The Chemical Abstracts Service (CAS) Registry Number is 35554-08-6. It is a cholinesterase inhibitor of molecular weight 299.29, and formula $C_{10}H_{17}N_7O_4$.

Signs, Symptoms, and Prognosis

Typical symptoms include numbness, muscle weakness, and, finally, respiratory paralysis. The human LD_{50} is 10 mg/kg by the oral route, but as low as 2.0 mg/kg by inhalation.

Treatment

There is no specific treatment for poisoning by saxitoxin. Supportive care, including respiratory support, is the treatment of choice.

TETRODOTOXIN

Description

Tetrodotoxin is found in the liver, gonads, intestines, and skin of many species of the fish order *Tetraodontidae*, including the globe fish, *Spheroides rubripes*. The Chemical Abstracts Service (CAS) Registry Number is 4368-28-9. The chemical name is octahydro-12-(hydroxymethyl)-2-imino-5,9:7,10a-dimethano -10aH-[1,3]dioxocino [6,5-d]pyrimidine-4,7,10,11,12-pentol, with molecular weight of 319.27, and elemental formula of $C_{11}H_{17}N_3O_8$.

Tetrodotoxin is naturally responsible for fugu (puffer fish) food poisoning. Personal importation of fugu into the United States is prohibited; however fugu chefs certified by the Ministry of Health and Welfare of Japan are allowed to import fugu into the U.S. When properly cleaned and prepared, the fugu flesh or musculature is edible and considered a delicacy by some Japanese. One meal sells for the equivalent of more than $400. Despite strict regulation, fugu causes approximately 50 deaths annually in Japan.

It has been claimed that tetrodotoxin is the active ingredient in zombie powder, used by Voodoo priests to induce a death-like trance. The zombie is given just enough tetrodotoxin to incapacitate him or her, then is revived and kept under control by the use of other drugs. However, claims concerning zombification are poorly documented.

Signs, Symptoms, and Prognosis

The human LD_{50} is about 334 mg/kg by ingestion. Typical symptoms from ingestion include numbness, tingling of the lips and inner mouth surfaces, weakness, paralysis of the limbs and chest muscles, and a drop in blood pressure have been reported within as little as 10 minutes after exposure. Death can occur within 30 minutes. The toxin is an ion-channel neurotoxin with the guanidinium moiety of tetrodotoxin lodging in the sodium channel of a nerve cell where it blocks transmission.

Treatment

There is no specific treatment for tetrodotoxin poisoning. Supportive care, including respiratory support, is the treatment of choice. Some Japanese physicians and hospitals claim to have expertise in keeping patients poisoned by puffer fish alive.

PRIONS

Prions are organisms which cause a variety of largely neurological diseases, such as scrapie, "mad cow" disease, chronic wasting disease, and Creutzfeldt-Jakob disease. While the effects are delayed, they are particularly devastating in that the brain, spinal tissue, and other nerve tissue is destroyed. Such tissues have essentially no recovery or repair mechanisms. Once the damage is done, death is the only outcome. While some cases are hereditary, most acquired cases in humans have resulted, it is assumed, from association with or consumption of infected nerve tissue.

Prions are infectious agents which do not have a nucleic acid genome; a protein alone is the infectious agent. A prion is defined as a small proteinaceous infectious particle which resists inactivation by procedures that would modify nucleic acids (DNA or RNA). The discovery that proteins alone could transmit an infectious disease has been a considerable surprise to the scientific community. Prion diseases are often called spongiform encephalopathies because of the post mortem appearance of the brain with large vacuoles in the cortex and cerebellum. Probably most mammalian species can develop these diseases. Specific examples include: scrapie in sheep; transmissible mink encephalopathy (TME) in mink; chronic wasting disease (CWD) in muledeer and elk; and bovine spongiform encephalopathy (BSE) in cattle. Humans are also susceptible to several prion diseases: Creutzfeld-Jakob Disease (CJD); Gerstmann-Straussler-Scheinker syndrome (GSS); Fatal Familial Insomnia (FFI); Kuru; and Alpers Syndrome.

Luckily, proteins can be easily denatured, or destroyed, by heating. Transmission of the disease is only possible from severely undercooked or raw nerve tissue; for this reason, most nations have prohibited the feeding of uncooked brain and other tissue to farm animals.

It is not known if any state or terrorist organization has studied or weaponized these diseases, but it would be more logical for terrorism than for military action due to the delayed effects. The psychological damage of a widespread outbreak of prion-caused disease could be immense. Deliberate contamination of a food supply with infected meat remains a possibility although, in most nations, sufficient safeguards are in place to protect the public. Likewise, most cooking processes would render the protein ineffective. However, post-distribution contamination, say in a school lunchroom, remains a possibility even though unlikely.

GENERAL COMMENTS ON MEDICAL TREATMENT

The impact of biological warfare may be a few patients with diarrhea or coughing, or a mass casualty situation. The first indication of a biological weapons attack will most likely be large numbers of patients arriving at a hospital or clinic with an illness. The routes of entry for biological weapon agents are the same as endemic diseases (*i.e.*, through inhalation, ingestion, or percutaneous inoculation). Biological agents are most likely to be delivered covertly and by aerosol. Other routes of entry are thought to be less important than inhalation, but may be potentially significant.

Aerosol Particles

Inhalation of agent aerosols, with resultant deposition of infectious or toxic particles within alveoli, provides a direct pathway to the systemic circulation. The natural process of breathing causes a continuing flux of biological agent to exposed individuals. The major risk is pulmonary retention of inhaled particles. Droplets as large as 20 microns can infect the upper respiratory tract, however, these relatively large particles generally are filtered by natural anatomic and physiological processes and only much smaller particles (ranging from 0.5 to 5 microns) reach the alveoli of the lungs efficiently.

Ingestion of contaminated food and water may occur during an aerosol bioweapons attack. Unwary consumption of such contaminated materials could result in disease. Direct contamination of food and water could be used as a means to disseminate infectious agents or toxins. This method of attack is most suitable for sabotage or terrorist activities, and might be used against limited targets such as water supplies or food supplies of a specific local area.

Percutaneous absorption is also possible although intact skin provides an excellent barrier for most, but not all, biological agents. However, mucous membranes and damaged skin constitute breaches in this normal barrier through which agents may readily pass.

The spread of diseases by releasing infected arthropods such as mosquitoes, ticks, or fleas is a possible terrorist scenario. These live vectors can be produced in large numbers and infected by allowing them to feed on infected animals, infected blood reservoirs, or artificially-produced sources of a biological warfare agent.

Preservation of toxins for extended periods and the protective influence of dust particles onto which microorganisms adsorb when spread by aerosols have been documented. Therefore, the potential exists for the delayed generation of secondary aerosols from contaminated surfaces. To a lesser extent, particles may adhere to an individual or to clothing, creating additional exposure hazards to the victim as well as to health care workers.

The spread of potential biological agents by person-to-person contact has also been documented. Humans, as unaware and highly effective carriers of a communicable agent, could readily become a source of dissemination (for example, plague or smallpox).

MANAGEMENT OF BIOLOGICAL WARFARE PATIENTS

Management of patients suffering from the effects of BW agents may include the need for isolation. Barrier nursing for patients suspected of suffering from exposure to biological agents will reduce the possibility of spreading the disease to health care providers and other patients. Specimens must be collected and submitted to the supporting laboratory for identification.

A biological weapons agent attack can produce a mass casualty situation. A major problem with a bioweapons mass casualty situation is that medical personnel are more susceptible to becoming a casualty to the agents. However, the ill patient may be the first indicator that a biological agent has been dispersed.

Biologically contaminated patients require some degree of decontamination. Contamination can be removed by use of a diluted disinfectant solution or a 0.5% chlorine solution.

Specific treatment is dependent upon the specific agent used. Patients should be treated for symptomatic presentation.

SUMMARY

There are a variety of pathogens and toxin producers that can be used as biological weapons. Most of these are naturally occurring bacteria, viruses, or fungi. However, there is evidence that at least some nations have developed enhanced pathogens by techniques like gene splicing. The possibilities are almost endless for the splicing of toxin-producing or vaccine-resistant DNA in normal organisms, some of which may be habituated to the human body. While ordinary anthrax and smallpox remain the most likely bioweapons, there is a possibility that dangerous and even fatal gene sequences have been spliced onto other bacteria and viruses. Medical treatments are the same as for the natural form of the disease.

10

DELIVERY SYSTEMS FOR BIOWEAPONS

PROBLEMS WITH USE OF BIOWEAPONS

Biological warfare can be waged against humans, livestock, or food crops. On the whole, the military usefulness of biological weapons is limited. Some of the intrinsic properties of the pathogens that have so far prevented the assimilation of bioweapons into mainstream military doctrines of most nations are difficulties in dispersion, unpredictability of the dispersion, and destruction or weakening of the agent by ambient conditions.

Technologies must be devised for the airborne release of biological warfare agents so that a relative high proportion of the agent will survive the separation from the delivery system. The separation of the agent from the delivery system by means of a small low explosive to break the container can generate enough heat and shock to kill a majority of the pathogens or inactivate the toxin. Therefore, the explosive charge must be designed in such a way that it creates a sufficient number of particles of the right size. Similarly, an aerosol generator also requires advanced design in order to produce sufficient particles of the right size. Such considerations are particularly important if a bioweapons user selects a non-contagious pathogen (*e.g.*, anthrax), which requires that each victim be individually exposed to a sufficient infective dose. Furthermore, biological warfare agents travel in the atmosphere as particles of microscopic clusters of individual pathogens or droplets of toxin that must be of the right size in order to penetrate the body and reach those parts where they will cause the onset of disease.

Because biological warfare agents travel as microscopic particles, they are dependent on environmental variables, such as air flow at different altitudes, air density, temperature, humidity, and presence or absence of clouds. These variables are beyond the control of the bioweapons user after the release of the agent.

If a bioweapons user selects a highly contagious agent (*e.g.*, smallpox) so that the disease is transmitted from victim to victim and propagated, a considerable risk exists that his own troops or civilian population will also fall victim to the disease. This means that such agents can only be considered for use in the deep rear of an enemy.

Ultraviolet rays in the sunlight and oxygen in the air may kill most of pathogens surviving the separation from the delivery system, so that bioweapon can only be effectively released during certain parts of the day (*e.g.*, at night) or under certain weather conditions (*e.g.*, cloud cover). Ultraviolet light, oxygen, and humidity may also affect the strength of toxins.

Most diseases require time for adverse effects (and symptoms) to develop. The delays may range from hours to days with respect to a single victim, and even longer before there are militarily significant effects in the target population. This gives the attacked party time to execute a retaliatory strike. However, the delays may make it more difficult to identify the bioweapons user, so that this characteristic may actually offer a strategic advantage or opportunities for clandestine operations to the user. Candidate biological warfare agents thus have to be selected on the grounds of a compromise between pathogenicity, survivability after release, and controllability.

Military biological weapon programs include lethal, incapacitating, and anti-crop agents. Biological warfare can be waged against animals and plants rather than against humans, and thus hit the target population's food sufficiency. Such attacks can be carried out as part of covert operations with little or no risk to the soldiers or civilian population of the bioweapons user. The great progress in biotechnology creates the potential to design biological warfare agents in order to increase their virulence, to overcome available medicines, to better resist the destructive influences from the environment during dispersion, or even to target particular genetic characteristics of certain population groups.

BIOLOGICAL ORGANISM DISSEMINATION

Biological agents are meant to be delivered using the same types of weapons as chemical warfare agents; the only difference is that the weapon is filled with an isolate of the biological agent rather than a chemical. Similarly, dispersal of a biological agent cloud is governed by many of the same phenomena as is the dispersal of chemical agents. Natural vectors and person-to-person spread are additional means of disseminating biological agents.

PARTICLE SIZE

The effect of particle size is key to the effectiveness of an attack with a biological warfare agent. Large particles have several drawbacks that generally make them inefficient transmitters of disease. These particles tend to settle out of the air relatively rapidly, limiting the time that they are effective. Once inhaled, large particles are most likely to impact on the mucous coat of the pharynx or nasal cavity. Although still capable of causing infection, the probability of infection decreases dramatically. Small particles remain airborne for a relatively long time; however, once they are inhaled, they are promptly exhaled. Small particles do not remain in the lung to cause infection. The particles most likely to cause infection are those of an intermediate size. These particles remain airborne, settling relatively slowly, but once they are inhaled they reach the alveoli in the lung and remain there with a large probability of causing infection. One hurdle to military use of biological weapons is that an aerosol of ideally-sized particles remains airborne for a relatively long period. It is subject to changes in the wind, which can potentially blow the cloud back at the attacker, and weather, which can wash the agent out of the atmosphere.

NATURAL VECTORS

Although most biological weapons were designed to function via inhalation, biological warfare agents can also enter the body through physical contact or ingestion. In addition, agents can be spread through natural vectors themselves, from contact with infected animals or from contact with contaminated food or water.

Some of the natural vectors are arthropods, which are hard-shelled invertebrates with segmented limbs, including:

- insects, including fleas, ticks, and mosquitoes
- crustaceans, including crabs, lobsters, shrimp, and barnacles
- arachnids, including spiders
- myriapods, including centipedes

Many agents also cause natural outbreaks of disease through these vectors. In fact, the use of natural vectors to spread biowarfare might be appealing to terrorist organizations. It would be difficult to differentiate this attack from a natural outbreak. It would also be almost impossible to prove that an attack even took place.

ARTILLERY SHELLS

Biological artillery shells or projectiles typically contain a reservoir filled with the biological agent surrounding an internal axial explosive charge called the burster. The nose of most such shells has a lifting lug screwed in. In order to fire the shell, the lifting lug is replaced with a fuse; in some cases the fuse is permanently affixed to the nose of the shell. Once the shell is fired, the fuse is armed. When the shell arrives at its target, the fuse detonates the explosive charge. The explosive charge must be of such a nature that there is not enough heat generated to destroy the agent. The burster shatters the shell casing and causes the biological agent to be dispersed as an aerosol. In the past, the U.S. used several munitions that have the same configuration and use identical shell bodies, bursters, and fuses as biological weapons, but contained different fills of flame- or smoke-generating chemicals.

A major problem with the use of biological agents in an artillery shell is keeping them viable during storage, transport, and firing. Generally, only shock- and heat-resistant items like anthrax spores are likely to be found in artillery shells.

LAND MINES

Biological agent land mines are not very likely since they remain at ambient conditions underground prior to use and use high explosive bursters. An exception was be for dissemination of spores or a toxin. Burial conditions would render many biological agents ineffective. Some agents such as anthrax spores might survive; if so, the mine would contain a reservoir filled with anthrax spores surrounding an explosive charge, similar to an artillery shell. The top of the land mine has a pressure plate; when the pressure plate is depressed, the fuse detonates the explosive charge or burster. The burster shatters the mine casing, blows away any soil covering the mine, and causes the biological agent to disperse as an aerosol. However, it is believed that the U.S. never fielded a biological land mine, and it is thought that no other nations did either. This is most likely due to the nature of land mine use in which the device remains hidden (often underground) for an extended period of time, limiting the number of viable organisms.

ROCKETS

A typical biological agent rocket includes, in order from the rear forward, a set of spring-loaded guidance fins, a rocket propellant, a reservoir filled with biological agent surrounding an axial explosive charge, and a fuse. Most systems use a launch tube to eject the rocket. When the rocket is fired, the propellant ignites and thrusts the rocket out of the firing tube. The guidance fins extend and the rocket flies to its target, while the fuse is armed in flight. When the rocket reaches its target, the fuse detonates the axial explosive charge or burster. The burster shatters the rocket casing and causes the biological agent to disperse as an aerosol. As with artillery shells, only a very hardy agent would survive the transit, bursting, and dissemination; this means that a spore-former like anthrax or a virus like smallpox or a toxin like T-2 is likely as a rocket-disseminated biological agent.

Iraq is believed to have launched rockets without bursters, like the Scud missile, containing biological agents. In this case, in which a ballistic missile simply carries the agent, a number of agents are possible; however, anthrax remains the most likely.

BOMBS

Biological agent bombs usually contain a reservoir filled with the agent surrounding an internal axial explosive charge. Biological bombs are generally stored without the tail fins and the detonating fuse. These items would be attached immediately before the bomb is loaded onto an aircraft. After the bomb is dropped, the fuse is armed. When the bomb arrives on its target, the fuse detonates the axial explosive charge or burster. The burster shatters the bomb casing and causes the biological agent to disperse as an aerosol. Again, only the most hardy agents (like the anthrax spores and some toxins) are likely to survive.

DISPERSAL OF BIOLOGICAL AGENTS

When a biological weapon detonates or a biological agent is released in an aerosolizer, it creates a primary cloud of either solid or liquid aerosols. This cloud then settles to the ground, depending upon characteristics of the aerosol and weather conditions, eventually landing on individuals, plants, equipment, and the ground. The contamination on the ground has a finite lifetime, but can injure people from contact with a secondary cloud of agent that results from being kicked up by people or vehicles. Eventually, the agent contamination disappears as the agent is destroyed by air, ultraviolet radiation, or weather conditions.

Factors that affect the hazard from the primary cloud are related to the local weather conditions. Factors which tend to diminish the hazard of the primary cloud include: variable wind directions, which causes dilution by redirection of the cloud; presence of wind velocities over 6 meters per second (13.4 miles per hour), which causes dilution by turbulence; presence of unstable air, which also causes dilution by turbulence; temperatures below 0° C (32° F), which result in less viability or even death of the organisms; temperatures above around 40° C (104° F) , which denature proteins and thus kill organisms; and precipitation, which washes out aerosol particles from the atmosphere. Factors which tend to increase the hazard of the primary cloud are generally the opposites of those previously cited: steady wind direction; wind velocity under 3 meters per second (10 feet per second); stable air (atmospheric inversion); temperatures between 20° C (68° F) and 35 ° C (95° F); and no precipitation. For a military operation, with the goal to cause as much damage as possible, these latter conditions are sought.

Winds vary almost constantly in force and direction, and are affected by every irregularity on the surface. Much work has been done on the effect of these irregularities of motion on the distribution of pollution through the atmosphere, and it is important to determine what can be learned about the probable distribution of chemical and biological agents from this work. As mentioned in Chapter One, F. Pasquill devised six categories of atmospheric turbulence that are generally accepted for making rough estimates of the spread of atmospheric pollutants; others have added a few categories for specialized studies. Of these, Pasquill category A is the most turbulent and is rare. His Category F is the most stable atmospheric state for which realistic predictions can be made. Examination of long-term meteorological records have shown that, in most places, states E, F, and those of greater stability than F are likely to occur at around midnight on about half the nights throughout the year. The longer the nights, the longer will be the period in which these degrees of stability will be observed; in mid-winter, it may even extend from 6:00 PM to 6:00 AM. For the purposes of this book, maintenance of this degree of stability for two hours would suffice; this degree of stability occurs in most locations during different times at all periods of the year. However, most of the information about the dissemination of pollutants in the atmosphere under different conditions of atmospheric stability has been derived from studies in open country; there is much less known about urban areas and areas of varying altitude. It is usual to predict the likely atmospheric stability from the wind speed, the state of the sky, and the time of the day. Over cities, atmospheric turbulence may be augmented by the extra surface roughness and by heating from industrial, commercial, and domestic processes. Release of aerosol particles at heights has been less studied and is much more complex. The cloud, and its particles, are moved by wind and other air currents forming a rough oval cross-section shape with the horizontal spread usually greater than the vertical. This was discussed in Chapter One.

Once the particles in the primary cloud settle to the ground, factors that tend to decrease the hazard due to ground contamination include: low ground temperature, which causes decreased viability of some agents; high wind velocity, which dilutes the agent; unstable air, which dilutes the agent; and heavy precipitation, which dilutes and washes the agent into the soil. Factors that tend to increase the hazard of ground contamination are related to the stability of the atmosphere. Factors that characterize stable air include low wind velocities and temperature inversions.

Persistency is the term used to describe the duration of an area's toxic contamination. Persistent agents are generally considered to be those for which contamination lasts a day or more. Non-persistent agents are generally considered to be those where contamination dissipates in a matter of minutes or hours. Chemical reactions that affect the persistence of biological agent in the environment include photochemical reactions from sunlight (especially UV radiation) and other reactions with compounds (often oxygen) present in the environment.

To calculate the precise lifetime of a biological warfare agent in the field can be very complex. This calculation requires knowledge of the parameters of the agent, plus climate information for the specific location, including temperature, humidity, rainfall, and solar flux, among others. Nevertheless, assumptions can be made for average values for humidity, rainfall, and solar flux, yielding approximate lifetimes that indicate the persistency under generalized climatic conditions.

EXPOSURE VERSUS DOSE

Most measurements of biological agents involve the amount of aerosol in the air. If there is a person at that location, this would be the person's exposure to the agent. However, the amount actually taken in by the person can be less than the exposure. As discussed in Chapter One, the chance that a particular dose will be received by a person when the air contains a toxic material is proportional to its concentration in the air and to the length of time of the exposure, Ct. For concentration expressed in mg/m^3 and time in minutes, the Ct is expressed in $mg\text{-}min/m^3$. The Ct is a measure of the intensity of exposure, and not of the dose that actually penetrates into the body. For infective agents, the term ID_{50} (Infective Dose 50) is used for the dose that has a 50% probability of producing a specified response to infection, considered in this book as signs or symptoms of disease or death.

TOXICITY, DOSE, AND EFFECTS

The LD_{50} (Lethal Dose 50) of a material is the dose that will kill 50% of the people receiving it. It may also be defined as the dose that has a 50% probability of killing any particular individual, and may be used for chemical and biological agents. The ID_{50} (Infective Dose 50) is the dose of a biological agent that has a 50% probability of producing any defined effect of infection. Likewise, the ID_5 indicates the dose likely to cause an effect in 5% of those exposed and the ID_{95} is the dose likely to affect 95% of exposed persons. Neither the LD_{50} nor the ID_{50} gives all the information needed to understand the relationship between dose and effect. As discussed in Chapter One, the additional information needed is provided by the slope of the probit line, which can be calculated from the data required for an accurate determination of an ID_{50}. The steepness of the slope indicates the factor by which the ID_{50} must be multiplied to give the probability of producing an infection in response to the agent. The steeper the slope, the more rapid the probability of an effect increases with increase in dose. If the probit slope is 0.5, the ratio of ID_{95} to ID_5 is very large, some 3.8×10^6; for a slope of 1, the ratio drops to 1.95×10^3; for a slope of 5, the ratio is 4.55; and for a slope of 20, the ratio is only 1.46. For most biological agents, the slope is highly unlikely to be very great; in fact, most probit slopes for biological agents are less than 1.

At low probit slopes common for biological agents, the probability of effect changes very slowly with dose, even in the neighborhood of the LD_{50}, so that it is often very difficult specify the LD_{50} with exactness. This has a bearing on the comparison of toxicities. Small differences between the ID_{50} values of two compounds will imply real differences in toxicity if both slopes are large, but very large differences in the ID_{50} may have no practical significance if both slopes are small. In the intermediate range of values of the slope, it is essential to consider the slope of the probit line and the ID_{50} together when making comparisons between effectiveness of biological agents. For example, there are two infective agents, A and B, being looked at. Agent A has an ID_{50} of 1 mg-min/m^3 and a probit slope of 1. Agent B has an ID_{50} of 5 mg-min/m^3 and a probit slope of 5. On the basis of ID_{50} values alone, agent A would appear to be 5 times as effective as B. But, using the probit slope ratios of ID_{95} to ID_5 given in Chart 1-1, the ID_{95} of A is 44 mg-min/m^3 (ID_{50} of 1×44.1 for slope of 1), while the ID_{95} of B is less than 11 mg-min/m^3 (ID_{50} of 5×10.65 for slope of 5). Therefore, when a high probability of effect is required, B is 4 times as effective as A. When the probit slopes are small (*e.g.*, 0.5 and 1.0), larger ratios of ID_{50} values can be misleading. Some very highly toxic chemical agents can have probit slopes between 5 and 10, but most biological agents have slopes less than or near unity. This means that a biological agent cloud may travel a considerable distance downwind before its dosage drops from an LCt_{50} to an LCt_5, but that exceedingly high concentrations are needed to give an ED_{95}.

SUMMARY

While there are many biological agents, ranging from viruses to bacteria to toxins, capable of causing disease and adverse effects, delivery and dissemination of many of these agents is very difficult. The requirements for an effective biological agent include being hardy or resistant enough to survive possible detonation or mechanical dispersion, being able to survive and remain viable in the environment, being difficult to identify or treat, being resistant to pharmaceuticals and vaccines, and being capable of causing incapacitation or lethality at relatively low concentrations.

11

BIOLOGICAL WEAPON DETECTION, PROTECTION, AND DECONTAMINATION

VIABILITY OF ORGANISMS

The term "viability," with respect to a biological organism, refers to the fact that the organism is capable of reproduction or replication. Most bacteria reproduce by cell division, in which one cell duplicates the genetic material in the cellular nucleus, forming a new exact copy of the nucleus. The two nuclei then move apart; a cell membrane forms between them, and two cells result. While this process takes place in a number of steps, with each step having different sensitivity to radiation or chemicals, for the purpose of this book it is enough to say that bacteria divide into two identical cells under most conditions. Rickettsial organisms are bacteria but cannot reproduce outside a host cell, similar to viruses. Viruses increase in number by a different process: the viral RNA or DNA unwinds; a mirror image of the structure is formed from available materials in the host cell, and a duplicate results. This process is known as replication rather than reproduction.

A common question often raised is "Is it alive?" A definition of life can be difficult, but one of the characteristics is the ability to duplicate. Thus, bacteria and viruses are capable of duplication; however, while bacteria are capable of reproduction as isolated organisms (assuming conditions such as moisture, temperature, and food are favorable), rickettsial organisms and viruses can only replicate inside the cells of a host. A virus lying on a tabletop cannot replicate. Some bacteria are capable of converting to a state similar to viruses in that they are not reproducing; this is known as spore formation. Spore formation usually occurs when conditions are not favorable for reproduction, such as high temperatures or dryness. The genetic material inside the spore can be thought of as "resting" or awaiting better conditions, which may be many years. When conditions get better, the bacterial spores return to their normal state and begin to reproduce again. Anthrax bacteria are among those bacteria forming these highly resistant spores.

Generally, most living organisms require water, an acceptable temperature range, and food; many, but not all, also require oxygen (these are known as aerobes). In order to reproduce, these conditions must be maintained. For those organisms capable of causing disease, called pathogens, these same requirements must be met.

To prevent reproduction, removal of moisture, food, or an unacceptable temperature are possibilities. To cause death of an organism, removal of the same factors can be done. For organisms requiring oxygen, called aerobes, removal of air can prevent reproduction and/or cause death. For those organisms capable of living without oxygen, called anaerobes, oxygen itself can cause death. In general, bacteria can be killed, but viruses can only be prevented from replicating, although it is possible to destroy their RNA or DNA by certain chemicals or radiation. Drugs that either kill or render the bacteria incapable of reproducing are called antibacterials. There are also a few chemicals capable of preventing viral replication; these are called antivirals. Those designed to kill fungi are called antifungals; to kill parasites, antiparasitics; and to kill mycobacteria, antimycobacterials. Together, any chemical designed either to kill a pathogen or to render it unable to replicate is called an antibiotic.

A material designed to render a pathogen unable to reproduce and multiply is usually called a biostat. These chemicals may not kill the pathogen, but since it cannot multiply, there is less chance of disease. A material designed to kill the pathogen is called a disinfectant.

Prophylaxis includes all the methods used to preserve health and prevent disease before it occurs. These encompass a wide variety of methods, from simple handwashing to the most modern pharmaceuticals. Other prophylactic means include wearing of gloves, wearing of face masks, use of disinfectants on surfaces and in potable water, prevention of aerosol suspensions by wet mopping rather than using a broom, isolation of contagious patients from others, prevention of potable water and sewage cross-connections, maintenance of proper temperatures for foods, immunization, *etc.*

Immunization is the process of forming antibodies to remain in the body by a limited exposure to the antigens of the pathogenic organism. When a pathogen enters the body, there are immune responses of the body to fight off infection. Most pathogens have specific chemical structures, called antigens, to which specific attack chemicals, called antibodies, will be drawn to destroy the pathogen. These antibodies may remain in the body for many years following exposure to the pathogen or the more limited exposure to the vaccine. Vaccine may contain a weakened form of the pathogen, called attenuated vaccine; the actual pathogen, called the live vaccine; or only a portion of the antigen known to cause the formation of the antibodies, usually a part of the protein coat of a virus or cell membrane of the bacterium. The presence of an antigen can be used to verify the presence of the pathogen in an infection or to verify the efficacy of the vaccine. In the latter case, a quantitative measure, called the titer, is made to ensure that enough of the antibodies are present to ward off the disease.

LABORATORY ANALYSES

Verification of a disease generally requires laboratory analysis. Most bacteria can be grown in a culture medium that supplies nutrients. Growth conditions are usually optimized by a high humidity and a temperature around 35° C (95° F). Unfortunately, it takes time for a bacterium to reproduce. Some bacteria duplicate within a few minutes, but others can take hours, often depending upon the temperature, availability of moisture and food, and absence of substances or other organisms harmful to the organism under study. In the laboratory, a moderate temperature is maintained, water and nutrients are available from the growth medium, and a cover is placed over the growth dish to prevent competing organisms from getting in. In most cases, at least twenty-four hours are needed for growth to reach a point where the bacterial colonies are visible to the unaided eye. Many bacteria can be grown on general culture media such as agar plates, but others require specific nutrients in the medium. For example, *Legionella* bacteria, the causative agent for Legionellosis, a form of pneumonia, requires a culture medium containing iron, amino acids, and charcoal (BCYE plates). Often, depending upon the culture medium, a specific strain of bacteria can be identified; for example, *Escherichia coli* and other coliform bacteria, when grown on the right medium, develop shiny gold-colored colonies while other bacteria appear black or another color. Following growth on the culture plate, a sample of the bacteria can be placed on a glass slide, stained with a dye, and examined under a microscope. One of the common dyes is gram stain. Some bacteria take up the dye and are called gram-positive. Others do not absorb the dye and are called gram-negative. This information, plus the appearance of the individual bacterium, whether rod-shaped, sphere-shaped, flagellated, in chains, in pairs, or in clumps, allow for identification of the type of bacteria. Because of the time delay in identifying the pathogens by growth of colonies, other methods have been developed.

Most analytical procedures include three steps: extraction, purification, and determination. The most significant recent improvement in the purification step is the use of solid-phase extraction. Test extracts are cleaned up before instrumental analysis, usually by thin layer or liquid chromatography, to remove co-extracted materials that often interfere with the determination of target analytes.

Polymerase Chain Reaction (PCR) analysis allows for the duplication of both bacterial and viral DNA. Specific unique gene sequences are looked for to verify the presence of a particular strain of bacteria or viruses. The method is much more rapid, allowing identification within a few hours; however, it suffers from the drawback that it cannot tell the difference between living organisms and dead ones. As long as the DNA is present, it will be amplified. For this reason, a growth test usually accompanies PCR analysis.

Indirect immunofluorescence assay (IFA) is a very dependable method of identifying a number of pathogens. In this method, a specific chemical is taken up by the pathogen; this chemical emits light at a particular wavelength when light is shined on it. The emitted light, called fluorescence, can be detected by a very sensitive light-sensitive diode and amplified by electronic circuits.

Thin layer chromatography (TLC), also known as flat bed chromatography or planar chromatography, is one of the most widely used separation techniques in analysis of toxins. This method employs the differential movement of certain materials, often proteins, when placed in a suitable solvent. The chemicals in a mixture are separated into individual unique components, which indicate the nature of the original material. The material being studied, the analyte, is placed on a thin sheet of plastic (the stationary phase) coated with specific chemicals. A solvent is applied and streaks (the mobile phase) appear as the components are separated. From the distinctive pattern, the analyte can be identified. Since 1990, it has been considered the method of choice to identify and quantify toxins, especially aflatoxins, at levels as low as one nanogram per gram (ng/g). The TLC method is also used to verify findings by newer, more rapid techniques.

Liquid chromatography (LC) is similar to TLC in many respects, including analyte application, stationary phase, and mobile phase. Liquid chromatography and TLC complement each other. For an analyst to use TLC for preliminary work to optimize LC separation conditions is not unusual. Liquid chromatography methods include normal-phase LC (NPLC); reversed-phase LC (RPLC) with pre- or before-column derivatization (BCD), that is, reactions with specific chemicals to form derivatives; RPLC followed by post-column derivatization (PCD); and RPLC with electrochemical detection.

Thin layer chromatography and LC methods for determining toxins are laborious and time consuming. Often, these techniques require extensive knowledge and experience of chromatographic techniques to solve separation and interference problems. Through advances in biotechnology, highly specific antibody-based tests are now commercially available that can identify and measure toxins in food in less than 10 minutes. These tests are based on the affinities of the monoclonal or polyclonal antibodies for toxins. The three types of immunochemical methods are radioimmunoassay (RIA) which uses a radioactive tracer; enzyme-linked immunosorbent assay (ELISA), which uses specific enzymes; and immunoaffinity column assay (ICA), which also uses specific enzymes but in a column rather than on a film. A number of bacteria and viruses can be verified using the immunochemical methods. For example, the Human Immunodeficiency Virus can be quickly identified using ELISA.

The presence of specific antibodies in the blood of an infected person, such as immunoglobins, such as Immunoglobin A (IgA), G (IgG), and M (IgM), can be used to identify specific strains of pathogens. One of the reasons that Acquired Immune Deficiency Syndrome (AIDS) is so dangerous is that that the Human Immunodeficiency Virus (HIV), which is the causative agent, attacks the immune system and prevents the normal immune responses of the T-Helper cells from functioning. But most pathogens cause the body to produce antibodies. Combined detection of multiple specific antibodies improves the specificity of the assays and provides better accuracy in diagnosis. For example, in Q Fever, IgM levels are helpful in the determination of a recent infection. In acute Q fever, patients will have IgG antibodies during later phases of the disease and IgM antibodies in both early and late phases. Increased IgG and IgA antibodies in the early phase are often indicative of Q fever endocarditis. There are many other diseases that can be verified by the laboratory analysis of antibodies. In fact, antibody measurement is one way of verifying that a vaccine is still effective.

There is a wide variety of test methods that can be employed to specifically identify a pathogen or toxin. Most commonly, the disease is suspected based upon symptoms, with later confirmation using laboratory methods to ensure the course of treatment is proper. All these methods are applicable to detection and verification of biological warfare agents for the naturally-occurring form of the pathogen.

SURETY

The term surety refers to the safe and proper handling and storage of biological agents. It includes many aspects, such as immunization of workers; isolation facilities, from entire facilities down to biohazard glove boxes and approved biohazard safety cabinets; documented information of the type, quantity, and nature of the agent; a chain of custody whereby every person who had access to the agent is identified and recorded; routine medical identification and evaluation of the workers; and records of all potential exposures. One of the most important aspects of biological agent surety is the maintenance of records, including work periods, immunization status, and potential exposures of all workers.

LABORATORY DETECTION OF BIOLOGICAL AGENTS

The detection and identification of biological agents is identical to those for routine disease surveillance. The methods discussed above, ELISA, RIA, TLC, gram staining, microscopic examination, and culture growth, can all be used to verify the presence and identity of a biological agent. However, just as for many normal diseases, the first step is usually to guess at the pathogen or toxin by examination of the patient's symptoms since all laboratory methods require at least some amount of time, ranging from perhaps an hour to as much as forty-eight hours. Treatment should begin immediately, before laboratory results are available. If the guess was wrong, a different treatment protocol may be used, but most pathogens respond to the broad spectrum antibiotics, so this may be a good place to start unless the symptoms presented clearly indicate a condition which does not respond to a particular treatment.

In many cases of bioweapons use, a complete clinical laboratory may not be available in a timely manner, or even at all. Therefore, field methods may have to be employed. If a person does not have access to military equipment, symptoms have been the best method of field identifying a biological agent.

MILITARY FIELD DETECTION OF BIOLOGICAL AGENTS

Since a person can't see, smell, or taste biological agents, how do you fight them? In many cases, a complete clinical laboratory may not be immediately available. In order to properly treat a medical condition, it is necessary to know what the biological agent is. In addition, to protect medical personnel, it may be necessary to decontaminate the victim before commencing treatment. The U.S. and other military forces have geared up defenses against these invisible killers when the threat of biological weapons became a reality during Operation Desert Storm and Operation Iraqi Freedom. Since the early 1990s, the U.S. military has fielded new protection equipment and detection systems, and more counter measures are in the works.

Remember that biological warfare and terrorism may encompass a variety of bacterial agents, viruses, and toxins. Lumping all biological warfare agents together as a single class is almost as dangerous as comparing biological and chemical agents directly; they can be very different. Depending on the situation, some are contagious, others are not. If a person is exposed to these agents, unprotected, they may be in trouble. But when they leave the contaminated area, they may not necessarily spread the disease in a contagious sense unless some of the agent remains on their clothing and is physically transferred to someone else.

Usually, biological agents are an aerosol threat. This can be thought of like the old perfume sprayer; that is placing a liquid in an atomizer and spraying it out in very, very small droplets. Most aerosols behave a lot like chemical vapors.

These deadly biological agents do have some natural enemies. Most biological agents don't weather well. There are a lot of things that make it difficult for these biological agents to exist. Ultraviolet light will destroy most agents; drying out will prevent duplication. There is some natural immunity to many of these diseases; thus, the same factors that keep us from having a continuous common cold will help out to some extent in the case of a biological attack.

Most biological agents can be dispersed one of two ways, by what's known as a "line source" or a "point source." Imagine you are at a location. Miles away, an enemy boat sailing along a river or coastline, or a train traveling a rail line, or a truck on a busy highway releases a spray. The enemy has determined the prevailing wind will carry the disease-laden aerosol in your direction. A terrorist might employ a quieter method of dispersion such as an aerosol sprayer from an aircraft. This "line source" can cover a sizable piece of geography.

A more commonly envisioned scenario, especially in a terrorist attack, involves a "point source." This is when the enemy launches a missile or other munition which, upon detonation, spews a biological agent directly on the victims. The area covered is much smaller than with a line source. Either way, the enemy is trying to take the organism or the toxin and disperse it into the atmosphere so that the targets then inhale it. The threat of most of these agents is airborne; just like the common cold, we get them via inhalation. Because these agents are very small particles, they can make it through our nasal passages, evading all of the normal protective measures we have to filter things out, and get down to the deep lung area, where they then cause disease.

Unlike a chemical agent attack, which might cause an almost immediate reaction, a biological attack would probably not cause a reaction until after an incubation period. It can take as long as forty-eight hours before a victim starts showing what are often nondescript, flu-like symptoms which then progress to whatever symptoms the specific agent would normally cause.

How best to defend yourself against an attack you may not even know has happened? The best defense comes from using a combination of immunization and physical protective measures. A number of vaccines exist to protect against biological threat agents. When vaccines are not available as a biological countermeasure, the answer is rapid detection, warning, reporting, and masking, if available. The military protective mask and even many commercial N95 or N100 particulate respirators are effective against every known biological agent, including those for which there are no vaccines. Defense officials are developing and fielding smaller, lighter, and simpler biological detectors on the battlefield and even in some high population density locations.

The Biological Integrated Detection System (BIDS) is a mobile laboratory that can detect four different agents simultaneously. It's mounted on the back of a heavy military vehicle such as a HUMVEE and is manned by four individuals. The U.S. Army deploys BIDS companies as corps or theater level assets to do biodetection for all the ground and air military forces. BIDS are point detectors, so it is necessary to wait for the cloud to come to the BIDS system. The same type of system is used by the U.S. Navy aboard ships and is called the Interim Biological Agent Detector (IBAD). A similar system has been placed in a number of large cities. After detecting what appears to be an unnatural agent, the systems first provide a warning, then determine specifically what the biological agent is. In a biological attack, the individual does not need to know immediately whether it is anthrax or plague, only that they need to put on protective equipment or evacuate the location.

U.S. defense officials also have acquired a long-range biological detector. It is a laser-scanning instrument that can look out about 50 kilometers when mounted on a helicopter. It cannot identify specific agents but looks for telltale, cigar-shaped plumes that come from someone laying down a line source. If the cloud can be seen some 50 kilometers off, that can give an amount of time to prepare for it or to go out and sample that cloud to ascertain exactly what it is. In the future, there will be detectors that are light enough and small enough to go on an unmanned drone aircraft that can fly through the cloud, find out what it is, and report back.

The U.S. military developed the "Portal Shield" device and deployed it in the Persian Gulf region in February 1998 during Operation Desert Thunder. It is about two-thirds the size of a typical office desk. It is fully modularized and self-contained, and it can detect eight different biological agents. It is a network sensor, using as many as eighteen sensors, depending on the geography, arrayed around a port or an airfield. The sensors are interconnected and allow for increased verification of agent presence. Using such an array system, the false positive rate is almost zero. In Bahrain, U.S. forces ran more than three thousand tests during twenty-four hour monitoring by the Portal Shield deployed there, with a false positive rate of less than one-half of one percent.

DECONTAMINATION

U.S. defense officials are also developing more user- and environmentally-friendly biological agent decontaminating materials. The equipment and materials commonly used for decontamination are similar to, or the same as, those for chemical agents. These materials are highly corrosive, especially to sensitive electronic equipment and people. Super tropical bleach (STB) is almost as hazardous in terms of being a caustic to human skin as some of the agents themselves. Officials are also evaluating what actually needs to be decontaminated after a biological attack. In a military operation, with highly mobile maneuvers, the goal is often to avoid contamination by simply going around it, or buttoning up and going through it. But if you can't move because you're stationed at a port or airport or in a city, officials need to know what key areas need to be decontaminated. The U.S. military is conducing studies to see how much decontamination is really needed on aircraft and in vehicles.

Almost all biological agents will not be in visible liquid or solid form; the hazard will be inhalation of minute aerosol particles, so detection by an individual is highly unlikely. If a visible solid or liquid is present, whether biological or chemical, for people not having access to military decontamination systems, the best methods are quite simple. The use of soap and water to wash hands and other body parts exposed to the agent is the quickest and best. Disinfectants and oxidants, such as isopropyl (rubbing) alcohol, household bleach, commercial sprays such as Lysol®, household hydrogen peroxide, and the like can be very helpful in decontaminating surfaces but not the skin, due to the potential for additional harm.

If a person thinks he or she was exposed to a biological agent, he or she should immediately leave the area, if possible, without contaminating others, remove external clothing by taking it off inside out to prevent additional exposure if there is liquid or solid agent on the clothing, wash with soap and water, and seek medical attention. A non-medically trained person should not take any medicines since he or she might make the condition worse or prevent the physician from making the correct diagnosis. If the skin is burning or itching, wash immediately but take care not to rub or abrade the skin. Do not attempt to neutralize any liquid or solid on the skin; either wash it off or use an absorbent tissue or sheet, again taking care not to rub the agent in or abrade the skin.

Part 3

CHEMICAL WEAPONS

12

INTRODUCTION TO CHEMICAL AGENTS

USE OF CHEMICAL WEAPONS IN ANCIENT TIMES

Noxious agents have been utilized in wars for centuries. It was reported by the ancient Greek, Solon of Athens, that an aqueduct from the Pleistus river was deliberately polluted with hellebore roots, a purgative, in 590 B.C. during the siege of Cirrha, to render the entrenched enemy unable to conduct battle. In 423 B.C., during the Peloponnesian War, Spartan allies took an Athenian-held fort by directing smoke through a hollowed-out beam into the fort. Incendiary devices and sulfur-based gases were blown by the wind onto the besieged Spartan city of Sphacteria by Demosthenes during the same war. Sparta used the toxic smoke generated by burning wood dipped in a mixture of tar and sulfur during at least one of its periodic wars with Athens. A primitive type of flamethrower was employed as early as the fifth century B.C.

It seems clear that the Chinese were the original masters of chemical warfare. It has been suggested that the pursuit of chemical weapons may have originated in the fumigation of dwellings to eliminate fleas, practiced by the Chinese as long ago as the seventh century B.C. Chinese writings contain hundreds of recipes for the production of poisonous or irritating smoke for use in war, and many accounts of their use. We know of the arsenic-containing "soul-hunting fog" and the irritating "five-league fog," made from low-burning gunpowder to which a variety of ingredients, including the excrement of wolves, was added to produce an irritating smoke. The use of a riot control agent, finely divided lime dispersed into the air, is described in a

Chinese account of the suppression of a peasant revolt in 178 A.D. Nor did they neglect delivery systems; descriptions of weapons with such poetic names as the "poison fog magic smoke eruptor" may be found in the artillery manuals of the Chinese army.

Writings of the Mohist sect in China, dating from the fourth century B.C., tell of the use of ox-hide bellows to pump smoke from furnaces in which balls of mustard and other toxic vegetable matter were being burned into tunnels being dug by a besieging army to discourage the diggers. The use of a toxic cacodyl (arsenic trioxide) smoke is also mentioned in early Chinese manuscripts. Gunpowder-like mixtures containing charcoal, sulfur, and saltpeter were known in China by 1044 A.D., used primarily as incendiaries, since the proportions were not right nor were they confined for a detonation to occur. From small fireworks, the Chinese developed bombs. By 1232, the Chinese had developed rockets and a weapon they called "heaven-shaking thunder," an iron bomb attached to a chain that could be lowered from the walls of a city to explode among attackers.

Choking smoke and caustic ash were used by the Roman cavalry at the Siege of Ambracia in Epirus in 187 B.C. But the inhabitants of the town of Ambracia dealt a setback to Roman soldiers trying to tunnel under their city wall. They filled a large jar with feathers, put fire in it, and attached a bronze cover perforated with numerous holes. After carrying the jar into the shaft and turning its mouth toward the enemy, they inserted a bellows in the bottom. By pumping the bellows, they caused such a large amount of disagreeable smoke that none of the Romans could endure it. As a result the Romans, despairing of success, made a truce and raised the siege. The Romans also practiced anti-crop chemical warfare, most famously after the defeat of Carthage, when all the fields were sown with salt to prevent resettlement by preventing the growth of crops.

The biggest use of chemicals in war in ancient times was in the area of flame weapons. The first use of fire in war probably followed shortly after the discovery of how to make it appear on command. By the fourth century B.C., a number of recipes existed for producing incendiary compositions, such as a mixture of pitch, sulfur, tow, granulated frankincense, and pine sawdust in sacks that were set alight. The most famous incendiary mixture is certainly the Greek fire of the Byzantine Empire. Greek fire is the modern name; the Byzantines called it "sea fire" and their Moslem enemies called it "Roman fire." The secret of this weapon is supposed to have been brought to the Byzantines by a refugee from Moslem-occupied Syria named Kallinikos (or Callinicus; the name is likely a pseudonym because it means "handsome winner," possibly a reference to the reward received for his invention). There was nothing particularly new about liquid fire; oil-based incendiaries had been in use for some time. But this material had the astonishing property of burning on contact with water. Used in combination with a spray device that allowed them to direct it against enemy ships, the Byzantines were able to utterly defeat the Arab fleet at the battle of Kyzikos in 678 A.D. The Byzantines had developed a decisive weapon, and they held it closely to keep their enemies from learning how to use or counter it. Greek fire would be credited with many victories, against the Moslems again in

717–718 A.D., and later against Russian attacks in 941 A.D. and 1043 A.D. Greek fire remained Byzantium's secret weapon against the Turks for five centuries. It seems that those in Syria must not have shared their science with the Turks or other Moslems. Later, the Turks themselves acquired it to conquer the Greek Empire in the fourteenth century. Greek fire was not only used during naval engagements, but was placed within pottery vessels and hurled though the air by siege engines. As a projectile, it was quite formidable and gave the Eastern Empire a distinct advantage over their Western adversaries in Rome. Known to the Arabs and Byzantines by the seventh century in a variety of mixtures, Greek fire was utilized to such a degree that it could also be considered a psychological weapon. The recipients of a barrage of Greek fire could not explain how the fire would ignite spontaneously, rise up, be projected downward, or why the Byzantine liquid flame could not be extinguished. The secret of Greek fire was held so closely that its exact composition has been lost to history. Those knowing the secret recipe often responded to inquiries by saying that the formulation had been revealed by an angel to the Constantines, and that any attempt to discover it would provoke the vengeance of God. It may even have been lost to the Byzantines because, when Constantinople fell into the hands of the Crusaders in 1204 A.D., Greek fire was not used.

While the Moslems did not have Greek fire, they did develop other oil-based incendiaries. In a siege of Mecca in 683 A.D., the Umayyads used catapults to hurl naphtha-based incendiary projectiles against the defenders, accidentally setting the cloth covering of the Ka'bah on fire. In 813 A.D., Baghdad would be essentially destroyed by naphtha barrels thrown into the city, and, in 1167 A.D., Cairo was destroyed by naphtha pots and bombs to deny it to the Crusaders. Distillation of petroleum to produce fractions like "white water naphtha," suited for incendiaries, was also known. A workshop for producing incendiary grenades, complete with a distilling furnace and a gasoline storage container, dating from the first half of the thirteenth century has been found in Hama, Syria. *The Book of Horsemanship and the Art of War*, written by Najm al-Din Ahda in 1285, tells how to build rockets with fire bomb warheads. The Moslems seem to have avoided the use of toxic additives in their flame weapons because of injunctions in the Koran against poisoning the air and water.

In the Middle Ages, before the introduction of gunpowder, smoke from burning straw or other material was employed, but its effectiveness is uncertain. Ancient armies attacking or defending fortified cities threw burning oil, molten lead, and fireballs from the parapets onto the troops below.

As the Middle Ages gave way to the Renaissance, gunpowder and firearms and artillery began to play a role in warfare and came to dominate the battlefield. They eventually reduced the utility of castle walls to the point where even the most advanced fortifications could be reduced and broken. The rise of gunpowder weapons also produced a little more interest in chemistry of weapons of warfare, but not much. Although there was no widespread interest in chemical warfare in the Renaissance, Leonardo da Vinci seems to have done some thinking on it. He offered a proposal to throw a powdered poison made of chalk, sulfide of arsenic, and powdered verdegris (poisonous copper acetate) onto enemy ships to cause asphyxiation. He also considered the issue of protection. In his notes, there is a description of a protective mask that would serve to shield the eyes, nose, mouth, and lungs from dust and smoke as a more effective protection against the use of his toxic powders than the damp cloth originally proposed to protect the users from retaliation.

The use of chemical weapons was largely considered improper. In fact, in August 1675, the French and Germans concluded the Strasbourg Agreement, which included an article banning the use of "perfidious and odious" toxic devices such as poisoned bullets.

Barrels of blinding quicklime were catapulted by the English fleet onto French vessels in the middle of the thirteenth century. "Stinking pots" and toxic bombs were used in large quantities by the forces of the Habsburg Holy Roman Emperors Ferdinand II and Ferdinand III, together with their Spanish cousin Philip IV during the Thirty Years War (1635–1648). The manufacture of more sophisticated devices based on arsenic, yellow arsenic, lead, white lead, red lead, verdigris, or antimony with the addition of belladonna, euphorbia, hellebore, aconite, nux vomica and poisons were mentioned in a military textbook in Germany in 1726; it is not known if any were actually used.

An English plan to smoke out the Russian garrison at Sebastopol using a lethal mixture involving 500 tons of sulfur and 200 tons of coke was considered during the Crimean War (1854-1855). Neither this plan, nor a subsequent one to use bombs containing Cadet's liquid, were executed. Cadet's liquid is a highly toxic heavy brown liquid which smells strongly of garlic and spontaneously bursts into flame when exposed to air; it is mainly cacodyl oxide, $[(CH_3)_2As)_2O]$ with other compounds such as dicacodyl, $[(CH_3)_2As)_2]$. At the same time, Sir Lyon Playfair proposed that the British use cyanide-filled shells to break the siege of Sebastopol. His proposal was rejected by the British admiralty because it was considered to be against the rules of warfare.

A plan to develop artillery shells filled with chlorine to be used against the Confederate forces in the U.S. Civil War (1861-1865) was seriously considered but rejected by the U.S. War Department and President Lincoln.

MODERN HISTORY OF CHEMICAL WEAPONS

Several nations in World War I, on both sides, began using munitions filled with irritants from almost the beginning of the war. The French first used shells filled with ethyl bromoacetate in August 1914, less than one month into the war, and chloroacetone was introduced into the French arsenal in November 1914. On October 27, 1914, the Germans at Neuve-Chappelle used the "Ni-Schrapnell" 105mm shell, which consisted of lead balls embedded in powdered o-dianisidine chlorosulfonate. The Germans first used gas on the Eastern Front against the Russians, on January 3, 1915. It was a tearing agent dispersed by artillery shells but was not successful. On January 31, 1915, at Boloimow, the Germans introduced 150 mm shells filled with

"T-Stoff," a mixture of brominated aromatics including xylyl bromide, xylylene bromide, and benzyl bromide. All these compounds are extreme irritants capable of severely limiting the effectiveness of unprotected troops. Scientists working with the military were quickly moving from irritants to lethal gases. For example, Professor Fritz Haber, chief of the German chemical warfare service during World War I, personally directed the first chlorine gas attack. Haber, a Nobel laureate renowned for his discovery of the process for synthesizing ammonia by the combination of nitrogen and hydrogen, is sometimes called the father of chemical warfare.

The first use of chlorine on the Western Front came on April 22, 1915, when German troops at Ypres discharged the gas from 5,730 cylinders on the line between Steenstraat on the Yser Canal, through Bixschoote and Langemark, to Poelcappelle, in Belgium during the Second Battle of Ypres. In the late afternoon, when weather conditions were favorable, the Germans released a 5-mile wide cloud of 168 tons of chlorine gas. The greenish-yellow cloud drifted over French and Algerian trenches, where it caused wide-spread panic and death. Being heavier than air, chlorine followed the ground's contours and sank into the trenches and shell holes that soldiers used as protection against shrapnel and bullets. This forced them to abandon their defenses in favor of higher ground. Those who did stay found it extremely difficult to fight with watery eyes, heaving stomachs, and burning lungs. It was thus more effective than artillery, which had to be very accurate to do any damage to a defensive position, or rifle fire, which required the enemy to expose himself. The French colonials, whose soldiers received full doses of the chlorine, gave way before the gas attack, leaving a breach in the Allied lines over four miles wide, which the Germans were quick to exploit. The First Canadian Division was asked to seal off the gap and hold the line that now was coming under attack, not only from additional gas and artillery, but also direct assaults by German infantry. In this the Canadians were largely successful. This first gas attack was successful, killing 5,000 soldiers and putting 1,500 more out of the war, and it did create fear and panic among the troops. On April 24, 1915, the Germans conducted a second chlorine gas attack at Ypres, this time against Canadian troops.

On May 31, 1915, an even more lethal attack was undertaken by the Germans, using a mixture of chlorine with some phosgene along 12 kilometers of the Russian front, at Bsura-Rumka. Some 12,000 bottles of gas were used, resulting in 9,000 casualties, including 6,000 deaths.

In July 1915, some 100,000 artillery shells containing benzyl bromide were fired by the Germans from 155mm guns in the Argonne. On December 19, 1915, the Germany army first used phosgene alone against British troops at Nieltje, Flanders, using some 88 tons of phosgene released from 4,000 cylinders. In March 1916, the first use of phosgene contained in artillery took place at Verdun, France; some 75 shells were used to devastating and deadly effect. In July 1916, shells filled with hydrogen cyanide were used during the Somme offensive. In March 1917, phosgene was spread from airplanes, creating vast and lethal concentrations of phosgene.

Mustard gas was first used also near Ypres on July 12, 1917. Mustard is the common name for 1,1-thiobis(2-chloroethane). Its chemical formula is $Cl\text{-}CH_2\text{-}CH_2\text{-}S\text{-}CH_2\text{-}CH_2\text{-}Cl$. Other names include H, yprite, sulfur mustard, and Kampstoff Lost, but the name mustard gas became more widely used because the impure agent is said to have an odor similar to that of mustard, garlic, or horseradish. When pure, it is both odorless and colorless. It was first synthesized by Frederick Guthrie in 1860, who reacted ethylene with chlorine and noticed the toxic effects it had on his own skin. Mustard causes severe burns to the skin, eyes, and respiratory tract, and is often fatal. Mustard caused more deaths in World War I than any other chemical agent. The first use of arsine-based "Clarks" was in September 1917; these were substances, causing nausea and vomiting, which gas mask filters could offer no protection against. By 1918, there was massive use by both sides of gas-filled shells. During these attacks, approximately 25% of projectiles used by one side or the other were chemical shells. It has been estimated that, if the conflict had continued for a year longer, it would have turned into a true chemical war.

According to Allied reports, the chemical agents used by the Germans in World War I included acrolein, allylisothiocyanate, arsenic trichloride, arsine, bromoacetone, bromoacetic ether, bromoethylmethylketone, benxyl bromide, xylyl bromide, toluyl bromide, bromine, carbon monoxide, phosgene, chloroacetone, chlorine, chloropicrin, cyanogen, mustard gas, dichloromethylether, dimethylsulfate, diphenylchloroarsine, diphenylfluoroarsine, ethyldichloroarsine, formaldehyde, hydrocyanic acid, hydrosulfuric acid, iodoacetic ether, iodoacetone, methylchlorsulfonic acid, monochlormethylchloroformate, nitrogen peroxide, phenylcarbylamine chloride, phosphine, phosphorus trichloride, sulfur dioxide, sulfur trioxides, and diphosgene.

Although improvements in gas masks and other means of protection considerably reduced total losses caused by gases, especially mustard gas, the death toll in World War I due to gas was 1,300,000, of which nearly 100,000 died in combat. Losses caused by other forms of weapons were put at 26,700,000, of which 6,800,000 died in combat. In many cases, survivors of gas attacks with varying degrees of injury subsequently fell victim to fatal infectious diseases. It should be noted that it was Russian troops who suffered the heaviest losses from gas attacks (11%, the average among the other belligerents was about 7%). Yet the next use of poison gas was by Russians in Russia. In 1920, chemical weapons were used during the Civil War between the White Russians and the Red Russians following the Russian Revolution.

German scientists, seeking better pesticides, first discovered Tabun. In December 1936, Dr. Gerhard Schrader of the I.G. Farbenindustrie laboratory in Leverkusen prepared Tabun (ethyl dimethylphosphoramidocyanidate, Trilon 83, or Agent GA) while working on development of new types of insecticides. He had been studying compounds such as acyl fluorides, sulfonyl fluorides, fluoroethanol derivatives, and fluoroacetic acid derivatives. The year before, he had prepared dimethylphosphoramidofluoridic acid, from which a number of pesticides might be produced as esters or salts. He began to investigate dimethylphosphoramides, which led to the preparation of Tabun. Schrader discovered that Tabun was a extremely powerful pesticide, requiring only 5 ppm to kill all the leaf lice he used in his first experiment. In January 1937, he was the first to see the effects of Tabun on human beings when he and an assistant began to experience meiosis, contraction of the pupils of the eyes, and shortness of breath when exposed to Tabun vapors in their laboratory. The Nazis had passed a law in 1935 requiring all inventions of potential military significance to be reported to the Ministry of War. A sample of Tabun was sent to the Chemical Warfare section of the Army Weapons Office at Berlin-Spandau in 1937. Shortly afterwards, Schrader was ordered to Berlin to give a demonstration.

The military quickly saw the great potential of Tabun, and it established a research laboratory at the town of Wuppertal-Eblerfeld. In 1939, a pilot plant for Tabun production was set up at Munster-Lager, near the German Army proving grounds at Raubkammer. In January 1940, the Germans began construction of the full-scale plant at Dyernfurth-am-Oder (now Brzeg Dolny, Poland). The plant did not reach full operation until June 1942 due to the corrosive nature of some of the intermediates and the toxicity of the Tabun. While the 3,000 workers were equipped with respirators and protective clothing, over 300 accidents occurred, and at least 10 workers were killed during the $2\frac{1}{2}$ years of the plant's operation. An I. G. Farbenindustrie subsidiary, Anorgana GmbH, operated the Tabun plant as well as the other chemical agent production plants in Germany. Another Farbenindustrie invention was Zyklon-B, a mixture containing hydrogen cyanide used by the Nazis to gas victims in the concentration camps.

In 1938, a second potent organophosphate nerve agent was discovered. This agent was named for its four discoverers: Schrader, Ambros, Rüdriger, and van der Linde, thus called Sarin (1-methylethyl methylphosphonofluoridate, Agent GB). In June 1939, the formula for Sarin was passed to the to Chemical Weapons section of the Army Weapons Office, along with a sample of the compound. All the synthetic routes for Sarin studied at that time required the use of hydrogen fluoride with accompanying severe corrosion problems. This necessitated the use of quartz- and silver-lined components. Pilot plants were constructed at Spandau, Münster Lager, and Dyernfurth. A production plant with a capacity of 500 tons per month was under construction at Falkenhagen, southeast of Berlin, near the end of World War II. Agent GB was the most expensive nerve agent selected for mass production by the Germans as part of their Grün 3 program because it required some 1058 tons of raw materials to produce only 100 tons of agent. A more efficient production method was later discovered, which required only 893 tons of raw materials to produce 100 tons of GB, but this is still large in comparison to the 356 tons of raw materials required to produce 100 tons of Tabun. GB produced by the second method is referred to in some documents of the period as Sarin-2. A similar compound is known as cyclosarin (cyclohexyl methylphosphonofluoridate; Agent GF); it has been suggested as a nerve agent, but there has apparently been no commercial production of it.

In August 1944, as the Soviet Red Army was approaching Silesia and the Western Allies began the race for the German border, the Nazis began destroying all the documents on the research and manufacture of Tabun and Sarin. By the end of 1944, Germany had produced 12,000 tons of Tabun with 2,000 tons loaded into projectiles and 10,000 tons loaded into aircraft bombs. These weapons were stored at Krappitz (Krapowice) in Upper Silesia as well as in abandoned mine shafts in Lausitz and

Saxony. Some stocks were also transported to Baveria in anticipation of a last-ditch defensive stand by the Nazis. In early 1945, Dyernfurth was to be destroyed, and the liquid nerve agents were simply poured into the Oder River. Although the plant was rigged with explosives, the Russians surrounded the plant before it could be destroyed. The Luftwaffe was then ordered to bomb the plant, but it also failed to destroy it. It is believed that the Soviets captured both the full-scale Tabun plant and the pilot Sarin plant intact. The Soviets later captured the nearly complete full-scale Sarin plant at Falkenhagen. Production at Dyernfurth resumed in 1946 under the Russians.

Richard Kuhn discovered Soman (1,2,2-trimethylpropyl methylphosphonofluoridate, Agent GD) in the spring of 1944, while working for the German Army on the toxicity of Tabun and Sarin. The documents detailing the discovery were buried in a mineshaft east of Berlin, but they were discovered by the Soviets. The Soviets produced and stockpiled Soman during the Cold War; it was the primary nerve agent of the U.S.S.R.

Scientists in other nations were aware of the German chemical weapons research, and quickly G agents were produced in the U.S. and U.K. In spite of the fact that both sides, the Axis and the Allies, in World War II possessed and prepared nerve agents, none were ever used in battle.

Scientists working at several chemical companies in the U.K, Germany, and Sweden independently discovered a class of organophosphate esters having substituted 2-aminoethanethiols for use as pesticides. Between 1951 and 1954, ICI researchers discovered Amiton (O,O-diethyl-S-[2-(diethylamino)ethyl] phosphorothiolate) and S-[2-(diethylamino)ethyl]-O-ethyl ethylphosphonothiolate. Schrader, the discoverer of Sarin, was now at Farbenfabriken Bayer AG, and prepared S-[2-(diethylamino)ethyl]-O-isopropyl methylphosphonothioate, while scientists at the Swedish Chemical Warfare Defense Laboratory prepared S-[2-(diethylamino)ethyl]-O-ethyl methylphosphonothiolate and S-[2-(dimethylamino)ethyl]-O-isopropyl methylphosphonothiolate. The British Chemical Weapons Laboratory at Porton Down began investigating these and related compounds. It informed the U.S. Chemical Weapons Laboratory at Edgewood, Maryland, of the results. The U.S. began a systematic investigation of the entire class of phosphonothiolate and fluorophosphonate chemicals. In 1958, the U.S. selected Agent VX (S-[2-[bis(1-methylethyl)amino]ethyl]-O-ethyl methylphosphonothiolate), for manufacture. Construction of a production plant began in 1959, and it operated from 1961 through 1968 at Rocky Mountain Arsenal (RMA) near Denver, Colorado. A Soviet team at the I. M. Sechenov Institute in Leningrad had already predicted the anticholinesterase activity of S-2-dialkylaminoethyl phosphono- and phosphorothiolates. The team had apparently learned the molecular formula ($C_{11}H_{26}NO_2PS$) for VX but without the chemical structure, from Soviet spies in the U.S. As a result, they developed Soviet V-gas with the same formula, but a different structure (S-[2-(diethylamino)ethyl]-O-ethyl isobutylphosphonothiolate).

The "G" nerve gases are volatile substances that do not readily penetrate intact skin, but toxicity significantly increases if the skin becomes permeable while dermal toxicity of the nonvolatile VX agent is high, even through intact skin. Agent GF can be taken up by inhalation or through skin contact. Sarin (Agent GB) has an especially rapid onset of action. On a weight basis it is less potent than VX, but one drop of the liquid on skin may be sufficient to cause death. Soman (Agent GD) is an extremely potent nerve agent. On a weight basis, Soman is less potent than VX, but more potent than Sarin or Tabun; it can be hazardous by any route of exposure, including dermal. Tabun has an especially rapid onset of action but, on a weight basis, is less potent than VX. VX is 100 times as toxic to humans as Sarin, persists much longer in the environment, and is better absorbed through the skin at normal ambient temperatures. VX is 300 times more lethal than Tabun on skin. On a weight basis, toxicity in descending order is VX > Soman = GF > Sarin > Tabun.

On December 2, 1943, German bombers attacked American tankers and munitions ships in Bari Harbor, off the southeast coast of Italy. They sank sixteen ships, partially destroyed four more, and set off at least two major explosions. The fires burned while hundreds of oil-soaked men were pulled out of the water. At first, many of the survivors seemed to be all right, though a few mentioned the odd smell of garlic. Soon they began showing symptoms of stinging eyes, skin lesions, and a variety of internal problems. Four survivors died later the first day, nine the next. By the end of a month, 83 men out of the 617 who made it to the hospital had died. One of the ships had held 100 tons of methyl bis-(beta-chloroethyl)amine hydrochloride, "nitrogen mustard" for short. It is related to, but different from, the mustard gas used in World War I. Later, the Army claimed the chemical agent was there as a deterrent. Most of the mustard gas burned off in the fires; the small part of it that was mixed with the floating oil was what had injured the sailors. This Bay of Bari incident produced the only mustard gas casualties in World War II, Americans who had been killed by American gas.

Between 1963 and 1967, Egypt used mustard bombs in the war with Yemen. In the 1983–1988 Iran-Iraq war, chemical weapons, chiefly mustard, were extensively used by Iraq against Iranian positions. At the end of that war, in 1988, Iraq, fearing a rebellion of the Kurds in its northern areas, killed over 5,000 people using mustard and other chemicals, including Sarin at the city of Halabjah. At the end of the first Persian Gulf War, in 1991, as part of the truce, Iraq agreed to destroy all chemical weapons and related technology, and to allow United Nations inspectors free access to all production and storage facilities. For almost seven years, U.N. inspectors were faced with uncooperative Iraqis, who hid almost all their stocks and facilities. Finally, in 1998, the U.N. withdrew weapons inspectors from Iraq because they were hindered in their work to the extent that nothing was being found. It was believed that new and upgraded facilities were built in Iraq for the production of chemical weapons. One of the reasons for Operation Iraqi Freedom was to find the chemical and biological weapons that had been reported by the Iraqis in 1991 so they could be destroyed, but neither the weapons nor proof of destruction have been found as of this writing.

The Aum Shinrikyo religious cult in Japan, in the summer of 1994, released Sarin gas in the city of Matsumoto, Japan. It had previously tried to spread botulism and anthrax but had failed, so it turned to chemical agents. No one died, but scores were permanently injured. This doomsday cult, in 1995, released Sarin gas from a trash-can in a Tokyo subway station. Five commuter trains were filled with the poison gas. The death toll was 12, but many more were seriously injured. The Sarin used was very impure, and the delivery system was ineffective; otherwise, many more would have died. The cult's production laboratory, near Mount Fuji, was discovered and destroyed by Japanese police.

U.S. CHEMICAL WEAPONS PROGRAM

The U.S. investigated a wide range of chemical agents prior to the Chemical Weapons Convention. The U.S. conducted most of its research and training in chemical weapons at Anniston Depot in Alabama, Aberdeen Proving Ground in Maryland, and Dugway Proving Ground in Utah. Production facilities were located at Rocky Mountain Arsenal (RMA) near Denver, Colorado, where a complete chemical manufacturing complex was used by the U.S. Army for the production of chemical weapons as well as conventional munitions between 1942 and 1982 and, later, at Pine Bluff Arsenal in Arkansas. Some RMA facilities were leased to Julius Hyman and Company in 1952, and subsequently to Shell Chemical Company for production of agricultural pesticides, continuing until 1982, when it was closed down. The U.S. Environmental Protection Agency (EPA) placed most of RMA on the Superfund National Priorities List (NPL) in 1987 and 1989; environmental cleanup is now almost complete. In 1992, Congress passed legislation designating RMA as a National Wildlife Refuge upon completion of the cleanup.

The U.S. began destroying its stockpiles of chemical weapons in 1993 and has agreed to complete the destruction by 2007. A number of possible methods were examined; incineration and chemical neutralization were decided upon as best available technologies. The U.S. Army has built demilitarization facilities at Johnston Atoll in the Pacific (incineration); at Anniston Depot in Alabama (chemical neutralization); at Tooele Army Depot in Utah (incineration); Pine Bluff Arsenal in Arkansas (chemical neutralization); Umatilla Army Depot in Oregon (chemical neutralization); Blue Grass Army Depot in Kentucky (chemical neutralization); Newport, Indiana (chemical neutralization); Pueblo Depot, Colorado (chemical neutralization); and Aberdeen Proving Ground in Maryland (chemical neutralization). The two methods employed are incineration in special incinerators capable of sustained high-temperatures and chemical neutralization followed by supercritical water oxidation. Lawsuits were filed seeking to prohibit the Army from using the incineration method but were unresolved at the time of publication.

The former method employs incinerators in which the chemical agent, which has been drained from the storage containers, rockets, or shells (these two after being sawed in two), is burned in a multiple-chamber incinerator at temperatures of over 1200° F. The stacks of the incinerator are equipped with both monitoring devices (so the operation can be shut down if there is any release) and alkaline scrubbing systems (to remove any trace amounts of agent or other acidic product). In the latter method, rockets or shells are sawed in two inside containment or the storage containers are drained (and rinsed); the agent drained out; the metal pieces sent to a "popping furnace" able to withstand the burning and possible detonation of propellants and bursters; and the liquid agent mixed with a highly alkaline solution made of sodium hydroxide. The neutralization is a form of hydrolysis, and the resulting mixture called a hydroxylate. Following the chemical neutralization, the hydroxylate is transported to a commercial facility in Deepwater, New Jersey, where it will be biologically treated. This final step was not used in earlier destruction, with the result that tons of still toxic "salts" were obtained when the hydroxlysate was dried in kilns. Nerve agents and mustard must be destroyed to at least 99.99999% efficiency. As of early 2003, the U.S. had destroyed about 25% of its inventory of chemical weapons.

U.S.S.R. CHEMICAL WEAPONS PROGRAM

The Soviets investigated and produced a wide variety of chemical weapons, including Agent GD, Agent V, mustard, hydrogen cyanide, and ricin. In fact, the Soviet chemical weapons program was probably the most extensive in the world. There were numerous research and production facilities around the former Soviet Union.

At the time that Russia became signatory to the Chemical Weapons Convention, it had some 44,000 tons of agents on hand. As of 2003, the Russians had destroyed only about 100 tons of mustard at its only facility, in Gorny. It was shut down for a period of time due to having been operated without proper environmental permits. It appears that it will take many years for the Russians to demilitarize their chemical stockpilc.

CHINESE CHEMICAL WEAPONS PROGRAM

China is widely reported to have active programs related to the development of chemical weapons, although essentially no details of these programs have appeared in the open literature. China is believed to have an advanced chemical warfare program that includes research and development, production, and weaponization capabilities. Its current inventory is believed to include the full range of traditional chemical agents. It also has a wide variety of delivery systems for chemical agents to include artillery rockets, aerial bombs, sprayers, and short-range ballistic missiles. Chinese forces have conducted defensive chemical weapons training and are prepared to operate in a contaminated environment. As China's program is further integrated into overall military operations, its doctrine, which is believed to be based in part on Soviet-era thinking, may reflect the incorporation of more advanced munitions for chemical weapons agent delivery.

On December 30, 1996, the Standing Committee of the National People's Congress China ratified the Chemical Weapons Convention (CWC). Previous, dual-use chemical-related transfers to Iran's chemical weapons program indicate that, at a minimum, China's chemical export controls were not operating effectively enough to ensure compliance with China's CWC obligation not to assist anyone in any way to acquire chemical weapons. In March 1997, Israeli authorities arrested an Israeli businessman, Nahum Manbar, for allegedly selling Chinese chemical weapon components to Iran.

On May 21, 1997, pursuant to the Chemical and Biological Weapons Control and Warfare Elimination Act of 1991, the U.S. government imposed trade sanctions on five Chinese individuals, two Chinese companies, and one Hong Kong company for knowingly and materially contributing to Iran's chemical weapons program. These individuals and companies were involved in the export of dual-use chemical precursors and chemical production equipment and technology. The Chinese companies were the Nanjing Chemical Industries Group (NCI) and the Jiangsu Yongli Chemical Engineering and Technology Import/Export Corp.

There are essentially no open source data on the subject of Chinese chemical weapons activities, and many legitimate research programs use similar, if not identical, equipment and facilities.

HAGUE CONVENTION OF 1899

The chief military nations of the world met at The Hague in 1899 to write rules for future warfare. There were four conventions and three declarations passed, which entered into force in September 1900. The conventions covered International Disputes, Laws and Customs of War on Land, Maritime Warfare, and Prohibitions on Launching of Projectiles and Explosives from Balloons. The three declarations covered Launching of Projectiles and Explosives from Balloons, Use of Projectiles for Diffusion of Asphyxiating or Deleterious Gases, and Use of Bullets Which Expand or Flatten Easily in the Human Body. Out of this meeting, the use of any projectile for disseminating asphyxiating or deleterious gases (chemical warfare agents) was banned. However, as soon as the Germans used mustard and chlorine in World War I, the Allies followed suit. It was apparent that the Hague Convention was without teeth.

GENEVA PROTOCOL OF 1925

The horrors of the continued suffering of thousands gassed in World War I led most of the nations of the world to meet in Geneva to establish stronger limits on war and military weapons. The resulting Protocol outlawed the use of chemical agents but primarily focused on phosgene, chlorine, mustard, and similar chemicals since nerve agents had not yet been developed. The Protocol prohibited all signatory nations from developing or using chemical weapons; however, its provisions were largely ignored by all the great powers, including Germany, Britain, the U.S., the U.S.S.R., and Japan.

CHEMICAL WEAPONS CONVENTION OF 1993

The 1993 convention strengthened the terms of the earlier treaties and protocols by adding precursor chemicals. Manufacture or possession of these precursors was not prohibited, but production figures, stockpiles, locations, and uses had to be disclosed. Unannounced inspections of industrial plants could be undertaken at any time by signatory states.

LIMITATIONS OF CHEMICAL AGENTS WEAPONS IN WAR

The use of chemical agents in warfare has a number of drawbacks. First, prevailing winds can change and send the agent in the wrong direction or even back onto the troops who released it. Second, the chemical agents can contaminate the area where troops might need to enter. Third, most agents are destroyed or weakened by high humidity, rain, oxygen in the air, and ultraviolet radiation in sunlight. Fourth, most militaries are generally prepared for operations in a chemically-contaminated environment and have protective equipment, even if of limited duration. Fifth, militaries know that their enemies also possess such weapons; mutual fear of retaliation exists. Most armies also have decontamination equipment and supplies available. For these reasons, the most likely scenario for use of chemical weapons is against unprotected civilians. Chemical weapons, although possessed by all sides in World War II, the Korean Conflict, the Viet Nam War, and the Persian Gulf War, were not used. The only use of chemical weapons has been against civilians, in such places as Thailand, Laos, Iraq, and the subways of Japan.

FUTURE USE OF CHEMICAL WEAPONS

The potential use of chemical weapons is a real threat, but their use by militaries in war is not very likely. Use by a terrorist organization against civilians is a much greater threat. As seen by the Aum Shinrikyo cult's use, death and lifelong injuries are possible. If the cult had a purer agent and a better dispersion method, thousands could have been killed and injured. A terrorist attack using chemical weapons could be undertaken using aerosol generation equipment similar to mosquito foggers or aerial spraying equipment. By the time that the extent of the attack is known, thousands could be casualties.

LETHAL CHEMICAL AGENT COUNTERMEASURES

There are a number of measures that can be taken to avoid exposure to lethal chemical agents: countermeasures begin with detection of the presence of a lethal chemical agent; once a person is aware that lethal chemical agents are present, they can wear various types of protective equipment to guard against exposure to the agent; decontamination of people and equipment can make it safe to remove the protective equipment; and, if detection, protection, or decontamination fail, there are a limited number of antidotes that can be administered. Each of these will be discussed in the following chapters.

CURRENT STATUS OF DEMILITARIZATION OF CHEMICAL WEAPONS

The United States and Russia, the two largest possessors of chemical weapons, are both behind schedule in demilitarization. The U.S. (about 20% destroyed) is ahead of Russia (about 2% destroyed), but neither nation appears able to complete destruction of the chemical agents until the period between 2008 and 2012, at the earliest. Russia was granted until December 2007 to reach 20%. South Korea requested and received permission under the Chemical Weapons Convention (administered by the Organization for the Prohibition of Chemical Weapons, OPCW), to also delay final destruction of its chemical weapons. South Korea has destroyed about 20% of its stockpile, but has until 2007 to reach 45%. If any nation cannot meet the revised April 2012 deadline for 100% destruction, the treaty will have to be amended.

13

EFFECTS OF CHEMICAL WEAPONS AND MEDICAL TREATMENTS

TYPES OF CHEMICAL WEAPONS

There is a variety of chemicals that have, or can, be used in warfare and/or terrorism. Most chemical agents fall into one or more of the following classes: blister agents, blood agents, choking agents, nerve agents, lacrimators, vomiting agents, irritants, and psychotropic compounds. However, other chemicals used on the battlefield include smokes and obscurants, incendiaries, and herbicides. We shall not consider explosives since these are not typically considered as chemical weapons even though they work based upon chemical principles.

BLISTER AGENTS

Blister agents can be either inhaled agents or contact agents. They cannot be smelled easily and are often used to harass rather than kill. The time most of them take to affect an individual depends on many factors, but they cause severe skin blisters, destroy the skin tissue, and most have a persistence time of one day to two months. The form of injury is particularly ugly. Modern blister agents are a Soviet development, based on improvements in the mustard gas used extensively during World War I. The effects are long-lasting, and the scars will never clear up.

For most blister agents, the mechanism of toxicity is an attack on dividing cells. Clinical effects may be seen in thirty minutes to one hour of exposure for some agents; others, like sulfur mustard, the effects may not appear for many hours. Typical symptoms and signs are blister formation in airways, with bloody sputum, which can result in sudden airway obstruction; nausea, vomiting, and diarrhea; apathy and lethargy; loss of white blood cells with loss of immune response; large fluid- or blood-filled blisters on the skin that heal with scarring; and conjunctivitis, sensitivity to bright light, pain, edema, and corneal perforation.

Medical management is first to decontaminate with water or a bleach/water mixture. Skin lesions are to be treated like a thermal burn. The eyes should be irrigated with copious amounts of water and treated like a corneal abrasion except do not use a tight patch. The patient may require intubation, assisted ventilation, and bronchodilators. The systemic treatments include preparing for infections, reverse isolation, and sterilization of the GI tract. The military field treatment is the same as for a burn, petrolatum gauze placed over the wound.

A number of protective measures against vesicants have been tried by various militaries: barrier creams, chlorine-containing socks and underwear, *etc*., but most have not been successful.

BLOOD AGENTS

Blood agents are inhaled chemicals that can often be smelled; however, most take effect in humans in less than five minutes and cause convulsions and suffocation because they interfere with the absorption of oxygen by the blood in the lungs. Some prevent the red blood cells from carrying oxygen to the cells, while others inhibit the mitochondria from utilizing the oxygen within the cell. Blood agents are almost always used to kill, not to harass, and most have a persistence time of approximately ten minutes. These agents are valued for the ability to act quickly in tactical situations requiring surprise.

The typical mechanism of toxicity for the modern blood agents is poisoning of the mitochondria of the cells so they are no longer able to use oxygen from the blood. The result is oxygen starvation.

The clinical effects appear very rapidly, as soon as 15 seconds after exposure with rapid breathing and gasping for breath. As soon as thirty seconds, convulsions begin. After two minutes, respiratory efforts cease and death ensues. A common sign is a cherry red color to the skin.

Medical management for most blood agents includes immediate administration of amyl nitrite, which is inhaled by the patient, followed by administration of 300 mg of sodium nitrite IV over two to four minutes, and 12.5 g of sodium thiosulfate IV. The military field treatment is to place a broken amyl nitrite capsule under the victim's nose and assist with breathing.

CHOKING AGENTS

Many choking agents can be smelled but can take as long as ten hours for severe adverse effect. These agents can be used to kill as well as cause coughing and suffocation. The persistence time for most is only approximately five minutes. These agents were among the first used in modern warfare, and phosgene was responsible for almost 80% of the fatalities in World War I. Chemical research is producing more efficient substitutes for these chemicals.

The typical mechanism of toxicity is a reaction of the agent with linings of the lungs, causing small holes where water leakage into the air sacs can occur. The result is delayed pulmonary edema (fluid buildup in the lungs).

The clinical effects typically occur between twenty minutes and twenty-four hours after the exposure and include irritation of the airways and delayed pulmonary edema. The most common sign is shortness of breath due to inability to oxygenate blood.

If pulmonary edema occurs within about four hours of exposure, the individual will most likely die despite the best medical efforts. However, medical management includes admission to the hospital with strict bed rest, administration of supplemental oxygen, as needed, and prepare for intubation since conditions can worsen precipitously. The military field treatment is to assist with breathing.

NERVE AGENTS

Some of the more common nerve agents are Tabun (GA), Sarin (GB), Soman (GD), CMPF, GP, VR-55, and VX; however, there are others. Nerve agents interfere with the transmission of electrical discharges across nerve synapses. In effect, they short-circuit the nerve system by preventing the formation of acetylcholine, which is used by the body at nerve synapses to shut off flow of electrons. These agents are also called anti-cholinesterase agents or cholinesterase inhibitors. They cannot be smelled easily and are very lethal, although they can be utilized at very low concentrations to harass rather than kill. Most of them take some five to ten minutes to take effect and cause death by convulsions, suffocation, and fluid buildup in the lungs. They can be designed to have persistence times from ten minutes to four months, depending upon climate and weather conditions.

For most nerve agents, the mechanism of toxicity is cholinesterase inhibition resulting in cholinergic over-stimulation, similar to pesticide intoxication. The clinical effects generally occur within minutes. These are pupil constriction, pain blurred vision, and bloodshot eyes; massive mucosal discharge; tightening of airways with large amount of secretions and shortness of breath; nausea, vomiting, and diarrhea; increased secretions; twitching, fatigue, weakness, and paralysis followed by loss of consciousness and seizures; and either slow or fast heart rate.

Medical management is to decontaminate the patient with copious amounts of water or bleach and water mixture; and assist ventilation. An antidote for most agents is atropine—2 mg IV every three to five minutes—given until ventilation is eased together with 1 gram of pralidoxime chloride (pryridine-2-aldoxime methyl chloride, 2 PAM chloride) IV over twenty minutes every hour, with a maximum of three doses. For symptoms, give 10 mg of diazepam (Valium®) IV, as needed. The military field treatment is to use the automatic ComboPen® of atropine and 2-PAM chloride, up to two doses.

LACRIMATORS

Lacrimators are also known as tear gases. Tear gas is an inhaled agent. Most have an odor, even if slight, and only take one minute to cause an effect. They are largely non-lethal, and, depending on the specific agent, can cause profuse crying, severe coughs, involuntary defecation, and vomiting. Although these agents are usually non-lethal, they are able to render soldiers totally helpless and are often used on the battlefield as a harassing agent. Most have a persistence time of thirty minutes or less.

Tear gases have been known since the middle of the nineteenth century. Chloropicrin (Agent PS) was first synthesized from picric acid (2,4,6-trinitrophenol) and calcium hypochlorite (chloride of lime) in 1848. During World War I, it was used both as an irritant and as a lethal chemical. Although its toxicity made chloropicrin a poor riot control agent, it continues to be used as a soil sterilant, a grain disinfectant, and an intermediate in chemical synthesis.

The use of tear gases in military operations predates the use of lethal chemical agents. Several belligerents in World War I use munitions filled with irritants. Ethyl bromoacetate was used in August 1914, chloroacetone in November 1914, and mixtures of xylyl bromide, xylylene bromide, and benzyl bromide in January 1915. Bromobenzylcyanide (Agent CA) was used by both the French and Americans towards the end of the war.

During World War I, the tear gas α-chloroacetophenone (Agent CN) was investigated in the U.S., but production plants were not finished by the end of the war. Police in many countries adopted Agent CN, which works primarily as an eye irritant, as a riot-control agent between World War I and World War II. During World War II, all belligerents manufactured CN in large quantities.

During the 1950s, one limitation of Agent CN became apparent; determined demonstrators could avoid much of the effect of Agent CN by closing their eyes. *o*-Chlorobenzylidine malononitrile (Agent CS) was adopted as an Agent CN replacement when it was found to have wider ranging and stronger effects. Agent CS tear gas takes effect almost immediately, causing severe burning and involuntary closing of the eyes, copious tearing, extreme burning in the nose, a tendency to breathe through the mouth, extreme burning in the throat, and coughing. In some cases there can be nausea and vomiting. Agent CS also causes a burning sensation on exposed parts of the body. A desirable feature of Agent CS as a riot-control agent is that recovery quickly follows when the affected person is no longer exposed; fresh air is a rapid antidote.

Agent CS and Agent CN are both solids that are used in riot control with pyrotechnics; a burning pyrotechnic candle disperses a solid aerosol of the agent. Alternatively, a liquid solution of Agent CN or Agent CS can be dispersed as an aerosol spray in products such as MACE®. In this application, a 1% (Agent CN) to 2% (Agent CS) solution in an inert solvent is sprayed through an aerosol nozzle.

A different kind of tear gas is oleoresin capsicum (Agent OC), also known as pepper spray, which is primarily an irritant of mucous membranes. It induces sensory neurons to release Substance P, which causes vasodilation, increased permeability, and altered neutrophil chemotaxis. Mucous membrane pain, cough, corneal abrasions, acute bronchospasm, and pulmonary edema result from exposure. Oleoresin capsicum is obtained from cayenne peppers. When sprayed in the face, oleoresin capsicum pepper spray will cause temporary blindness and restricted breathing. The effects of the pepper spray last for thirty minutes to one hour. This pepper spray is relatively non-toxic and nonflammable. There has been one reported death where pepper spray possibly contributed to the death of an asthmatic. Several nations are using Agent OC for riot control. Since oleoresin capsicum pepper spray is an inflammatory agent, unlike MACE®, it is effective against those who feel no pain such as psychotics, drunks, or drug abusers. Interestingly, while the tear gases can be used as riot-control agents, their use in war is forbidden by a strict interpretation of the Chemical Weapons Convention.

VOMITING AGENTS

Vomiting agent are inhaled agents. They cannot be smelled easily and most take only one minute to be effective. Most are non-lethal and cause vomiting, headaches, coughing, and nausea. Most have a persistence time of approximately thirty minutes. These agents are basically very strong tear gases. They are a favorite chemical weapon for clearing enemy troops out of congested or built-up areas like urban environments.

Certain arsenic-containing compounds have been identified as causing severe irritation and vomiting. They can be thought of as more toxic tear gases. Diphenylchloroarsine (Agent DA), 10-chloro-5, 10-dihydrophenarsazine (Agent DM, Adamsite), and diphenylcyanoarsine (Agent DC) are three closely-related compounds developed near the end of World War I. All three compounds are solids when pure, and must be used as aerosols; this requires some means of aerosol generation, so these agents are typically used in either thermal grenades or smoke generators. The effects of all three are similar. There is severe irritation of the eyes, nose, and throat. If the agent is inhaled for a few minutes, tightness of the chest and headache are experienced. The headache develops into general nausea, which can result in vomiting in approximately three minutes. At concentrations expected to occur in combat conditions, fatalities are not expected; however, these compounds can be fatal at higher concentrations.

Agent DA, the first in the series, was used a few times by German troops in 1917. Agent DA was considered a significant development because it penetrated the activated carbon gas mask filters used in World War I. Its irritant behavior was considered more important than its lethality. Agent DA was used in combination with phosgene and diphosgene; Agent DA caused victims to remove their masks to sneeze, cough, or vomit, rendering them vulnerable to the toxic effects of the other agents. Agent DA alone saw some use as a riot-control agent up to the 1930s.

Agent DC was used by the Germans in 1918. It was intended to combine the vomiting potential of Agent DA with the lethality of cyanide. However, Agent DC did not prove particularly lethal in tests.

Agent DM had been discovered by German scientists in 1913 but was never used by the military. It was independently discovered by an American and a British team, both at the beginning of 1918. Agent DM was produced, but not used, by the Americans at the very end of the war. It may have been used by the Italian army. It was produced by many nations for use as a riot-control agent until it was superseded by α-chloroacetophenone (Agent CN) and similar tear agents. It was also found to be effective as a pesticide, which kept it in production for years. By 1920, gas mask filters had been improved to protect against aerosol particles, which may account for the termination of this line of development.

By mutual agreement, Agents DA and DC are not to be used by any of the NATO members. Use of Agents DA, DC, and DM against civilians was banned by the Western nations by the 1930s because of the toxicity of their arsenic-based by-products. There is documentation of limited Soviet production of Agent DA in 1936. Agent DC was a standard agent, called Red Number One, in the arsenal of the Japanese Imperial Forces between 1931 and 1945. Adamsite was prepared by all the belligerent states during World War II. Smoke generators filled with a mixture of Adamsite and a-chloroacetophenone were also developed. By 1967, protection had advanced to the point that the value of Agents DA, DC, and DM in combat was considered questionable.

CHEMISTRY OF OTHER MILITARY COMPOUNDS

In addition to the highly toxic lethal chemical warfare agents, there are a number of other chemical compounds of military interest. These compounds are used in a wide variety of military contexts: riot control, irritants, psychogenics, herbicides, and flame- and smoke-generating chemicals. Riot-control agents such as tear gas are widely used to control civilian populations. The ideal riot-control agent has very low toxicity, with effects that can quickly incapacitate an individual but are rapidly reversed when the individual leaves the area. Arsenical irritants are a class of more toxic riot-control agents that contain arsenic; these agents, which are no longer in use by the U.S., cause intense irritation and vomiting. Psychogenic agents are nonlethal incapacitating agents. They cause disorientation with intense visual and auditory hallucinations, similar to the effects from using LSD. Herbicides are used to kill plants and trees; they were used extensively during the Viet Nam War to deny jungle plants as cover for the Viet Cong. The herbicides are being used to destroy drug crops in Colombia, Afghanistan, and elsewhere. They could also be used to destroy food crops. Flame- and smoke-generating chemicals are used to burn structures and to hide troop movements from opposing forces.

In addition, there are recent developments of non-toxic incapacitating chemical weapons such as the following: a very light aqueous foam that can be used to engulf an individual to reduce his or her visual acuity, and a dense, sticky foam which, when sprayed on the lower extremities of a person, makes it impossible for him or her to walk or stand.

PSYCHOGENICS OR INCAPACITATING AGENTS

An incapacitating agent renders its victims physically dysfunctional and is considered a separate category by itself. Not as much is known about this type of agent and a great deal of research is still required, but consideration will be given to two categories which are better known: CNS depressants (anticholinergics) and Central Nervous System (CNS) stimulants (*e.g.*, LSD). Although cannabinols (from marijuana) and psylocibin (from certain mushrooms) have been considered chemical agents, their effective dose is too high to be useful as agents for use in the field. CNS depressants produce their effects by interfering with transmission of information across central synapses. An example of this type of agent is 3-quinuclidinyl benzilate (Agent BZ), which blocks the muscarinic action of acetylcholine, both peripherally and centrally. In the central nervous system anticholinergic compounds disrupt the high integrative functions of memory, problem solving, attention and comprehension. Relatively high doses produce toxic delirium, which destroys the ability to perform any military task. CNS stimulants cause excessive nervous activity by facilitating transmission of impulses. The effect is to flood the cortex and other higher regulatory centers with too much information, making concentration difficult and causing indecisiveness and inability to act in a sustained purposeful manner. A well known drug that acts in this way is D-lysergic acid diethylamide (LSD); similar effects are sometimes produced by large doses of amphetamines.

Agent BZ was weaponized in bomblets with a pyrotechnic mixture. The pyrotechnic mixture was ignited, which produces a solid aerosol of the high-melting Agent BZ. Production of Agent BZ in the U.S. began in 1962 at Pine Bluff Arsenal, Arkansas, and lasted through the late 60s. Between 1988 and 1990, all Agent BZ munitions were destroyed at Pine Bluff. No Agent BZ munitions remain in the U.S. stockpile.

In 1938, Albert Hofmann and Adolph Stoll in Germany first synthesized LSD along with other derivatives of lysergic acid. It was not until 1943 that the hallucinogenic properties of this substance were accidentally discovered by Hofmann, and subsequently confirmed by self-experimentation. During the 1950s, LSD was a subject of considerable interest as a potential chemical warfare agent. In both the U.S. and Britain, test subjects voluntarily ingested LSD in an effort to determine whether LSD could serve as a non-lethal incapacitant. Other tests were done involuntarily; Project MK-ULTRA tested LSD (along with a large number of other behavior-modifying and psychoactive drugs) on unwitting subjects. At least two fatalities were tied directly to these testing programs; it is probable that more deaths are indirectly related. LSD was considered for military use but rejected as ineffective and too dangerously unpredictable. The action of LSD on the nervous system results from it being a potent serotonin antagonist.

In 2002, Russian police, attempting to free hostages held by members of a Chechen terrorist group, filled a theater with a powerful form of the muscle relaxant fentanyl. This was the first reported use for such operations, although the drug is commonly used in medical surgery procedures. There was some evidence in the scientific literature that the drug could cause seizures and tracheal constriction with respiratory depression, even at doses as low as 0.006 mg/kg. A number of deaths, of the terrorists and hostages, resulted, probably due to the restricted air flow in the crowded building. A more dangerous form, α-methylfentanyl, has been reported in the medical literature to cause a number of deaths. Since fentanyl and its derivatives are not considered likely as either warfare or terrorist agents, they are not included in the tabular data that follows. Also not included are a number of other plant and animal toxins, which are unlikely to be used as chemical agents; for example, cobra and other snake venom have been used to kill individuals but are not suspected as being used in a mass casualty situation due to difficulties in delivery of the toxin.

HERBICIDES

The most famous of the military herbicides was Agent Orange, but there are a number of other chemicals which have been used to destroy foliage in order to allow for better visibility. Most of the herbicides affect the plant metabolism, either at the beginning of the growth cycle, or in the mature plant, so that the plant is unable to utilize nutrients.

The common herbicides used by the U.S. military during the Viet Nam War have been called the "Rainbow Agents" because many of the code names were those of colors (Agent Orange, Agent Orange II or "Super Orange," Agent Purple, Agent Pink, Agent Green, Agent White, Agent Blue); others were Dinoxol, Trinoxol, Diquat bromide, Bromacil, Tandex, Monuron, Diuron, and Dalapon). Agents Orange, "Super Orange," Purple, and Dinoxol were different mixtures of 2,4-D and 2,4,5-T with Agent Purple almost identical to Agent Orange. Agents Pink, Green, and Trinoxol were different concentrations of 2,4,5-T. Agent White was a mixture of Picloram (4-amino-3,5,6-trichloropicolinic acid) and 2,4-D. Agent Blue was the only arsenic compound, cacodylic acid or dimethylarsinic acid. Dalapon is 2,2-dichloroproprionic acid. All the others are the common or trade name of the product; for example, monuron is N-para-chlorophenyl-n',n'-dimethyl urea; and bromacil is 5-bromo-3-sec-butyl-6-methyluracil. The bromacil and monuron formulations were usually applied at much higher concentrations (15–30 kilograms active ingredient (a.i.) per hectare (kg/ha)) than the others, except for Agents Orange and Purple (15–50 kg/ha). Since the majority of the military defoliants were either solutions or mixtures of 2,4-D (2,4-dichlorophenoxy-acetic acid esters and salts) and/or 2,4,5-T (2,4,5-tricholorphenoxyacetic acid esters), or cacodylic acid only Agents Orange, Green, and Blue will be included in the tabular information found at the end of this chapter.

INCENDIARIES

Incendiaries are chemical weapons intended to start fires. All are self-igniting when used in bombs and artillery shells. Some contain phosphorus, which ignites when in contact with oxygen in the air.

Two major types of metal incendiaries exist, those that are magnesium based and those of the thermite/thermate type. Most types generally are encountered only in the military or industrial setting.

Magnesium, a silvery white metal of atomic weight 24.32, ignites at 632° C (1170 ∞F) and burns at 1982° C (3600° F), with magnesium oxide (MgO) as its primary combustion product. In a highly exothermic reaction, metallic magnesium can ignite to produce magnesium dihydroxide ($Mg(OH)_2$) and hydrogen. Magnesium is used in either powdered or solid form as an incendiary agent for both illumination and antipersonnel purposes. Various alloys of magnesium (*e.g.*, aluminum/zinc/magnesium alloy found in the U.S. M126 illumination round) are mechanically sturdier but also can be ignited easily.

Thermite is a mixture of powdered or granular aluminum and powdered iron oxide. When combined with other substances, such as binders, the material is termed a "thermate." All such materials react vigorously when heated to the combustion temperature of aluminum. This reaction produces aluminum oxide, elemental iron, and sufficient heat to melt the iron. The reaction temperature is approximately 2200° C (4000° F). Since the burning temperature of these chemicals is so very high, standard hazardous-materials clothing is not protective at all; even some firefighter bunker gear may be ineffective.

Napalm—from the fact that one of the components using in the 1940s was sodium (Na) palmitate (from palm oil)—contains gasoline. The use of gasoline alone can be washed off people and equipment with a spray of water; likewise, the gasoline can be washed off by diving into water. Mixing gasoline and a thickening agent, formerly a surfactant such as sodium palmitate but today an aluminum-based surfactant forms what is called jellied gasoline, which adheres to objects. Napalm can be employed in bombs, artillery shells, and by a flamethrower. The current U.S. DoD definition for napalm is powdered aluminum soap or similar compound used to gelatinize oil or gasoline for use in napalm bombs or flame throwers; or the resultant gelatinized substance.

SMOKES AND OBSCURANTS

Battlefield obscurants are intended to conceal troops and equipment. The earliest forms were dust raised by vehicles or smoke from burning fuels. The Iraqi military, in opposing Operation Iraqi Freedom forces, used burning crude oil in trenches to produce thick black clouds of smokes in hopes of obscuring portions of the city of Baghdad. In the years before and just after World War I, Navy vessels were able to hide behind smoke clouds made of the exhaust of the coal- and oil-burning boilers, if the wind was in the right direction. Later, red and white phosphorus were found to produce clouds that were opaque to visible light. Fog oil, a substance similar to motor oil without the additives, can be sprayed onto a heated surface such as a vehicle manifold or in a smoke generator. The fog oil forms an aerosol cloud of minute droplets, which blocks visible light. Several of the world's navies used chlorosulfonic acid, which forms a dense cloud over high humidity; however, it is highly corrosive to people and equipment, so use was very limited. Another chemical which had limited use as a smoke was titanium tetrachloride, but it, too, is toxic and corrosive, forming hydrochloric acid with water in the air. Some smokes are used for signaling devices; these typically contain organic dyes.

With increasing use of electronic devices such as Forward-Looking Infrared (FLIR), a wide amount of the spectrum had to be blocked. It was found that a mixture of fog oil with small particles of graphite effectively blocked infrared (IR) and visible light. With the addition of radar to the battlefield, a research project began in the U.S. and the former U.S.S.R. to find an effective tri-spectral smoke that would block visible, radio, and IR radiation. A number of materials were tried, including fog oil, graphite, plus copper or aluminum particles, but none were completely successful.

Toxicity of Common Smoke Agents—Petroleum Smokes

The petroleum smokes, fog oil and diesel fuel (DF-2), contain highly variable mixes of hydrocarbons, which have not been well characterized and may vary from manufacturer to manufacturer and from lot to lot. The disseminated smokes contain essentially the same class fractions as the bulk materials from which they are generated, although the vapor phases of the smokes are generally richer in the more volatile components. The smoke particles contain the less volatile components and show a slight increase in the proportion of aromatics to aliphatics than present in the bulk material. This enrichment may be due to oxidation of some of the aliphatic components. Fog oil contains approximately 50% by weight of aromatics, while diesel fuel has only about half this amount. The mass median diameter of the aerosol particles is generally between 1 and 1.5 mm, depending on concentration, and are within the deep lung respirable range for humans and laboratory animals.

Continuous skin exposure to petroleum products can produce severe dermal irritation and other dermatoses. The oils and fuels break down the protective defenses of the skin barrier with resultant inflammation, eczemas and infection. Diesel fuel produces more severe dermal toxicity than the lubricating oils that constitute fog oil smoke. Neither petroleum product elicits skin sensitization responses. Dermal toxicity reactions are not observed when the oils are promptly removed from skin surfaces with soap and water and when proper protective clothing is worn.

Inhalation of petroleum smokes does not produce immediate toxicity in animals. Concentrations in the range of 5 to 10 g per m^3 at 3 to 4 hours exposure produce approximately 50% mortality in mice, rats and guinea pigs. For shorter exposure periods, higher concentrations are required to produce mortality. Mortality in acute studies is due to vascular injury, acute inflammatory lesions, epithelial necrosis of the lung, and bronchopneumonia. Surviving animals showed recovery from some acute toxicity responses at 14 days post-exposure. Repeated daily exposures over four- to nine-week periods at various concentrations, daily exposure duration, and frequency of exposures per week demonstrated that the lung was the main target organ for the petroleum smoke's toxicity in rats. Focal accumulations of free pulmonary cells were observed associated with thickening and hypercellularity of alveolar walls, and end expiratory volume and functional residual capacity were increased. These effects persisted for up to two weeks after the end of exposure. The severity of the biologic effects in the diesel fuel exposures are highly correlated with the number of weekly exposures; three days per week exposure produced more severe toxic reactions than one weekly treatment. Six-hour daily exposures at concentrations of 1.3 and 2.0 g per m^3 also produced greater toxic effects than two-hour daily exposures at 4.0 or 6.0 g per m^3. In fog oil experiments, concentration was seen to be the most important determinant of toxic response (1.5 versus 0.5 g per m^3) rather than exposure duration (1 or 3.5 hours per day) or exposure frequency (2 or 4 days per week).

Longer term exposures of 13 weeks to lower concentrations of the petroleum smokes demonstrated reduced toxicity. Although the frequency of weekly exposures were different for the diesel fuel and fog oil (2 versus 4 weekly exposures), 3.5- to 4-hour daily exposures at 1.5 g per m^3 for each smoke gave similar results. At the end of the exposure period, histologic sections of lung showed marked hypercellularity in the alveolar region and septa. Lung wet weights were increased indicating an edematous reaction. In diesel fuel-exposed rats, all lesions were reversible after an 8-week recovery period. The fog oil-treated rats exhibited multifocal granuloma formation 4 weeks after the end of exposure. The development of granulomas after cessation of exposure may be a nonspecific response of the lung to foreign particles when its clearance mechanisms are overwhelmed. At lower concentrations of the petroleum smokes (0.25 and 0.5 g per m^3), mild to moderate hypercellularity was observed after the end of exposure but tissue damage was not present.

Epidemiologic studies of non-military workers who are exposed to lubricating oils and oil aerosols have shown increased incidences of skin squamous cell carcinomas but not increased respiratory tumors due to exposure to lubricating oil aerosols. The major health hazard associated with highly refined mineral oils such as fog oils purchased after April 1986 was lipoid pneumonia. Efforts to protect those who may be occupationally exposed is expressed as an 8-hour time weighted average (TWA) exposure limit of 5 mg per m^3 (for respirable fraction) in the 2003 American Conference of Governmental Industrial Hygienists (ACGIH) "Threshold Limit Values And Biological Exposure Indices." The evidence for diesel fuel carcinogenicity from laboratory studies is uncertain.

Troops and, in particular, operators and instructors may be exposed to fog oil smoke concentrations ranging from 1 mg per m^3 downwind of a fog oil generator to 1000 mg per m^3 in the immediate area. Estimated exposure times may vary from less than 1 hour per day to 6 hour per day over several days. Exposure frequencies may vary from once per week to several times per week. The helicopter fog oil smoke system can reportedly lay a smoke screen that measures 5 by 7 kilometers.

Exposure to diesel fuel smoke from the U.S. military Vehicle Engine Exhaust Smoke System (VEESS) may range from less than 1 mg per m^3 downwind from the system to an estimated range of 1300 to 35,000 mg per m^3 within the immediate area of the source. Exposures to diesel fuel smoke are estimated to be of less than 30 minutes duration; however, this could occur several times per day.

Toxicity of Common Smoke Agents —HC (Hexachloroethylene) Smoke

The major component in HC smoke is zinc chloride ($ZnCl_2$) particles with organochlorine compounds in the gaseous phase comprising approximately 10% of the total smoke. Traces of lead and cadmium are also present in the particulate phase. The initial mass median particle diameter is generally around 0.3 mm, which increases with concentration and time after detonation.

$ZnCl_2$ is an extremely corrosive material and has induced severe corneal damage in laboratory animals. Severe skin irritation has been produced in rabbits and in humans, particularly at sites of recent injury.

There are only a few laboratory studies on the inhalation toxicity of HC smoke. Earlier reports did not specify the concentration and duration of exposures producing mortality. Incidents of accidental human exposure and laboratory studies have shown that lethal concentrations are in the range of 4 to 5 g per m^3 or greater for single 30 minute exposures. The typical symptoms and clinical signs of exposure to high concentrations are indicative of severe irritation, inflammation, epithelial necrosis and parenchymal infiltration. Manifestations of toxicity can be delayed at lower concentrations and in some individuals may be slowly resolved. Volunteers exposed to atmospheres containing 120 mg per m^3 complained of nose, throat and chest irritation, cough, and nausea after 2 minutes. After 2 minutes at 80 mg per m^3, most subjects had slight nausea, few coughed, and all could smell the smoke.

Although information on pulmonary and skin absorption of zinc in humans remains unavailable, during a previous experiment, 14 pre-terminal patients were given an intravenous infusion of radiolabelled zinc to determine distribution and organ uptake at autopsy. Between 1 and 174 days post exposure, skeletal muscle and the liver showed maximum uptake followed by spleen, lung, pancreas and prostate uptake. Additional studies in laboratory animals have shown identical patterns of absorption and distribution.

Troops may be exposed to the combustion products of HC smoke pots, grenades, and artillery rounds. Exposure concentrations may vary from 7 mg per m^3 downwind to greater than 100 mg per m^3 within 100 meters of the source. Smoke pot exposures and personnel exposed should be similar to those using the fog oil as they are used in conjunction with each other. Smoke grenade concentrations may range from less than 3 mg per m^3 to 50 mg per m^3, depending on personnel location. Duration of exposure is typically expected to be less than 3 minutes. Exposures to artillery projectiles may range from 45 mg per m^3 to 1840 mg per m^3; however, exposure from these rounds would probably be at low concentrations as troops are not usually located near impact areas.

The most toxic smoke in the current U.S. military inventory is HC smoke (M5 and M4 smoke pots and M8 smoke grenade); the substitution of terpthalic acid in training grenades is a recent development to limit soldiers' exposure to HC.

Toxicity of Common Smoke Agents—Phosphorus Smokes

Phosphorus munitions contain either red or white phosphorus in various matrices: felt, butyl rubber or polymer epoxy binder. The composition of the phosphorus smokes are very similar, being composed primarily of polyphosphoric acids with trace levels (less than one%) of organic compounds. Particle size distributions usually center around 1.0 mm with a narrow range.

Transient eye irritation has been noted in animals exposed for 8 minutes a day. In humans, the primary symptom is upper respiratory irritation. The minimum harassing concentration of phosphorus smokes is about 0.6 to 0.7 g per m^3 for 2-4 minutes. Respiratory distress, nasal discharge, coughing, and soreness and irritation of the throat have been noted. The LC_{50} in rats for single or five daily one-hour exposures ranges from 1.6 to 2.3 g per m^3. Since five daily 4-hour exposures to approximately 1.0 mg per m^3 did not produce deaths, concentration is probably the major determinant of lethality.

Repeated exposures of rats for brief daily periods (15 minutes) to 0.5 or 1.0 g per m^3 led to the production of laryngitis and tracheitis with some animals at the highest concentration displaying interstitial pneumonia after 6 weeks of exposure. These histopathologic changes were also present after 13 weeks of exposure and were not resolved four weeks after the end of exposure. A concentration of 0.2 g per m^3 had no effect. Rats given repeated daily exposures for longer periods (2 hours) to concentrations in the range of 0.4 to 1.2 g per m^3 displayed terminal bronchiolar fibrosis after 4 weeks of exposure. This lesion increased in incidence and severity with increasing concentration and weeks of exposure. The histologic changes did not regress during an eight-week recovery period after 13 weeks of exposure. There were indications that pulmonary defense mechanisms are compromised after 13 weeks of exposure. In similar studies, a decline in pulmonary bactericidal activity in laboratory rats exposed to the aerosol was displayed. Mutagenic tests, both *in vivo* and *in vitro*, did not provide evidence of a genotoxic effect for phosphorus smokes.

The U.S. military's M8 phosphorus grenade concentrations may range from 6 mg per m^3 to greater than 100 mg per m^3. Duration of exposure is expected to be less than 15 minutes in most situations. Concentrations from artillery projectiles may range from 28 mg per m^3 to 550 mg per m^3; however, the exposure would probably be at low concentrations as troops are not normally located near impact areas.

Toxicity of Common Smoke Agents—Brass

The disseminated brass smoke is an irregular flake and contains an alloy of approximately 75% copper and 25% zinc. The mass median aerodynamic diameter (MMAD) is usually between 2.1 and 2.3 mm. This material is not a skin irritant and is only a mild eye irritant after 24 hours. Acute exposures for four hours to concentrations up to 0.2 g per m^3 produced acute inflammatory responses in the respiratory tract, which are reversible. The effects of repeated inhalation exposures are under investigation.

The U.S. military's M76 grenade concentrations may range from 10 mg per m^3 to greater than 1000 mg per m^3. Duration of exposures is expected to be less than 15 minutes.

Toxicity of Common Smoke Agents—Dyes

Dye mixtures in the U.S. and NATO M18 colored smoke grenades, yellow, green, red and violet, gave positive results in the Ames Salmonella reversion assay. One of the major components in the older, now obsolete, yellow grenade dye mix, dibenzochrysenedione (Vat Yellow 4), gave positive results in a National Cancer Institute (NCI) carcinogenic bioassay. A major component of the red grenade and part of the dye mix in the violet grenade, 1-methylaminoanthraquinone. undergoes oxidative demethylation during grenade detonation to produce small quantities of 2-aminoanthroquinone which is a suspected animal carcinogen. The major component in the green grenade, 1,4-di-p-toluidino-9,10-anthraquinone (Solvent Green 3) was negative in the Ames bioassay.

The yellow dye now used in the M18 grenade, Solvent Yellow 33, was negative in the standard Ames test strains and an *in vivo* cytogenetic test but was positive in a mouse lymphoma assay. Repeated inhalation exposure to 234 mg per m^3 (MMAD: 3-5 mm) produced mild toxic effects in rats. The green grenade contains 30% of this dye and 70% of Solvent Green 3. This mixture gave the same mutagenic results as the Solvent Yellow 33. The green dye is retained in rat lung and elicits a mild inflammatory response. Additional studies are needed to better characterize the toxicity of colored dyes used in smoke grenades; in addition, new, less toxic materials are being considered as replacements for those in the U.S. military inventory.

M18 grenade concentrations may range from 1 mg per m^3 to 14 mg per m^3. Duration of exposures is expected to be less than 10 minutes.

Toxicity of Common Smoke Agents—Titanium Dioxide

Generally classified as a "nuisance dust," titanium dioxide (TiO_2) is used in numerous non-military applications to include paints, plastics, and floor coverings. Occupational exposure to TiO_2 has generally shown no evidence of serious adverse pulmonary effects. The 2003 American Conference of Governmental Industrial Hygienists (ACGIH) TWA-TLV for titanium dioxide is 10 mg per m^3.

Toxicity of Common Smoke Agents—Graphite

Also classified as a "nuisance dust," graphite is composed predominantly of carbon with a few trace impurities totaling less than 1% by weight. Two synthetic graphite powders are currently in use by the U.S. military (Micro-260 and KS-2).

Exposures, in occupational settings, typically show no adverse effects (during a working lifetime) if exposures are kept below 10 mg per m^3. Exposures greater than this may compromise clearing mechanisms, resulting in the accumulation of dust-laden macrophages and the proliferation of Type II pneumocytes.

The 2003 American Conference of Governmental Industrial Hygicnists (ACGIH) TWA-TLV for graphite exposure (except fibers) is 2 mg per m^3 for the respirable fraction.

DETAILS OF COMMON BLISTER AGENTS

Agent HD	
Agent Type	Blister Agent
Common Name	Distilled Mustard
Chemical Name	Bis(chloroethyl) sulfide
Formula	$(ClCH_2CH_2)_2S$
Code Name	HD
Molecular Weight	159.08
State at 20° C	Colorless to pale yellow liquid
Vapor Density (air = 1)	5.4
Liquid Density (g/cc)	1,268 @ 25° C
Freezing Point (° C)	14.45
Boiling Point (° C)	217
Vapor Pressure (mg/m^2)	0.072 @ 20° C
Volatility (mg/m^3)	610 @ 20° C
Flash Point	105° C; ignited by large explosive charges
Decomposition Temperature (° C)	149-177
Heat of Vaporization (° C)	94
Odor	Garlic-like
Median Lethal Dosage (mg-min./m^3)	1,500 by inhalation; 10,000 by skin exposure
Median Incapacitating Dosage (mg-min./m^3)	200 by eye effect; 2000 by skin effect
Rate of Detoxification	Very low, cumulative
Eye and Skin Toxicity	Eyes very susceptible, skin less so
Rate of Action	Delayed hours to days
Physiological Action	Blisters; destroys tissues, injures blood vessels
Protection Required	Protective mask and clothing
Stability	Stable in steel
Decontamination	Bleach, fire, DS2, M258 kit
Means of Detection in Field	M256A and M18A2 kits
Typical Use	Delayed action casualty agent

Figure 13.1 Agent HD Characteristics

Agent HN-1	
Agent Type	Blister Agent
Common Name	Nitrogen Mustard
Chemical Name	Bis(chloroethyl) ethyl amine
Formula	$(ClCH_2CH_2)_2NC_2H_5$
Code Name	HN-1
Molecular Weight	170.08
State at 20° C	Dark liquid
Vapor Density (air = 1)	5.9
Liquid Density (g/cc)	1.09 @ 25° C
Freezing Point (° C)	-34
Boiling Point (° C)	194
Vapor Pressure (mg/m^2)	0.24 @ 25° C
Volatility (mg/m^3)	1,520 @ 20° C
Flash Point	High enough not to interfere with military use
Decomposition Temperature (° C)	Decomposes before boiling point is reached
Heat of Vaporization (° C)	77
Odor	Fishy or musty
Median Lethal Dosage (mg-min./m^3)	1,500 by inhalation; 20,000 by exposure
Median Incapacitating Dosage (mg-min./m^3)	200 by eye effect; 9,000 by skin effect
Rate of Detoxification	Not detoxified, cumulative
Eye and Skin Toxicity	Eyes susceptible to low concentration; less toxic to skin
Rate of Action	Delayed action, 12 hours or longer
Physiological Action	Blisters; affect respiratory tract; destroys tissues, injures blood vessels
Protection Required	Protective mask and clothing
Stability	Adequate for use as agent
Decontamination	Bleach, fire, DS2, M258 kit
Means of Detection in Field	M256A and M18A2 kits
Typical Use	Delayed action casualty agent

Figure 13.2 Agent HN-1 Characteristics

Agent HN-2	
Agent Type	Blister Agent
Common Name	Nitrogen Mustard
Chemical Name	Bis(chloroethyl) methyl amine
Formula	$(ClCH_2CH_2)_2NCH_3$
Code Name	HN-2
Molecular Weight	156.07
State at 20° C	Dark Liquid
Vapor Density (air=1)	5.4
Liquid Density (g/cc)	1.15 @ 20° C
Freezing Point (° C)	-60 to -65
Boiling Point (° C)	75 @ 15 mm Hg
Vapor Pressure (mg/m^2)	0.29 @ 20° C
Volatility (mg/m^3)	3,580 @ 25° C
Flash Point	High enough not to interfere with military use
Decomposition Temperature (° C)	Below boiling point; polymerizes with heat generation
Heat of Vaporization (° C)	78.8
Odor	Soapy in low concentrations; fruity in high concentrations
Median Lethal Dosage (mg-min./m^3)	3,000 by inhalation
Median Incapacitating Dosage (mg-min./m^3)	Less than HN-1; more than HN-3; 100 for eye effect
Rate of Detoxification	Not detoxified-cumulative
Eye and Skin Toxicity	Toxic to eyes; blister skin
Rate of Action Skin	Effect delayed 12 hours or longer
Physiological Action	Similar to Distilled Mustard, Bronchopneumonia may occur after 24 hours
Protection Required	Protective mask and clothing
Stability	Unstable
Decontamination	Bleach, fire, DS2, M258 kit
Means of Detection in Field	M256A and M18A2 kits
Typical Use	Delayed action casualty agent

Figure 13.3 Agent HN-2 Characteristics

Agent HN-3	
Agent Type	Blister Agent
Common Name	Nitrogen Mustard
Chemical Name	Triethylamine
Formula	$N(CH_2CH_2)_3$
Code Name	HN-3
Molecular Weight	204.54
State at 20 °C	Dark Liquid
Vapor Density (air = 1)	7.1
Liquid Density (g/cc)	1.24 @ 25° C
Freezing Point (° C)	-3.7
Boiling Point (° C)	256
Vapor Pressure (mg/m^2)	0.0109 @ 25° C
Volatility (mg/m^3)	121 @ 25° C
Flash Point	High enough not to interfere with military use
Decomposition Temperature (° C)	Below boiling point
Heat of Vaporization (° C)	74
Odor	None, if pure
Median Lethal Dosage (mg-min./m^3)	1,500 by inhalation; 10,000 by skin exposure (estimated)
Median Incapacitating Dosage (mg-min./m^3)	200 by eye effect; 2500 by skin effect
Rate of Detoxification	Not detoxified-cumulative
Eye and Skin Toxicity	Eyes very susceptible, skin less so
Rate of Action	Serious effects same as for HD; minor effect sooner
Physiological Action	Similar to HN-2
Protection Required	Protective mask and clothing
Stability	Stable
Decontamination	Bleach, fire, DS2, M258 kit
Means of Detection in Field	M256A and M18A2 kits
Typical Use	Delayed action casualty agent

Figure 13.4 Agent HN-3 Characteristics

Agent CX	
Agent Type	Blister Agent
Common Name	---
Chemical Name	Phosgene oxime dichloroformate
Formula	CCl_2NOH
Code Name	CX
Molecular Weight	113.94
State at 20° C	Colorless solid or liquid
Vapor Density (air = 1)	---
Liquid Density (g/cc)	---
Freezing Point (° C)	39 to 40
Boiling Point (° C)	53 to 54 at 28 mm Hg
Vapor Pressure (mg/m^2)	High
Volatility (mg/m^3)	---
Flash Point	---
Decomposition Temperature (° C)	Decomposes slowly at ambient temperatures
Heat of Vaporization (° C)	---
Odor	Sharp, penetrating
Median Lethal Dosage (mg-min./m^3)	---
Median Incapacitating Dosage (mg-min./m^3)	---
Rate of Detoxification	---
Eye and Skin Toxicity	Powerful irritant to eyes and nose
Rate of Action	Immediate effects on contact
Physiological Action	Violently irritates mucus membrane of eyes and nose
Protection Required	Protective mask and clothing
Stability	Decomposes slowly
Decontamination	None is entirely effective; wash
Means of Detection in Field	M256A, M151A2N,M18A2
Typical Use	Delayed action casualty agent

Figure 13.5 Agent CX Characteristics

Agent L	
Agent Type	Blister Agent
Common Name	Lewisite
Chemical Name	Bis(chloroethyl) chloroarsine
Formula	$(ClCH_2CH)_2As_3Cl$
Code Name	L
Molecular Weight	207.35
State at 20 °C	Dark oily liquid
Vapor Density (air = 1)	7.1
Liquid Density (g/cc)	1.89 @ 20° C
Freezing Point (° C)	-18
Boiling Point (° C)	190
Vapor Pressure (mg/m^2)	0.394 @ 20° C
Volatility (mg/m^3)	4,480 @ 20° C
Flash Point	None
Decomposition Temperature (° C)	> 100
Heat of Vaporization (° C)	58 (from 190 to 0° C)
Odor	Variable, but may resemble geranium flowers
Median Lethal Dosage ($mg\text{-}min./m^3$)	1,200 to 1,500 by inhalation; 100,000 by skin exposure
Median Incapacitating Dosage ($mg\text{-}min./m^3$)	Below 300 by eye effect; over 1,500 by skin effect
Rate of Detoxification	Not detoxified
Eye and Skin Toxicity	1,500 $mg/min/m^2$ exposure severely damages cornea; skin less susceptible
Rate of Action	Rapid
Physiological Action	Similar to HD plus may cause systemic poisoning
Protection Required	Protective mask and clothing
Stability	Stable in steel and glass
Decontamination	Bleach, fire, DS2, caustic soda M258 kit
Means of Detection in Field	M18A2 kit
Typical Use	Moderately delayed casualty agent

Figure 13.6 Agent L Characteristics

Agent HL	
Agent Type	Blister Agent
Common Name	Mustard Lewisite
Chemical Name	---
Formula	Mixture of mustard and lewisite
Code Name	HL
Molecular Weight	186.4
State at 20° C	Dark oily liquid
Vapor Density (air = 1)	6.5
Liquid Density (g/cc)	1.66 @ 20° C
Freezing Point ($^{\circ}$ C)	25.4
Boiling Point ($^{\circ}$ C)	< 19
Vapor Pressure (mg/m^2)	0.248 @ 20° C
Volatility (mg/m^3)	2,730 @ 20° C
Flash Point	High enough not to interfere with military use
Decomposition Temperature ($^{\circ}$ C)	> 100
Heat of Vaporization ($^{\circ}$ C)	58 (from 190 to 0° C)
Indeterminate value	Odor Garlic-like
Median Lethal Dosage (mg-min./m^3)	1,500 by inhalation; 10,000+ by skin exposure
Median Incapacitating Dosage (mg-min./m^3)	200 by eye effect; 1,500 to 2000 by skin effect
Rate of Detoxification	Not detoxified
Eye and Skin Toxicity	Very high
Rate of Action	Prompt stinging; delayed (about 12 hours) for blistering
Physiological Action	Similar to HD, but may cause systemic poisoning
Protection Required	Protective mask and clothing
Stability	Stable in lacquered steel
Decontamination	Bleach, fire, DS2, caustic soda, M258 kit
Means of Detection in Field	M18A2
Typical Use	Delayed-action casualty agent

Figure 13.7 Agent HL Characteristics

Agent PD	
Agent Type	Blister Agent
Common Name	---
Chemical Name	Phenyldichloroarsine
Formula	$C_6H_5AsCl_2$
Code Name	PD
Molecular Weight	222.91
Statc at 20° C	Colorless liquid
Vapor Density (air = 1)	7.7
Liquid Density (g/cc)	1.65 @ 20° C
Freezing Point (° C)	-20
Boiling Point (° C)	252 to 255
Vapor Pressure (mg/m^2)	0.033 @ 25° C
Volatility (mg/m^3)	39 @ 20° C
Flash Point	High enough not to interfere with military use
Decomposition Temperature (° C)	Stable to boiling point
Heat of Vaporization (° C)	69
Odor	None
Median Lethal Dosage (mg-min./m^3)	2,600 by inhalation
Median Incapacitating Dosage (mg-min./m^3)	16 as vomiting agent; 1,800 as blistering agent
Rate of Detoxification	Probably rapid
Eye and Skin Toxicity	633 mg-min./m^3 produces eye casualty; less toxic to skin
Rate of Action	Immediate eye effect; skin effects 1/2 to 1 hour
Physiological Action	Irritates, causes nausea and vomiting, blistering
Protection Required	Protective mask and clothing
Stability	Stable
Decontamination	Bleach, DS2, caustic soda; M258 kit
Means of Detection in Field	M18A2 kit
Typical Use	Delayed casualty agent

Figure 13.8 Agent PD Characteristics

Agent ED	
Agent Type	Blister Agent
Common Name	---
Chemical Name	Ethyldichloroarsine
Formula	$C_2H_5AsCl_2$
Code Name	ED
Molecular Weight	174.88
State at 20° C	Colorless liquid
Vapor Density (air = 1)	6.0
Liquid Density (g/cc)	1.66 @ 20° C
Freezing Point (° C)	-65
Boiling Point (° C)	156
Vapor Pressure (mg/m^2)	2.09 @ 20° C
Volatility (mg/m^3)	20,000 @ 20° C
Flash Point	High enough not to interfere with military use
Decomposition Temperature (° C)	Stable to boiling point
Heat of Vaporization (° C)	52.5
Odor	Fruity but biting; irritating
Median Lethal Dosage (mg-min./m^3)	3,000 to 5,000 by inhalation;100,000 by skin effect
Median Incapacitating Dosage (mg-min./m^3)	5 to 10 by inhalation
Rate of Detoxification	Rapid
Eye and Skin Toxicity Vapor	Harmful only on long exposure; liquid blisters less than L
Rate of Action	Immediate irritation; delayed blistering
Physiological Action	Damages respiratory tract, affects eyes, blisters can cause death
Protection Required	Protective mask and clothing
Stability	Stable in steel
Decontamination	None needed in field; bleach caustic soda, or DS2 in closed spaces, M258 kit
Means of Detection in Field	M18A2 kit
Typical Use	Delayed action casualty agent

Figure 13.9 Agent ED Characteristics

Agent MD	
Agent Type	Blister Agent
Common Name	--
Chemical Name	Methyldichloroarsine
Formula	CH_3AsCl_2
Code Name	MD
Molecular Weight	160.86
State at 20° C	Colorless liquid
Vapor Density (air = 1)	5.5
Liquid Density (g/cc)	1.83 @ 20° C
Freezing Point (° C)	-55
Boiling Point (° C)	133
Vapor Pressure (mg/m^2)	7.76 @ 20° C
Volatility (mg/m^3)	74,900 @ 20° C
Flash Point	High enough not to interfere with military use
Decomposition Temperature (° C)	Stable to boiling point
Heat of Vaporization (° C)	49
Odor	None
Median Lethal Dosage (mg-min./m^3)	3,000 to 5,000 (estimated)
Median Incapacitating Dosage (mg-min./m^3)	25 by inhalation
Rate of Detoxification	Rapid
Eye and Skin Toxicity	Cornea damage possible; blisters less than HD
Rate of Action	Rapid
Physiological Action	Irritates respiratory tract, injures lungs and eyes, causes systemic poisoning
Protection Required	Protective mask and clothing
Stability	Stable in steel
Decontamination	Bleach, caustic soda, DS2; M258 kit
Means of Detection in Field	M18A2 kit
Typical Use	Delayed action casualty agent

Figure 13.10 Agent MD Characteristics

Dimethyl Sulfate	
Agent Type	Blister Agent
Common Name	DMSO4
Chemical Name	Dimethyl Sulfate
Formula	$(CH_3)_2\ SO_4$
Code Name	--
Molecular Weight	126.13
State at 20° C	Colorless oily liquid
Vapor Density (air = 1)	4.35
Liquid Density (g/cc)	1.3322 @ 20° C
Freezing Point (° C)	-27
Boiling Point (° C)	188 (with decomposition)
Vapor Pressure (mm Hg)	0.677 @ 20° C
Volatility	Vaporizes @ 50° C
Flash Point (°C)	--
Decomposition Temperature (° C)	188
Heat of Vaporization (° C)	--
Odor	Essentially odorless
Median Lethal Dosage (mg/m^3)	500, for 10 minutes, for resting male
Median Incapacitating Dosage (mg/m^3)	10 for resting male
Rate of Detoxification	Moderate
Eye and Skin Toxicity	High
Rate of Action	Delayed
Physiological Action	Respiratory failure following mucosal inflammation, edema of major airways; cessation of breath; death may follow
Protection Required	Protective mask and clothing
Stability	Stable in steel at ordinary temp.
Decontamination	Bleach slurry, dilute alkali, or DS2, steam and ammonia in confined area, M258 kit.
Means of Detection in Field	--
Typical Use	Delayed action casualty agent

Figure 13.11 Dimethyl Sulfate Characteristics

DETAILS OF COMMON BLOOD AGENTS

Agent HCN	
Agent Type	Blood Agent
Common Name	Cyanide
Chemical Name	Hydrogen cyanide
Formula	HCN
Code Name	AC
Molecular Weight	27.02
State at 20° C	Colorless liquid, easily vaporized
Vapor Density (air = 1)	0.93
Liquid Density (g/cc)	0.687 @ 10° C
Freezing Point (° C)	-13
Boiling Point (° C)	25.7
Vapor Pressure (mg/m^2)	742 @ 25° C
Volatility (mg/m^3)	1,080,000 @ 25° C
Flash Point	0° C, ignited 50% of time when disseminated from artillery shell.
Decomposition Temperature (° C)	65.5
Heat of Vaporization (° C)	233
Odor	Bitter almonds
Median Lethal Dosage (mg-min./m^3)	Varies widely with concentrations
Median Incapacitating Dosage (mg-min./m^3)	Varies widely with concentrations
Rate of Detoxification	Rapid; 0.017 mg/kg/min.
Eye and Skin Toxicity	Moderate
Rate of Action	Very rapid
Physiological Action	Interferes with use of oxygen by body tissues; accelerated rate of breathing
Protection Required	Protective mask and clothing
Stability	Stable if pure, can inflame on shell explosion
Decontamination	None needed in field
Means of Detection in Field	M18A2 kit
Typical Use	Quick-acting casualty agent

Figure 13.12 Agent HCN Characteristics

Agent CK	
Agent Type	Blood Agent
Common Name	Cyanogen
Chemical Name	Cyanogen chloride
Formula	CNCl
Code Name	CK
Molecular Weight	61.48
State at 20° C	Colorless gas
Vapor Density (air = 1)	2.1
Liquid Density (g/cc)	1.18 @ 10° C
Freezing Point (° C)	-6.9
Boiling Point (° C)	12.8
Vapor Pressure (mg/m^2)	1,000 @ 25° C
Volatility (mg/m^3)	2,600,000 @ 12.8° C
Flash Point	None
Decomposition Temperature (° C)	> 100
Heat of Vaporization (° C)	103
Odor	Bitter almonds
Median Lethal Dosage (mg-min./m^3)	11,000
Median Incapacitating Dosage (mg-min./m^3)	7,500
Rate of Detoxification	Rapid; 0.02 to 0.01 mg/kg/min.
Eye and Skin Toxicity	Low, lacrimatory and irritating
Rate of Action	Rapid
Physiological Action	Chokes, irritates, causes slow breathing rate
Protection	Required protective mask
Stability	Tends to polymerize; may explode
Decontamination	None needed in field
Means of Detection in Field	M18A2 and M256A
Typical Use	Quick acting casualty agent

Figure 13.13 Agent CK Characteristics

Agent SA	
Agent Type	Blood Agent
Common Name	Arsine
Chemical Name	Arsenic trihydride
Formula	AsH_3
Code Name	SA
Molecular Weight	77.93
State at 20° C	Colorless gas
Vapor Density (air = 1)	2.69
Liquid Density (g/cc)	1.34 @ 20° C
Freezing Point (° C)	-16
Boiling Point (° C)	-62.5
Vapor Pressure (mg/m^2)	11,100 @ 20° C
Volatility (mg/m^3)	30,900,000 @ 0° C
Flash Point	Below shell detonation temp; mixtures with air may explode spontaneously
Decomposition Temperature (° C)	> 280
Heat of Vaporization (° C)	53.7
Odor	Mild garlic-like
Median Lethal Dosage (mg-min./m^3)	5,000
Median Incapacitating Dosage (mg-min./m^3)	2,500
Rate of Detoxification	Low
Eye and Skin Toxicity	None
Rate of Action	Delayed action to two hours to as much as eleven days
Physiological Action	Damages blood, liver, and kidneys
Protection Required	Protective mask
Stability	Not stable in uncoated metal containers
Decontamination	None needed
Means of Detection in Field	None
Typical Use	Delayed action casualty agent

Figure 13.14 Agent SA Characteristics

DETAILS OF COMMON CHOKING AGENTS

Agent CG	
Agent Type	Choking Agent
Common Name	Phosgene
Chemical Name	Carbonyl chloride
Formula	$COCl_2$
Code Name	CG
Molecular Weight	98.92
State at 20° C	Colorless gas
Vapor Density (air = 1)	3.4
Liquid Density (g/cc)	1.37 @ 20 °C
Freezing Point (° C)	-128
Boiling Point (° C)	7.6
Vapor Pressure (mg/m^2)	1.173 @ 20° C
Volatility (mg/m^3)	4,300,000 @ 7.6° C
Flash Point	None
Decomposition Temperature (° C)	800
Heat of Vaporization (° C)	59
Odor	Similar to new mown hay, green corn
Median Lethal Dosage (mg-min./m^3)	3,200
Median Incapacitating Dosage (mg-min./m^3)	1,600
Rate of Detoxification	Not detoxified-cumulative
Eye and Skin Toxicity	None
Rate of Action	Immediate to three hours
Physiological Action	Damages and floods lungs
Protection Required	Protective mask
Stability	Stable in steel, if dry
Decontamination	None needed in field, aeration in closed spaces
Means of Detection in Field	M18A2 kit, odor
Typical Use	Delayed or immediate action, casualty agent

Figure 13.15 Agent CG Characteristics

Agent DP	
Agent Type	Choking Agent
Common Name	Diphosgene
Chemical Name	Diphosgene
Formula	$ClCOOCCl_3$
Code Name	DP
Molecular Weight	197.85
State at 20° C	Colorless liquid
Vapor Density (air = 1)	6.8
Liquid Density (g/cc)	1.65 @ 20° C
Freezing Point (° C)	-57
Boiling Point (° C)	127 to 128
Vapor Pressure (mg/m^2)	4.2 @ 20° C
Volatility (mg/m^3)	45,000 @ 20° C
Flash Point	None
Decomposition Temperature (° C)	300 to 350
Heat of Vaporization (° C)	None
Odor	Similar to new mown hay or green corn
Median Lethal Dosage (mg-min./m^3)	3,200
Median Incapacitating Dosage (mg-min./m^3)	1,600
Rate of Detoxification	Not detoxified, cumulative
Eye and Skin	Toxicity slightly lacrimatory
Rate of Action	Immediate to three hours depending on concentration
Physiological	Action Damages and floods lungs
Protection Required	Protective mask
Stability	Unstable, tends to convert to Agent CG
Decontamination	None needed in field, aeration in closed spaces
Means of Detection in Field	Odor
Typical Use	Delayed or immediate action casualty agent

Figure 13.16 Agent DP Characteristics

DETAILS OF COMMON NERVE AGENTS

Agent GA	
Agent Type	Nerve Agent
Common Name	Tabun
Chemical Name	Ethyldimethylphosphoramidocyanide
Formula	$(CH_3)_2$ $NP(O)(C_2H_5)(O)(CN)$
Code Name	GA
Molecular Weight	162.3
State at 20° C	Colorless to brown liquid
Vapor Density (air = 1)	5.63
Liquid Density (g/cc)	1.073 @ 25° C
Freezing Point (° C)	-50
Boiling Point (° C)	240
Vapor Pressure (mg/m^2)	0.07 @ 25° C
Volatility (mg/m^3)	610 @ 25° C
Flash Point	78° C
Decomposition Temperature (° C)	150
Heat of Vaporization (° C)	79.56
Odor	Faintly fruity, none when pure
Median Lethal Dosage (mg-min./m^3)	400 for resting male
Median Incapacitating Dosage (mg-min./m^3)	300 for resting male
Rate of Detoxification	Slight but definite
Eye and Skin Toxicity	Very high
Rate of Action	Very rapid
Physiological Action	Cessation of breath, death may follow
Protection Required	Protective mask and clothing
Stability	Stable in steel at ordinary temp.
Decontamination	Bleach slurry, dilute alkali, or DS2, steam and ammonia in confined area, M258 kit.
Means of Detection in Field	M256A and M18A2 kits
Typical Use	Quick action casualty agent

Figure 13.17 Agent GA Characteristics

Agent GB	
Agent Type	Nerve Agent
Common Name	Sarin
Chemical Name	(1-methylethyl) methylphosphonofluoride
Formula	$(CH_3)_2CHO\ (CH_3)FPO$
Code Name	GB
Molecular Weight	140.10
State at 20° C	Colorless liquid
Vapor Density (air = 1)	4.86
Liquid Density (g/cc)	1.0887 @ 25° C
Freezing Point (° C)	-56
Boiling Point (° C)	158
Vapor Pressure (mg/m^2)	2.9 @ 25° C
Volatility (mg/m^3)	22,000 @ 25° C
Flash Point	Nonflammable
Decomposition Temperature (° C)	150
Heat of Vaporization (° C)	80
Odor	Almost none when pure
Median Lethal Dosage ($mg\text{-}min./m^3$)	100 for resting male
Median Incapacitating Dosage ($mg\text{-}min./m^3$)	75 for resting male
Rate of Detoxification	None; cumulative toxin
Eye and Skin Toxicity	Very high
Rate of Action	Very rapid
Physiological Action	Cessation of breath and death may follow
Protection Required	Protective mask and clothing
Stability	Stable when pure
Decontamination	In confined area steam and ammonia; hot soapy water
Means of Detection in Field	M256A and M18A2 kits
Typical Use	Quick action casualty agent

Figure 13.18 Agent GB Characteristics

Agent GD	
Agent Type	Nerve Agent
Common Name	Soman
Chemical Name	1,2,2-trimethylpropyl methylphosphonofluoride
Formula	$(CH_3)_3CCH(CH_3)OPF(O)CH_3$
Code Name	GD
Molecular Weight	182.178
State at 20° C	Colorless liquid
Vapor Density (air = 1)	6.33
Liquid Density (g/cc)	1.0222 @ 22° C
Freezing Point (° C)	-42
Boiling Point (° C)	198
Vapor Pressure	0.92 mm Hg @ 20° C
Volatility (mg/m^3)	3,900 @ 25° C
Flash Point	High enough not to interfere with military use
Decomposition Temperature (° C)	130
Heat of Vaporization (° C)	72.4
Odor	Fruity, camphor odor when pure
Aqueous Solubility (grams per liter)	34 @ 0° C
Median Lethal Dosage (mg-min./m^3)	100 - 400 for resting male
Median Incapacitating Dosage (mg-min./m^3)	75 - 300 for resting male
Rate of Detoxification	Low ; essentially cumulative toxin
Eye and Skin Toxicity	Very high
Rate of Action	Very rapid
Physiological Action	Cessation of breath and death may follow
Protection Required	Protective mask and clothing
Stability	Stable when pure
Decontamination	Bleach slurry, dilute alkali, hot soapy water, M258 kit, Na_2CO_3, NaOH, or KOH
Means of Detection in Field	M256A and M18A2 kits
Typical Use	Quick action casualty agent

Figure 13.19 Agent GD Characteristics

Agent GF	
Agent Type	Nerve Agent
Common Name	Cyclosarin
Chemical Name	Cyclohexyl methylphosphonofluoride
Formula	(C_6H_{10}) CHO (CH_3)FPO
Code Name	GF
Molecular Weight	191.9
State at 20° C	Colorless liquid
Vapor Density (air = 1)	About 5
Liquid Density (g/cc)	1.06 @ 20° C
Freezing Point (° C)	--
Boiling Point (° C)	--
Vapor Pressure (mg/m^2)	@ 25° C
Volatility (mg/m3)	@ 25° C
Flash Point	--
Decomposition Temperature (° C)	--
Heat of Vaporization (° C)	--
Odor	Odorless
Median Lethal Dosage ($mg\text{-}min./m^3$)	--
Median Incapacitating Dosage ($mg\text{-}min./m^3$)	--
Rate of Detoxification	Low; essentially cumulative toxin
Eye and Skin Toxicity	Very high
Rate of Action	Rapid
Physiological Action	Produces casualties when inhaled or absorbed
Protection Required	Protective mask and clothing
Stability	Relatively stable at room temperature
Decontamination	STB slurry or DS2 solution; hot soapy water, M258 kit
Means of Detection in Field	M256A and M18A2 kits
Typical Use	Quick acting casualty agent

Figure 13.20 Agent GF Characteristics

Agent V	
Agent Type	Nerve Agent (Russian)
Common Name	V-Gas
Chemical Name	S-(2-(diethylamino)ethyl)-O-2-methylphosphonothioate
Formula	$C_{11}H_{26}NO_2PS$
Code Name	V-Gas
Molecular Weight	267.38
State at 20° C	Colorless liquid
Vapor Density (air = 1)	9.2
Liquid Density (g/cc)	1.0083 @ 20° C
Freezing Point (° C)	< -51
Boiling Point (° C)	298
Vapor Pressure (mg/m^2)	0.0007 @ 25° C
Volatility (mg/m3)	10.5 @ 25° C
Flash Point	159° C
Decomposition Temperature (° C)	Half-life of 36 hours @ 150° C
Heat of Vaporization (° C)	78.2 @ 25° C
Odor	Odorless
Median Lethal Dosage (mg-min./m^3)	100
Median Incapacitating Dosage (mg-min./m^3)	50
Rate of Detoxification	Low; essentially cumulative toxin
Eye and Skin Toxicity	Very high
Rate of Action	Rapid
Physiological Action	Produces casualties when inhaled or absorbed
Protection Required	Protective mask and clothing
Stability	Relatively stable at room temperature
Decontamination	STB slurry or DS2 solution; hot soapy water, M258 kit
Means of Detection in Field	M256A and M18A2 kits
Typical Use	Quick acting, persistent casualty agent

Figure 13.21 Agent V Characteristics

Agcnt VX	
Agent Type	Nerve Agent (U.S. and NATO)
Common Name	VX
Chemical Name	S-(2-(bis(1-methylethyl)amino)ethyl)-O-ethylmethylphosphonothioate
Formula	$C_{11}H_{26}NO_2PS$
Code Name	VX
Molcular Weight	267.38
State at 20° C	Colorless liquid
Vapor Density (air = 1)	9.2
Liquid Density (g/cc)	1.0083 @ 20° C
Freezing Point (° C)	< -51
Boiling Point (° C)	298
Vapor Pressure (mg/m^2)	0.0007 @ 25° C
Volatility (mg/m3)	10.5 @ 25° C
Flash Point	159° C
Decomposition Temperature (° C)	Half-life of 36 hours @ 150° C
Heat of Vaporization (° C)	78.2 @ 25° C
Odor	Odorless
Median Lethal Dosage (mg-min./m^3)	100
Median Incapacitating Dosage (mg-min./m^3)	50
Rate of Detoxification	Low; essentially cumulative toxin
Eye and Skin Toxicity	Very high
Rate of Action	Rapid
Physiological Action	Produces casualties when inhaled or absorbed
Protection Required	Protective mask and clothing
Stability	Relatively stable at room temperature
Decontamination	STB slurry or DS2 solution; hot soapy water, M258 kit
Means of Detection in Field	M256A and M18A2 kits
Typical Use	Quick acting, persistent casualty agent

Figure 13.22 Agent VX Characteristics

DETAILS OF COMMON TEAR GASES

Agent CN	
Agent Type	Tear Agent
Common Name	Tear gas
Chemical Name	-Chloroacetophenone
Formula	$C_6H_5COCH_2Cl$
Code Name	CN
Molecular Weight	154.59
State at 20° C	Solid
Vapor Density (air = 1)	5.3
Density (g/cc)	1.318 @ 20° C
Freezing Point (° C)	54
Boiling Point (° C)	248
Vapor Pressure (mg/m^2)	0.0041 @ 20° C
Volatility (mg/m^3)	343 @ 20° C
Flash Point	High enough not to interfere with military use
Decomposition Temperature (° C)	Stable to boiling point
Heat of Vaporization (° C)	98
Odor	Apple blossoms when diluted; choking
Median Lethal Dosage (mg-min./m^3)	11,000 - 14,000
Median Incapacitating Dosage (mg-min./m^3)	80
Rate of Detoxification	Rapid
Eye and Skin Toxicity	Temporary severe eye irritation; mild skin irritation
Rate of Action	Instantaneous
Physiological Action	Lachrymatory; irritates respiratory tract
Protection Required	Protective mask
Stability	Stable
Decontamination	Aeration in open; soda ash solution or alcoholic caustic soda in closed spaces
Means of Detection in Field	M-nitrobenzene and alkali in white-band tube or detector kit
Typical Use	Military training and riot control agent

Figure 13.23 Agent CN Characteristics

Agent CNC	
Agent Type	Tear Agent
Common Name	---
Chemical Name	Chloroacetophenone in chloroform
Formula	$C_6H_5COCH_2Cl$ in CHOCl
Symbol	CNC
Molecular Weight	128.17 on basis of components
State at 20° L	Liquid
Vapor Density (air = 1)	4.4
Liquid Density (g/cc)	1.40 @ 20° C
Freezing Point (° C)	0.23
Boiling Point (° C)	Variable 60 to 247
Vapor Pressure (mg/m^2)	127 @ 20° C
Volatility (mg/m^3)	Indeterminate
Flash Point	None
Decomposition Temperature (° C)	Stable to boiling point
Heat of Vaporization (° C)	Not applicable
Odor	Chloroform
Median Lethal Dosage (mg-min./m^3)	11,000 (estimated)
Median Incapacitating Dosage (mg-min./m^3)	80
Rate of Detoxification	Rapid
Eye and Skin Toxicity	Temporary severe eye irritation; mild skin irritation
Rate of Action	Instantaneous
Physiological Action	Lachrymatory, irritates respiratory tract
Protection Required	Protective mask
Stability	Adequate for riot-control use
Decontamination	Aeration in open; soda ash solution or alcoholic caustic soda in closed spaces
Means of Detection in Field	M-nitrobenzene and alkali in white-band tube or detector kit
Typical Use	Military training and riot-control agent

Figure 13.24 Agent CNC Characteristics

Agent CNS	
Agent Type	Tear Agent
Common Name	--
Chemical Name	Chloroacetophenone in chloroform
Formula	--
Code Name	CNS
Molecular Weight	141.78 on basis of components
State at 20° C	Liquid
Vapor Density (air = 1)	Approximately 5
Liquid Density (g/cc)	1.47 @ 20° C
Freezing Point (° C)	2
Boiling Point (° C)	Variable between 60 and 247
Vapor Pressure (mg/m^2)	78 @ 20° C
Volatility (mg/m^3)	610,000 @ 20° C (includes chloroform solvent)
Flash Point	None
Decomposition Temperature (° C)	Stable to boiling point
Heat of Vaporization (° C)	Not applicable
Odor	Flypaper-like; chloroform
Median Lethal Dosage ($mg\text{-}min./m^3$)	11,400
Median Incapacitating Dosage ($mg\text{-}min./m^3$)	60
Rate of Detoxification	Slow
Eye and Skin Toxicity	Irritating; not toxic
Rate of Action	Instantaneous
Physiological Action	Acts as vomiting and choking agent as well as tear agent
Protection Required	Protective mask
Stability	Adequate for tear gas use
Decontamination	None needed in field; hot solution of soda ash and sodium sulfite in enclosed spaces
Means of Detection in Field	CN test, as alkaline sulfite in blue-band tube of detector kits
Typical Use	Former military training and riot-control agent

Figure 13.25 Agent CNS Characteristics

Agent CNB	
Agent Type	Tear Agent
Common Name	---
Chemical Name	Chloroacetophenone in benzene and carbon tetrachloride
Formula	---
Code Name	CNB
Molecular Weight	Average 119.7 on basis of components
State at 20° C	Liquid
Vapor Density (air = 1)	Approximately 4
Liquid Density (g/cc)	1.14 @ 20° C
Freezing Point (° C)	-7 to -30
Boiling Point (° C)	Variable between 75 and 247
Vapor Pressure (mg/m^2)	Variable; mostly solvent vapors
Volatility (mg/m^3)	Indeterminate
Flash Point	< 4.44
Decomposition Temperature (° C)	> 247
Heat of Vaporization (° C)	Not applicable
Odor	Benzene
Median Lethal Dosage (mg-min./m^3)	11,000 (estimated)
Median Incapacitating Dosage (mg-min./m^3)	80
Rate of Detoxification	Rapid unless large amounts of solvent is inhaled
Eye and Skin Toxicity	Temporary severe eye irritation; mild skin irritation
Rate of Action	Instantaneous
Physiological Action	Powerfully lachrymatory
Protection Required	Protection mask
Stability	Adequate
Decontamination	Aeration in open; soda ash solution or alcoholic caustic soda in closed spaces
Means of Detection in Field	M-nitrobenzene and alkali in white-band tube of detector kit
Typical Use	Former military training and riot-control agent

Figure 13.26 Agent CNB Characteristics

Agent CA	
Agent Type	Tear Agent
Common Name	---
Chemical Name	Bromobenzyl cyanide
Formula	$BrC_6H_4CH_2CN$
Code Name	CA
Molecular Weight	196
State at 20° C	Liquid
Vapor Density (air = 1)	6.7
Liquid Density (g/cc)	1.47 @ 25° C
Freezing Point (° C)	25.5
Boiling Point (° C)	Decomposes at 242
Vapor Pressure (mg/m^2)	0.011 @ 20° C
Volatility (mg/m^3)	115 @ 20° C
Flash Point	None
Decomposition Temperature (° C)	60 to 242
Heat of Vaporization (° C)	55.7
Odor	Sour or decayed fruit
Median Lethal Dosage (mg-min./m^3)	8,000 to 11,000 (estimated)
Median Incapacitating Dosage (mg-min./m^3)	30
Rate of Detoxification	Rapid at low doses
Eye and Skin Toxicity	Irritating; not toxic
Rate of Action	Instantaneous
Physiological Action	Irritates eye and respiratory passages
Protection Required	Protective mask
Stability	Fairly stable in glass, lead, or enamel
Decontamination	20% alcoholic caustic
Means of Detection in Field	M-nitrobenzene and alkali in white-band tube of detector kit
Typical Use	Former military training and riot-control agent

Figure 13.27 Agent CA Characteristics

Agent CS	
Agent Type	Tear Agent
Common Name	---
Chemical Name	o-chlorobenzylmalononitrile
Formula	$ClC_6H_4CHC(CN)_2$
Code Name	CS
Molecular Weight	188.5
State at 20° C	Colorless solid
Vapor Density (air = 1)	---
Density (g/cc)	1.04 @ 20° C
Freezing Point (° C)	93 to 95
Boiling Point (° C)	310 to 315(with decomposition)
Vapor Pressure (mg/m^2)	---
Volatility (mg/m^3)	0.71 @ 25° C
Flash Point	197° C
Decomposition Temperature (° C)	---
Heat of Vaporization (° C)	53.6
Odor	Pepper-like
Median Lethal Dosage (mg-min./m^3)	61,000
Median Incapacitating Dosage (mg-min./m^3)	10 to 20
Rate of Detoxification	Rapid; sublethal in five to ten minutes
Eye and Skin Toxicity	Highly irritating; not toxic
Rate of Action	Instantaneous
Physiological Action	Highly irritating; but not toxic
Protection Required	Protective mask and clothing
Stability	Stable
Decontamination	Water, 5% sodium bisulfite, and water rinse
Means of Detection in Field	None
Typical Use	Military training and riot-control agent

Figure 13.28 Agent CS Characteristics

Agent OC	
Agent Type	Tear Agent
Common Name	Pepper spray; Oleoresin capsicum
Chemical Names	N-[(4-Hydroxy-3-methoxyphenyl)methyl]-8-methyl-6-nonenamide
Formula	$C_{18}H_{27}NO_3$
Code Name	OC
Molecular Weight	305.42
State at 20° C	Solid
Vapor Density (air = 1)	---
Density (g/cc)	Around 1.7 @ 20° C
Freezing Point (° C)	65
Boiling Point (° C)	210
Vapor Pressure (mg/m^2)	--
Volatility (mg/m^3)	-
Flash Point	--
Decomposition Temperature (° C)	--
Heat of Vaporization (° C)	--
Odor	Red pepper-like
Median Lethal Dosage ($mg\text{-}min./m^3$)	Probable oral Lethal Dose 0.5 -5 g/kg
Median Incapacitating Dosage ($mg\text{-}min./m^3$)	--
Rate of Detoxification	Rapid
Eye and Skin Toxicity	Highly irritating; not toxic
Rate of Action	Instantaneous
Physiological Action	Highly irritating; but low toxicity
Protection Required	Protective mask and clothing
Stability	Stable
Decontamination	Water, 5% sodium bisulfite, and water rinse
Means of Detection in Field	None
Typical Use	Riot-control agent

Figure 13.29 Agent OC Characteristics

Agent PS	
Agent Type	Tear Agent
Common Name	---
Chemical Name	Chloropicrin
Formula	CCl_3NO_2
Code Name	PS
Molecular Weight	164.39
State at 20° C	Slightly oily liquid with intense odor
Vapor Density (air = 1)	5.7
Density (g/cc)	1.656 @ 20° C
Freezing Point (° C)	Between -64 and -67
Boiling Point (° C)	112
Vapor Pressure (mm Hg)	16.91 @ 20° C
Volatility (mg/m^3)	-
Flash Point	Non-flammable
Decomposition Temperature (° C)	--
Heat of Vaporization (° C)	--
Odor	Pepper-like
Water Solubility (grams per liter)	2.272 at 0° C; 1.621 at 25° C
Median Lethal Dosage (mg/m^3)	2000 (in 10 minutes)
Median Incapacitating Dosage (mg/m^3)	Around 9
Rate of Detoxification	Rapid
Eye and Skin Toxicity	Highly irritating
Rate of Action	Instantaneous
Physiological Action	Highly irritating; but not toxic at low concentrations
Protection Required	Protective mask and clothing
Stability	Stable
Decontamination	Water, 5% sodium bisulfite, and water rinse
Means of Detection in Field	None
Typical Use	Riot-control agent

Figure 13.30 Agent PS Characteristics

DETAILS OF COMMON VOMITING AGENTS

Agent DA	
Agent Type	Vomiting Agent
Common Name	---
Chemical Name	Diphenylchloroarsine
Formula	$(C_6H_5)_2AsCl$
Code Name	DA
Molecular Weight	264.5
State at 20° C	White to brown solid
Vapor Density (air = 1)	Forms little amount of vapor
Liquid Density (g/cc)	1.387 @ 50° C
Freezing Point (° C)	41 to 44.5
Boiling Point (° C)	333
Vapor Pressure (mg/m^2)	0.0036 @ 45° C
Volatility (mg/m^3)	48 @ 45° C
Flash Point	350° C
Decomposition Temperature (° C)	300
Heat of Vaporization (° C)	56.6
Odor	None
Median Lethal Dosage (mg-min./m^3)	15,000 estimated
Median Incapacitating Dosage (mg-min./m^3)	12 over a ten-minute period
Rate of Detoxification	Rapid
Eye and Skin	Toxicity irritating; not toxic
Rate of Action	Very rapid
Physiological Action	Cold-like symptoms, plus headache, vomiting, nausea
Protection Required	Protective mask
Stability	Stable if pure
Decontamination	None needed in field, caustic soda or chlorine in enclosed spaces
Means of Detection in Field	None
Typical Use	Former military training and riot-control agent

Figure 13.31 Agent DA Characteristics

Agent DM	
Agent Type	Vomiting Agent
Common Name	Adamsite
Chemical Name	10-chloro-5,10-dihydrophenarsazine
Formula	$C_6H_4(AsCl)(NH)C_6H_4$
Code Name	DM
Molecular Weight	277.57
State at 20° C	Yellow to green solid
Vapor Density (air = 1)	Forms very little vapor
Liquid Density (g/cc)	1.65 (solid) @ 20° C
Freezing Point (° C)	195
Boiling Point (° C)	410
Vapor Pressure (mg/m^2)	Negligible
Volatility (mg/m^3)	Negligible
Flash Point	None
Decomposition Temperature (° C)	Above boiling point
Heat of Vaporization (° C)	80
Odor	None
Median Lethal Dosage (mg-min./m^3)	15,000
Median Incapacitating Dosage (mg-min./m^3)	22 for 1-minute exposure; 8 for 60-minute exposure
Rate of Detoxification	Rapid for low doses
Eye and Skin Toxicity	Irritating; relatively nontoxic
Rate of Action	Very rapid
Physiological Action	Cold-like symptoms, plus headache, vomiting, nausea
Protection Required	Protective mask
Stability	Stable in glass or steel
Decontamination	None needed in field; bleach or DS2 in confined spaces
Means of Detection in Field	None
Typical Use	Former military training and riot-control agent

Figure 13.32 Agent DM Characteristics

Agent DC	
Agent Type	Vomiting Agent
Common Name	---
Chemical Name	Diphenylcyanoarsine
Formula	$(C_6H_5)_2AsCN$
Code Name	DC
Molecular Weight	255.0
State at 20° C	White to pink solid
Vapor Density (air = 1)	Forms very little vapor
Liquid Density (g/cc)	1.3338 @ 35° C
Freezing Point (° C)	31.5 to 35
Boiling Point (° C)	350
Vapor Pressure (mg/m^2)	0.0002 @ 20° C
Volatility (mg/m^3)	2.8 @ 20° C
Flash Point	Low
Decomposition Temperature (° C)	300 (25% decomposes)
Heat of Vaporization (° C)	71.1
Odor	Bitter almond-garlic mixture
Median Lethal Dosage (mg-min./m^3)	10,000 (estimated)
Median Incapacitating Dosage (mg-min./m^3)	30 for 30-second exposure; 20 for 5-minute exposure
Rate of Detoxification	Rapid
Eye and Skin Toxicity	Irritating; not toxic
Rate of Action	More rapid than DM or DA
Physiological Action	Cold-like symptoms, plus headache, vomiting, nausea
Protection Required	Protective mask
Stability	Stable
Decontamination	None needed in field, alkali solution or DS2 in closed spaces
Means of Detection in Field	None
Typical Use	Former training and riot control agent

Figure 13.33 Agent DC Characteristics

DETAILS OF COMMON INCAPACITATING AGENTS

Agent BZ	
Agent Type	Incapacitating Agent
Common Name	BZ
Chemical Name	α-Hydroxy-α-phenylbenzeneacetic acid, 1-azabicyclo[2,2,2]-oct-3-yl ester
Formula	$C_{21}H_{23}NO_3$
Code Name	BZ
Molecular Weight	337.4
State at 20° C	Crystalline solid
Vapor Density (air = 1)	11.6
Density (g/cc)	Bulk 0.51
Freezing Point (° C)	167.5
Boiling Point (° C)	412
Vapor Pressure (mg/m^2)	0.03 @ 70° C
Volatility (mg/m^3)	0.5 @ 70° C
Flash Point	246° C
Decomposition Temperature (° C)	Begins at 170
Heat of Vaporization (° C)	62.9
Odor	Odorless
Median Lethal Dosage (mg-min./m^3)	---
Median Incapacitating Dosage (mg-min./m^3)	---
Rate of Detoxification	---
Eye and Skin Toxicity	---
Rate of Action	Delayed from one to four hours
Physiological Action	Fast heartbeat; dizziness, vomiting, dry mouth, blurred vision, stupor, increased random activity
Protection Required	Protective mask
Stability	Adequate for military use
Decontamination	Wash with soap and water; shake or brush; hypochlorite or caustic alcoholic solutions; detergent vetting solutions
Means of Detection in Field	None
Typical Use	Former delayed action and temporarily incapacitating agent

Figure 13.34 Agent BZ Characteristics

LSD	
Agent Type	Incapacitating Agent
Common Name	Acid
Chemical Name	9,10-didehydro-N,N-diethyl-6- methyl-8b-ergoline-8-carboxamide; lysergic acid diethylamide
Formula	$C_{20}H_{25}N_3O$
Code Name	--
Molecular Weight	323.42
State at 20° C	Crystalline solid
Vapor Density (air = 1)	--
Density (g/cc)	--
Freezing Point (° C)	Between 80 - 85
Boiling Point (° C)	--
Vapor Pressure (mg/m^2)	Very low @ 70° C
Volatility (mg/m^3)	-
Flash Point	--
Decomposition Temperature (° C)	--
Heat of Vaporization (° C)	--
Odor	Odorless
Median Lethal Dosage (mg-min./m^3)	--
Median Incapacitating Dosage (mg-min./m^3)	--
Rate of Detoxification	--
Eye and Skin Toxicity	--
Rate of Action	Delayed from one to four hours
Physiological Action	Fast heartbeat; dizziness, vomiting, dry mouth, blurred vision, stupor, increased random activity
Protection Required	Protective mask
Stability	Adequate for military use
Decontamination	Wash with soap and water; shake or brush; hypochlorite or caustic alcoholic solutions; detergent vetting solutions
Means of Detection in Field	None
Typical Use	Former delayed action and temporarily incapacitating agent

Figure 13.35 LSD Characteristics

DETAILS OF COMMON MILITARY HERBICIDES

Agent Blue	
Agent Type	Herbicide
Common Name	Cacodylic acid
Chemical Name	Dimethylarsinic acid
Formula	$C_2H_7AsO_2$
Code Name	Blue
Molecular Weight	138.0
State at 20° C	White solid; water solutions colored blue
Vapor Density (air = 1)	--
Density (g/cc)	--
Freezing Point (° C)	195
Boiling Point (° C)	200
Vapor Pressure (mg/m^2)	Very low @ 70° C
Volatility (mg/m^3)	Very low @ 70° C
Flash Point	--
Decomposition Temperature (° C)	--
Heat of Vaporization (° C)	--
Odor	Odorless
Median Lethal Dosage	In rats, 1350 mg/kg.
Median Incapacitating Dosage (mg-min./m^3)	Not applicable; usually applied at 3 - 8 kg/ha.
Rate of Detoxification	---
Eye and Skin Toxicity	---
Rate of Action	Rapid
Physiological Action	Affects plant metabolism
Protection Required	Protective mask
Stability	Adequate for military use
Decontamination	Wash with soap and water
Means of Detection in Field	None
Typical Use	Military defoliant

Figure 13.36 Agent Blue Characteristics

Agent Green	
Agent Type	Herbicide
Common Name	2,4,5-T
Chemical Name	2,4,5-trichlorophenoxyacetic acid and esters
Formula	$C_8H_5Cl_3O_3$ (2,4,5-T)
Code Name	Green
Molecular Weight	255.49 (2,4,5-T)
State at 20° C	Colorless to tan solid; solutions may be green
Vapor Density (air = 1)	--
Density (g/cc)	1.80 @ 20° C
Freezing Point (° C)	153
Boiling Point (° C)	Decomposes
Vapor Pressure	Less than 0.01 mPa @ 20° C
Volatility (mg/m^3)	@ 70° C
Flash Point	--
Decomposition Temperature (° C)	--
Heat of Vaporization (° C)	--
Odor	Odorless
Median Lethal Dosage (mg-min./m^3)	In rats, 500 mg/kg.
Median Incapacitating Dosage (mg-min./m^3)	Not applicable
Rate of Detoxification	---
Eye and Skin Toxicity	---
Rate of Action	Moderate
Physiological Action	Alters plant metabolism (growth accelerator)
Protection Required	Protective mask
Stability	Adequate for military use
Decontamination	Wash with soap and water
Means of Detection in Field	None
Typical Use	Military defoliant

Figure 13.37 Agent Green Characteristics

Agent Orange	
Agent Type	Herbicide
Common Name	Mixture of 2,4-D and 2,4,5-T
Chemical Name	2,4-dichlorophenoxyacetic acid and esters; and 2,4,5-trichlorophenoxyacetic acid and esters
Formula	---
Code Name	Orange
Molecular Weight	--
State at 20° C	Colorless solid; water solutions may be orange
Vapor Density (air = 1)	--
Density (g/cc)	Around 1.8
Freezing Point (° C)	Depends upon formulation
Boiling Point (° C)	Depends upon formulation
Vapor Pressure	Less than 0.01 mPa @ 20° C
Volatility (mg/m^3)	Very low @ 70° C
Flash Point	--
Decomposition Temperature (° C)	--
Heat of Vaporization (° C)	--
Odor	Odorless
Median Lethal Dosage (mg-min./m^3)	In rats, around 400 mg/kg active ingredient
Median Incapacitating Dosage (mg-min./m^3)	Not applicable; applied at 15 - 50 kg/ha of active ingredient
Rate of Detoxification	---
Eye and Skin Toxicity	---
Rate of Action	Moderate
Physiological Action	Metabolic disrupter in plants (growth accelerator)
Protection Required	Protective mask
Stability	Adequate for military use
Decontamination	Wash with soap and water
Means of Detection in Field	None
Typical Use	Former military defoliant

Figure 13.38 Agent Orange Characteristics

DETAILS OF COMMON BATTLEFIELD SMOKES AND OBSCURANTS

Fog Oil	
Agent Type	Smoke
Common Name	Fog oil; White oil
Chemical Name	Mixture of short-chain hydrocarbons
Formula	--
Code Name	--
Molecular Weight	Variable
State at 20° C	Oily Liquid
Vapor Density (air = 1)	--
Density (g/cc)	Around 0.8 (light) to 0.9 (heavy)
Freezing Point (° C)	--
Boiling Point (° C)	Around 360
Vapor Pressure (mg/m^2)	@ 70° C
Volatility (mg/m^3)	@ 70° C
Flash Point (Closed Cup)	135° C
Decomposition Temperature (° C)	--
Heat of Vaporization (° C)	--
Odor	Odorless
Median Lethal Dosage (mg-min./m^3)	--
Median Incapacitating Dosage (mg-min./m^3)	--
Rate of Detoxification	--
Eye and Skin Toxicity	Low
Rate of Action	Not applicable
Physiological Action	Not applicable
Protection Required	Protective mask
Stability	Adequate for military use
Decontamination	Wash with soap and water
Means of Detection in Field	None
Typical Use	Military smoke screen

Figure 13.39 Fog Oil Characteristics

Agent RP	
Agent Type	Smoke
Common Name	Red phosphorus
Chemical Name	Phosphorus
Formula	P_6
Code Name	RP
Molecular Weight	123.9
State at 20° C	Rcd solid
Vapor Density (air = 1)	4.77
Density (g/cc)	1.85
Freezing Point (° C)	44
Boiling Point (° C)	280
Vapor Pressure (mg/m^2)	0.026 @ 20° C
Volatility (mg/m^3)	@ 70° C
Flash Point (° C)	Autoignition at 260
Decomposition Temperature (° C)	--
Heat of Vaporization (@ 20° C)	49.8 X 10^6 J/mole
Odor	Slight odor like sulfur oxides or garlic
Median Lethal Dosage (mg-min./m^3)	--
Median Incapacitating Dosage (mg-min./m^3)	--
Rate of Detoxification	--
Eye and Skin Toxicity	--
Rate of Action	--
Physiological Action	--
Protection Required	Protective mask
Stability	Adequate for military use
Decontamination	Wash with soap and water
Means of Detection in Field	None
Typical Use	Military battlefield obscurant (typically vehicles)

Figure 13.40 Agent RP Characteristics

Agent WP	
Agent Type	Smoke
Common Name	White Phosphorus
Chemical Name	Phosphorus
Formula	P
Code Name	WP
Molecular Weight	123.90
State at 20° C	White solid
Vapor Density (air = 1)	4.42
Density (g/cc)	1.85
Freezing Point (° C)	44
Boiling Point (° C)	280
Vapor Pressure (mm Hg)	0.026 @ 20° C
Volatility (mg/m^3)	@ 70° C
Flash Point (° C)(Open Cup)	260; autoignition at 30° C
Decomposition Temperature (° C)	800
Heat of Vaporization (at 20° C)	49.8×10^6 J/mole
Odor	Slight odor like sulfur oxides or garlic
Median Lethal Dosage ($mg\text{-}min./m^3$)	--
Median Incapacitating Dosage ($mg\text{-}min./m^3$)	--
Rate of Detoxification	--
Eye and Skin Toxicity	--
Rate of Action	Rapid
Physiological Action	Burning of skin, eyes, and mucous membranes
Protection Required	Protective mask
Stability	Adequate for military use
Decontamination	Wash with soap and water
Means of Detection in Field	None
Typical Use	Battlefield obscurant

Figure 13.41 Agent WP Characteristics

Graphite Fibers	
Agent Type	Smoke
Common Name	Graphite
Chemical Name	Graphite
Formula	C
Code Name	---
Molecular Weight	12.01
State at 20° C	Black solid
Vapor Density (air = 1)	Non volatile
Density (g/cc)	2.25
Freezing Point (° C)	3550
Boiling Point (° C)	Above 4827
Vapor Pressure (mg/m^2)	@ 70° C
Volatility (mg/m^3)	@ 70° C
Flash Point	Autoignition temperature in flowing air of 452 - 518° C
Decomposition Temperature (° C)	--
Heat of Vaporization (° C)	--
Odor	Odorless
Median Lethal Dosage (mg-min./m^3)	--
Median Incapacitating Dosage (mg-min./m^3)	--
Rate of Detoxification	--
Eye and Skin Toxicity	--
Rate of Action	Not applicable
Physiological Action	Essentially none
Protection Required	Protective mask
Stability	Adequate for military use
Decontamination	Wash with soap and water
Means of Detection in Field	None
Typical Use	Battlefield obscurant

Figure 13.42 Graphite Fibers Characteristics

SUMMARY

There is a large number of other chemicals, many of these common industrial materials, which could be used by terrorists or saboteurs. The so-called opportunistic chemicals include pesticides, chlorine, ethylene and propylene oxides, nitriles, solvents, fertilizers, ethers, *etc*. The types are too many to list here. Explosives have not been included since these are not typically considered as chemical agents. Explosive materials include RDX, trinitrotoluene (TNT), nitroglycerin, ammonium perchlorate, a number of other nitrites and nitrates, hydroxyammonium nitrate, and many, many others. The chemicals listed above are most of the military-relevant chemical agents that have been, or could be, used in war or a terrorist action.

14

DELIVERY SYSTEMS FOR CHEMICAL WEAPONS

TYPES OF CHEMICAL WEAPON DELIVERY SYSTEMS

Chemical warfare agents in the U.S. stockpile have typically been stored either in the chemical weapons themselves or in bulk storage containers, often one-ton cylinders. Chemical weapons were generally designed with two objectives in mind: compatibility with existing weapons systems, *i.e.*, chemical artillery shells could have been fired from the same guns as conventional artillery shells, chemical bombs could have been dropped from the same airplanes as conventional bombs, *etc.*, and efficient generation of an aerosol of chemical weapons agent at the target

Artillery Shells

Chemical artillery shells or projectiles typically contain a reservoir filled with the chemical agent surrounding an internal axial explosive charge called the burster. The nose of most such shells has a lifting lug screwed in. In order to fire the shell, the lifting lug is replaced with a fuse; in some cases the fuse is permanently affixed to the nose of the shell. Once the shell is fired, the fuse is armed. When the shell arrives at its target, the fuse detonates the explosive charge. The explosive charge must be such a nature that there is not enough heat generated to destroy the agent. The burster shatters the shell casing and causes the chemical agent to be dispersed as an aerosol. In the past, the U.S. used several munitions that have the same configuration and use identical shell bodies, bursters, and fuses as chemical weapons, but contain different fills of flame or smoke-generating chemicals.

Land Mines

Chemical agent land mines, such as the U.S. M23 unit, usually contain a reservoir filled with chemical agent surrounding an explosive charge, similar to the artillery shell. The top of the land mine has a pressure plate; when the pressure plate is depressed, the fuse detonates the explosive charge, or burster. The burster shatters the mine casing, blows away any soil covering the mine, and causes the chemical agent to disperse as an aerosol. Unlike biological agents, most chemical agents can remain effective in land mines buried in the ground for extended periods of time.

Rockets

A typical chemical agent rocket, like the former U.S. M55 4½-inch unit, includes, in order from the rear forward, a set of spring-loaded guidance fins, an M28 double-based rocket propellant, a reservoir filled with chemical agent surrounding an axial explosive charge, and a fuse. Most systems used a launch tube to eject the rocket. When the rocket was fired, the propellant ignited and thrust the rocket out of the firing tube. The guidance fins extended and the rocket flew to its target, while the fuse was armed in flight. When the rocket reached its target, the fuse detonated the axial explosive charge, or burster. The burster shattered the rocket casing and caused the chemical agent to disperse as an aerosol. The M55 rocket was probably the most hazardous chemical weapon in the U.S. stockpile from a safety standpoint because the manufacturing process, coupled with an aging of the inventory of these weapons, effectively prevented the separation of the propellant and explosive from the chemical agent fills; this complicated the destruction of rockets as compared to all other chemical munitions. As the rocket propellant ages, it can become unstable; in fact, all M55 chemical rockets today contain 40 year-old propellant. In addition, the rocket was made of light aluminum rather than the heavy gauge steel of other chemical weapons. Sarin-containing rockets posed a particular hazard because the small amount of fluorides ion in the aging Sarin increased the corrosion of the aluminum body of the rocket. Sarin-containing M55 rockets account for the largest proportion of leaking chemical munitions in the U.S. stockpile. The M55 rockets are destroyed by carefully sawing the rocket body in half, pouring out the chemical agent, which is sent to neutralization facilities, while the explosive and propellant are burned in an explosives furnace. The M55 rockets are in the process of being destroyed, or demilitarized, but it appears the U.S. will not meet the 2007 deadline for complete destruction.

Bombs

Chemical bombs usually contain a reservoir filled with the chemical agent surrounding an internal axial explosive charge. Chemical bombs are generally stored without the tail fins and the detonating fuse. These items are attached immediately before the bomb is loaded onto an aircraft. After the bomb is dropped, the fuse is armed. When the bomb arrives on its target, the fuse detonates the axial explosive charge, or burster. The burster shatters the bomb casing and causes the chemical agent to disperse as an aerosol.

BULK STORAGE ITEMS

Chemical agents that are not stored in weapons are stored in bulk containers. These containers are similar to containers in general use in the chemical industry, for water treatment plants using chlorine. The most common device is the so-called "ton unit," a steel pressure cylinder capable of holding about 2,000 pounds of chemical agent. This looks similar to a horizontal propane gas cylinder with protective flanges at each end. It is probably the least hazardous way to store chemical agent because it does not contain any explosive or energetic components and is made with thick walls.

DISPERSAL OF CHEMICAL AGENTS

When a chemical weapon detonates, it creates a primary cloud of either solid or liquid aerosols. This cloud then settles to the ground, depending upon characteristics of the aerosol and weather conditions, eventually landing on individuals, plants, equipment, and the ground. The contamination on the ground has a finite lifetime, but can injure people from direct contact or from contact with a secondary cloud of agent that results from evaporation of the ground contamination. Eventually, the agent contamination disappears as the agent decomposes or is diluted below toxic levels by physical action.

Factors that affect the hazard from the primary cloud are related to the local weather conditions. Factors which tend to diminish the hazard of the primary cloud include variable wind directions, which causes dilution by redirection of the cloud; presence of wind velocities over six meters per second (about 13 miles per hour), which causes dilution by turbulence; presence of unstable air, which also causes dilution by turbulence; temperatures below 0° C (32° F), which result in less evaporation from the liquid or solid aerosol particles so aerosol particles settle to the ground more quickly than the agent vapor; and precipitation, which hydrolyzes the chemical and washes out both aerosol particles and vapor from the atmosphere. Factors which tend to increase the hazard of the primary cloud are generally the opposites of those previously cited: steady wind direction; wind velocity under 3 meters per second (about 6 miles per hour); stable air (atmospheric inversion); temperatures above 20° C (68° F); and no precipitation. For a military operation, with the goal to cause as much damage as possible, these latter conditions are considered the most optimal.

Once the primary cloud settles to the ground, factors that tend to decrease the hazard due to ground contamination include high ground temperature, which causes decomposition of some agents; high wind velocity, which dilutes the agent; unstable air, which dilutes the agent; and heavy precipitation, which dilutes, hydrolyzes, and washes the agent into the soil. Factors that tend to increase the hazard of ground contamination are related to the stability of the atmosphere. Factors that characterize stable air include low wind velocities and temperature inversions.

Persistency is the term used to describe the duration of an area's toxic contamination. Persistent agents are generally considered to be those for which contamination lasts a day or more; these are of low volatility. Non-persistent agents are generally considered to be those where contamination dissipates in a matter of minutes or hours; these are volatile. In some cases, thickeners are added to non-persistent agents to extend their persistence. Eventually, agent contamination will disappear as the agents decompose or are diluted below toxic levels by physical action. Chemical reactions that affect the persistence of agents in the environment include hydrolysis by water in the environment, photochemical reactions from sunlight (especially UV radiation), thermochemical decomposition, and other reactions with compounds present in the environment. Physical transformations of chemical agents are dominated by volatilization, which is a function of the vapor pressure of the agent.

To calculate the precise lifetime of a chemical warfare agent in the field can be very complex. This calculation requires knowledge of the parameters of the agent, plus climate information for the specific location, including temperature, humidity, rainfall, and solar flux, among others. Nevertheless, assumptions can be made for average values for humidity, rainfall, and solar flux, yielding approximate lifetimes that indicate the persistency under generalized climatic conditions.

EXPOSURE VERSUS DOSE

Most measurements of chemical agents involve the amount of aerosol and vapor in the air. If there is a person at that location, this would be the person's exposure to the agent. However, the amount actually taken in by the person can be less than the exposure. The chance that a particular dose will be received by a person when the air contains a toxic substance will be proportional to its concentration in the air and to the length of time of the exposure. The product of concentration and time is referred to as the concentration $\times$ time (Ct). If the agent concentration is expressed in mg/m^3 and time is in minutes, then the Ct is expressed in mg-min/m^3. The Ct is a measure of the intensity of exposure and not of the dose that actually penetrates the body.

TOXICITY, DOSE, AND EFFECTS

The LD_{50} (Lethal Dose 50) of a chemical is the dose that will kill 50% of the subjects receiving it. It may also be defined as the dose that has a 50% probability of killing any particular individual. Correspondingly, the ED_{50} (Effective Dose 50) is the dose that has a 50% probability of producing any defined effect (usually incapacitation), not necessarily death. The ED_5 indicates the dose likely to cause an effect in 5% of those exposed and the ED_{95} is the dose likely to affect 95% of exposed persons. Neither the LD_{50} nor the ED_{50} gives all the information needed to understand the relationship between dose and effect. The additional information needed is provided by the slope of the probit line, as discussed in Chapter One. Most chemical agents generally don't travel very far, but they usually require only low concentration to cause high fatality levels.

SUMMARY

While there are many chemical agents, ranging from blood agents to vesicants to nerve agents, capable of causing serious harmful and even fatal effects, delivery and dissemination of many of these agents is difficult. The requirements for an effective chemical agent include remaining unaffected by possible detonation or mechanical dispersion, being difficult to identify or treat, being persistent in the environment, and being capable of causing incapacitation or lethality at relatively low concentrations.

15

CHEMICAL AGENT SAFETY, PROTECTION, DETECTION, AND DECONTAMINATION

BINARY CHEMICAL WEAPONS

Due to concerns about the eventual demilitarization of chemical agent-filled munitions, it was decided that it would be simpler to destroy agents if the final, most toxic form had not been made. In addition, there was some concern that aging chemical agents might be less effective when used. Therefore, the U.S. decided to manufacture chemical weapons in which two or more components were not mixed until just before use. In theory, during the flight of a rocket, artillery shell, or bomb, a small burster would allow for two chemicals to mix. The final agent would then be fresh. If the weapon were never to be used, the less toxic components could be drained out and destroyed. Binary chemical weapons mix two separate, relatively non-toxic chemicals in flight to create a toxic chemical agent.

Agent GB Binary

Methylphosphonyl difluoride (Agent DF) is initially located in one canister, while a mixture of isopropyl alcohol and isopropyl amine (Agent OPA) is located in a separate canister. When the weapon is fired or otherwise delivered, a disk between the canisters ruptures, and the two components react in flight to produce Agent GB.

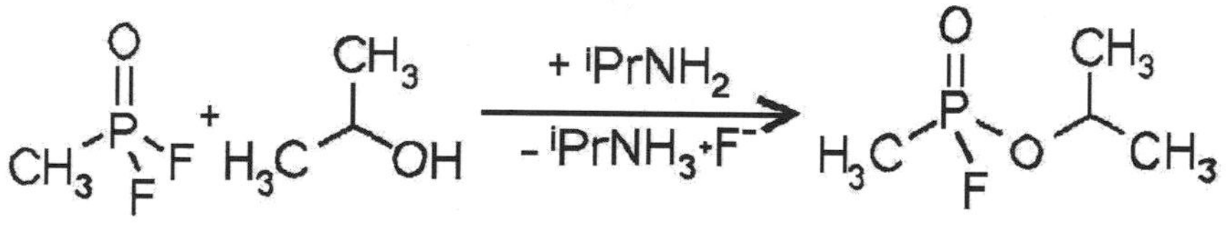

Figure 15.1 Agent GB Binary Reaction

Agent GD Binary

Methylphosphonyl difluoride (Agent DF) is initially located in one canister, while a mixture of pinacolyl alcohol and an amine is located in a separate canister. When the weapon is fired or otherwise delivered, a disk between the canisters ruptures, and the two components react in flight to produce GD.

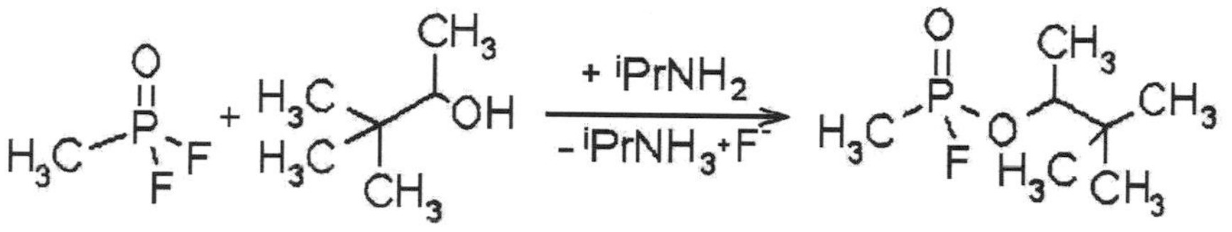

Figure 15.2 Agent GD Binary Reaction

Agent VX Binary

O-Ethyl-O-2-diisopropylaminoethyl methylphosphonite (Agent QL) is initially located in one canister, while elemental sulfur is located in a separate canister. When the weapon is fired or otherwise delivered, a disk between the canisters ruptures, and the two components react in flight to produce VX.

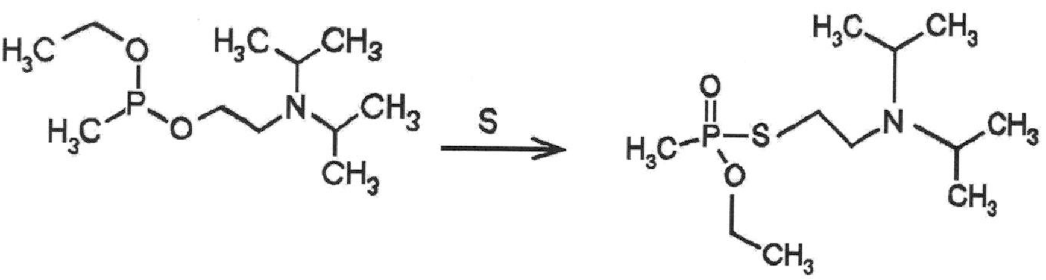

Figure 15.3 Agent VX Binary Reaction

DECONTAMINATION OF PERSONNEL

If the skin becomes contaminated with a chemical warfare agent, it is critical to remove the agent as quickly as possible. The first step is to flush the eyes with copious amounts of *clean*, that is agent uncontaminated, water. The next step is to decontaminate the skin. The U.S. military has two types of technology for this purpose.

M258A1 kits contain two sets of wipes, DECON-1 wipes wetted with 72% ethanol, 10% phenol, 5% sodium hydroxide, 0.2% ammonia, and 12.8% water are used first, followed by the DECON-2 wipes impregnated with chloramine-B and sealed glass ampoules filled with 45% ethanol, 5% zinc chloride and the 50% water. The ampoule is crushed into the wipe, which is then used to clean the affected area.

The newer M291 kit contains six packets, each of which contains an applicator bag filled with Ambergard® J XE-555 decontaminant resin. The dry resin is rubbed over the contaminated area and the resin absorbs any agent. If no decontamination kit is available, the use of soap and water is the best alternative, taking care that the runoff and rinse water may still contain some agent.

DECONTAMINATION OF EQUIPMENT

All types of equipment, from vehicles to weapons, must be decontaminated. For some of the more volatile chemical agents, only aeration and ventilation are required for decontamination. For liquid agents with lower vapor pressures, treatment with decontamination solution is required. In World War I and World War II, chlorine-containing bleaches were commonly used as decontamination solutions, and are still today. Most chemical agents can be rendered less harmful by the use of oxidizing agents (*e.g.*, bleach) and alkaline agents (*e.g.*, sodium or ammonium hydroxide). Examples of these include:

- 2–6% aqueous sodium hypochlorite solution (household bleach); in the event that there are no other kits available, the use of household bleach right out of the bottle is acceptable
- 7% aqueous slurry of calcium hypochlorite (HTH®, commonly used to disinfect swimming pools)
- 7–70% aqueous slurries of calcium hypochlorite and calcium oxide, known as supertropical bleach (STB)
- solid mixture of calcium hypochlorite and magnesium oxide (Dutch powder)
- 0.5% aqueous calcium hypochlorite buffered with sodium dihydrogen phosphate and detergent (ASH)
- 0.5% aqueous calcium hypochlorite buffered with sodium citrate/citric acid and detergent (SLASH)

- British anti-Lewisite (dimercaprol; 2,3-dimercaptopropanol; BAL), used in World War II to neutralize mustard agents; the same material can be used today. BAL can also be taken internally as an effective chelating agent for heavy metals (*e.g.*, lead, uranium)

Other decontamination solutions, which may be used if others are not available, include:

- aqueous solutions of sodium hydroxide (NaOH) or potassium hydroxide (KOH)
- diluted alkali solutions, including ammonia (NH_4OH)
- steam and ammonia or hot, soapy water for confined areas
- fire or extreme heat which can be used to decontaminate heat resistant solid surfaces

Some of the systems currently used by the U.S. military for equipment decontamination include:

- The M280 kit for use on personal equipment (individual weapons, radios, *etc.*) uses the same materials as the M258A1 personal decontamination kit: a set of towelettes wetted with 72% ethanol, 10% phenol, 5% sodium hydroxide, 0.2% ammonia, and 12.8% water, which are used first, followed by a second set of towelettes impregnated with chloramine B and wetted with 45% ethanol, 5% zinc chloride and the 50% water.
- The M13 portable decontamination apparatus, for use on vehicles and equipment, uses decontamination solution number 2 (DS2): 70% diethylenetriamine, 28% ethylene glycol, monomethyl ether, and 2% sodium hydroxide. DS2 is less corrosive to metals than are the bleach-based decontamination solutions, but it can damage painted surfaces and plastics (while they are both decontaminants, STB and DS-2 should never be mixed together since heat is generated and a fire can result).
- The M295 Individual Equipment Decontamination Kit uses the Ambergard® J XE-555 decontaminant sorptive resin as the M291 kit for personal use.
- The M17 transportable decontamination system for equipment, vehicles, and personnel is capable of delivering water at temperatures up to 120° C (248° F) and pressures up to 689 kPa (100 pounds per square inch (psi)).

The U.S.S.R. developed a wheeled vehicle having a modified jet engine mounted on the truck bed. Fed into the engine was a solution of hypochlorite and sodium hydroxide. Tanks and other vehicles could drive through the washing solution and be decontaminated very quickly. The U.S. Army is studying a similar system.

CHEMICAL AGENT DETECTION, VERIFICATION, AND IDENTIFICATION

The ideal technology for detecting the presence of a chemical agent, verifying that an agent is present, and uniquely identifying that agent would accomplish all three sufficiently quickly for threatened personnel to put on appropriate protective equipment before being affected by the agent. Unfortunately, the current state of technology is such that all three tasks cannot be accomplished within that relatively short length of time. Therefore, the challenge is usually broken down into several components: individual detection, point detection, and subsequent identification and verification.

DETECTION OF CHEMICAL WARFARE AGENTS BY INDIVIDUAL SOLDIERS

Individual detection technologies provide the first warning of the presence of a chemical agent. These technologies must be simple, light, and portable, so that they can be issued to many individuals, and fast enough to give an immediate warning of the presence of a chemical agent. The first warning of a chemical attack on the battlefield is often direct observation by well-trained troops of the presence of agent clouds or liquid fallout, as well as characteristic odors of certain of the chemical warfare agents. During World War I, the smell of mustard or new-mown hay was a clue to the presence of chemical weapons; unfortunately, this meant that some amount of the agent was inhaled. Today, many of the agents are odorless and must be detected by other means. A soldier will use a test paper (or other method) to verify the actual presence of an agent.

Detection Paper

The U.S. Army and U.S. Marine Corps issue M8 and M9 detection papers, either as sheets or booklets of a number of sheets or an adhesive-backed tape. These papers are impregnated with agent-soluble pigments or dyes. The M8 paper gives immediate, qualitative verification of the presence of liquid V and G nerve agents and H blister agents. When a sheet of M8 paper is brought in contact with liquid nerve or blister agents, the agents react with chemicals in the paper to produce agent specific color changes. There are three sensitive indicator dyes suspended in the paper matrix. The paper is blotted on a suspected liquid agent and observed for a color change; there is a color chart inside the front cover of the booklet for comparison. The chemical reaction between the M8 paper and chemical agent creates a pH-dependent color change on the M8 paper. V-type nerve agents turn the M8 paper dark green, G-type nerve agents turn it yellow, and blister agents (H) turn it red. However, M8 paper cannot be used to detect chemical agents in water or aerosol agents in the air, and it does not detect vapors. It is best suited for non-porous materials. The M9E1 tape, which is similar but can be worn on the uniform since it has an adhesive back, detects the presence of liquid V and G nerve agents and H and L blister agents. Detection papers are not as reliable as other means of detection because they depend on the liquid agent contacting the surface of the paper; neither type of paper detects traces of chemical agent vapors. Some solvents, standard decontaminating solutions, and motor oils cause false-positive reactions by the M8 paper. Extremely high temperatures, scuffs, certain types of organic liquids, and decontamination solution type two (DS-2) cause false-positive reactions by M9 paper.

Detection Tubes

In these systems, a small manual air pump draws air through a tube impregnated with an indicator that absorbs the agent in the air. The tube may require the addition of a developer for verification, or may be self-indicating. In a civilian setting, several commercially-available industrial tubes might be used; these include those for detecting cyanide, fluorides, and amines, depending upon the agent suspected.

Detection Kits

The M256A1 Chemical Agent Detector Kit is a portable chemical agent detector kit that can detect and identify nerve, blister, or blood agents as vapors. It is typically used to determine when it is safe to unmask after a chemical agent attack. A test disk contains a glass ampoule with compounds that react with an agent to give a color change. The ampoule is crushed, the activated test disk is exposed to the ambient air, and the disk is compared to a color chart to determine if an agent is present.

The M18A2 Chemical Agent Detector Kit uses both detector tubes and paper tickets to detect and identify dangerous concentrations of lethal chemical agents as vapors in the air, as well as liquid chemical agent contamination on exposed surfaces. Agents detected are CK, H, HN-1, HN-3, CX, AC, CG, L, ethyl dichloroarsine, methyl dichloroarsine, the G nerve agents, and VX. Each kit consists of 12 disposable sampler-detectors, one booklet of M8 paper, and a set of instruction cards attached by a cord to a plastic carrying case. Each sampler-detector contains a square impregnated spot for blister agents, a circular test spot for blood agents, a star test spot for nerve agents, and a Lewisite-detecting tablet and rubbing tab. There are eight glass ampoules, six containing reagents for testing and two for a chemical heater to vaporize agents at low temperatures. When the ampoules are crushed between the fingers, formed channels in the plastic sheets direct the flow of liquid reagent to wet the test spots. Each test spot or detecting tablet develops a distinctive color that indicates whether a chemical agent is or is not present in the air.

At present, there are no packaged military detection kit systems for the arsenical vomiting agents, tear gases, or incapacitating agents. All of the individual detection devices are simple, very sensitive, and give a very rapid response. The major drawback to the simplicity is that these devices generally can not identify the specific agent present; the primary drawback to the high sensitivity is that these devices can give false positive readings. Therefore, separate systems are needed to verify the presence of an agent and to identify the specific agent.

POINT DETECTION DEVICES

The next step in complexity and reliability comes with point detection devices. These devices can provide the first warning of the presence of a chemical agent, and are generally used at a single location for a longer period of time. These technologies must be portable, but they do not need to be as light or as simple as detection devices issued to individuals; the devices do not need to move as often and are operated by personnel with specific training. The point detection devices are required to be fast, so that they give an immediate warning of the presence of a chemical agent.

Battlefield Devices

The hand-held Improved Chemical Agent Monitor or ICAM, is also known as CAM2^{TM}, and is based on ion mobility spectrometry (IMS). Air is drawn into the IMS unit and is ionized by a weakly radioactive source, such as polonium-210. Chemical agent molecules in the vapor phase form low-mobility ionic clusters. The time of flight of the ionic clusters in a drift tube is measured relative to a reference cluster (since different ions move at different speeds with the heaviest taking longer to pass a certain distance). A computer examines the pattern of the time of flight, determines the level of chemical agent present, and indicates on a display or by an alarm the level of hazard.

The Automatic Chemical Agent Alarm (ACADA) system is another detector based on the GID-3^{TM} IMS. This system is a larger, but still portable, IMS system with a communications interface to support battlefield automation systems. ACADA replaced the earlier M8A1 Automatic Chemical Agent Alarm System, which was used for years up to the 1991 Gulf War. The M8A1 is also a portable chemical agent alarm based on IMS. The M8 family of detectors have been in use for more than thirty years, but they suffer from frequent false positives, even from vehicle exhaust; they are no longer in use by active forces.

The U.S. military M22 system is an enhanced automatic detector that uses ion mobility spectrometry to identify nerve and blister agents. It has been reported to be some 99% effective with much reduced false positive results than predecessor systems.

The Joint Service Lightweight Standoff Chemical Agent Detector (JSLSCAD) is a state-of-the-art detection system designed to provide U.S. forces with enhanced capability in detecting chemical warfare agents. It is a lightweight, passive, and fully automatic detection system that scans the surrounding atmosphere for chemical warfare agent vapors. It furnishes continuous 360-degree coverage from a variety of tactical and reconnaissance platforms at distances up to 5 kilometers. It is a second-generation system that significantly improves on the capabilities of the currently-fielded M21 Remote Sensing Chemical Agent Alarm, discussed below. The JSLSCAD provides the U.S. military with enhanced early warning to avoid chemically-contaminated battle areas. When avoidance is not possible, the JSLSCAD will give personnel extra time for the soldiers to their don Mission Oriented Protective Posture (MOPP) gear.

The JCAD (Joint Chemical Agent Detector) is a multi-mission chemical agent point detection system, currently in development for the U.S. military. JCAD will fulfill the missions to detect, identify, quantify, and report the presence of nerve, blister, and blood agents, and even toxic industrial chemicals. The JCAD units can be placed as stand alone or interfaced as a network around base perimeters and inside buildings. JCAD has the ability to accumulate and report meiosis level cumulative concentrations of one chemical agent while still providing a rapid alarm response to high concentration exposure from a different agent while storing up to 72 hours of cumulative dosages. JCAD is will eventually replace all current U.S. military chemical point detection systems. It is designed for personnel (worn on load-bearing equipment (LBE), aircraft interior, ground vehicle (wheeled and tracked), shipboard, fixed site/advanced warning and equipment survey applications.

The U.S. Navy uses a Shipboard ACADA man-portable point detection system to detect all classic nerve and blister agents as well as other chemical warfare agent vapors. The system can easily be upgraded to detect new agents. Designed and patented by the U.S. Navy, it is designed to operate in a shipboard environment and to detect agents at low concentrations in real time while ignoring the presence of common vapor interferents. The system has both visible and audible alarms. The ability of the system to disregard common shipboard interferents, thus minimizing false alarms, distinguishes it from other systems. Vapor analysis is achieved by the use of two ion-mobility spectroscopy (IMS) cells; a radioactive source, sealed inside each cell, is used as an ionizer.

All of the point detection devices are very sensitive and give a very rapid response. Again, the drawback to this sensitivity is that these devices can give false positive readings. Therefore, separate systems are needed to verify the presence of an agent and to identify the specific agent.

Even though such electronic instruments are available, in some combat situations, other means are employed. In the second Persian Gulf War, called Operation Iraqi Freedom, the U.S. Marine Corps began tending chickens in the Kuwaiti staging areas. It has been shown by a number of studies that chickens are particularly sensitive to chemical nerve agents. While this might seem a little bizarre in today's world, birds and rabbits were used as sentinels for the presence of chemical agents in World War I.

LONG RANGE STAND-OFF DEVICES

The M21 Remote Sensing Chemical Agent Alarm (RSCAAL) is based on a passive infrared (IR) detector. An incoming IR signal is compared against known agent spectra; when a match is detected a display lights and an alarm sounds. The display also indicates in which of the seven fields of view, spread over a 60° arc, the agent was detected, so the direction of the attack can be determined. This also allows the operator to track a moving agent cloud. The M21 is capable of detecting nerve and blister agents in the vapor phase up to 5,000 meters away; however, it must have a direct line of sight to the agent cloud.

CHEMICAL AGENT DEPOT SYSTEMS

Two detection technologies are used at chemical warfare agent manufacturing, storage, and demilitarization facilities. The first of these systems is similar to the point detection system in that it is automated, rapid, and sensitive. The second system is used to verify the results obtained with the first, but it requires sampling and subsequent analysis at an on-site laboratory.

The Miniature automatic Continuous Agent Monitoring System or MINICAMS™ is an automated near real time gas chromatograph (GC). An air sample is drawn through a pre-concentrator loop filled with an adsorbent. Periodically, the system is switched so that a carrier gas stream flows through the pre-concentrator loop as it is heated, carrying the adsorbed sample into the gas chromatograph (GC). While the sample is being analyzed on the GC, the pre-concentrator loop collects the next sample. The GC separates the chemical compounds in the sample based on differential partitioning between the carrier gas and a stationary phase (small coated beads) in the GC column. The MINICAMS™ detects the presence of an agent with a Flame Photometric Detector (FPD), which specifically detects the chemiluminescent reactions in a hydrogen/air flame of compounds containing sulfur and phosphorus. The entire cycle from sample collection to detection typically requires 3 to 5 minutes. MINICAMS™ is a refinement of the larger Automatic Continuous Agent Monitoring System (ACAMS) detector; ACAMS is still in use in some locations.

The Depot Area Air Monitoring System (DAAMS) is used to confirm the detection of an agent by MINICAMS™. Large air samples are drawn continuously through the DAAMS tube, which contains a correspondingly large amount of adsorbent. At either a pre-determined time, or when a confirmation of a MINICAMS™ is required, the DAAMS tube is physically transported to the laboratory, where the sample is desorbed into a laboratory gas chromatograph.

Identification and verification of chemical warfare agents can be conducted in the field using mobile systems, or samples can be collected for subsequent laboratory analyses.

IDENTIFICATION AND VERIFICATION

Identification and verification of chemical warfare agents can be conducted in the field using mobile systems, or samples can be collected for subsequent laboratory analyses.

The M93A1 Fox Nuclear-Biological-Chemical Reconnaissance System (NBCRS) Vehicle

The M93A1 Fox, adapted from a German design, is a self-contained mobile system equipped with the M21 Remote Sensing Chemical Agent Alarm, a passive infrared detection device, and the MM-1 Mobile Mass Spectrometer built on a truck body. The Fox has remote sampling capability by use of an articulated arm as well as remote sampling ports. The mobile mass spectrometer is capable of identifying and verifying chemical warfare agents in samples. The mobile instruments used in the Fox are a little less sensitive and selective than the instruments used in a fixed laboratory because they are limited in their size, weight, and power consumption relative to the laboratory instrumentation, and must be built to withstand a higher level of vibration. However, the lower sensitivity and selectivity are offset by the fact that the Fox can be driven to the site of a chemical attack, so that no time is required for the sample to be shipped from the site to the laboratory. Since the number of false positives can be fairly high, all results are considered presumptive until verified at better equipped laboratories. Ultimately, samples are sent to the U.S. Army Edgewood Chemical and Biological Center in Maryland. In addition, suspected biological agents may also be sent to the U.S. Army Medical Research Institute for Infectious Diseases (USAMRIID) at Fort Detrick, MD, or the Naval Medical Research Center at Bethesda, MD. The Fox and its instrumentation can detect about 60 standard military chemical agents but has the capability of adding an industrial computer chip to expand the number to 115 chemicals, including toxic industrial and petroleum products. Typically, in the field, the Fox is programmed to look for 10 persistent and nonpersistent agents plus hydrocarbons, especially petroleum products.

Sample Collection in the Field

The most certain method to positively identify a chemical warfare agent and to verify its presence is analysis in an analytical laboratory. Once an individual or point detection device sounds an alarm, and the appropriate protective equipment is donned, physical samples can be collected and sent to the laboratory. In addition to its use for immediate detection, the M18A2 kit can also be used to collect samples of unidentified chemical agents for submission to a laboratory for identification.

The drawback to sampling and subsequent analysis is that it takes a significant amount of time to ship a sample to the laboratory; at best, days will elapse from the time the sample is taken until the results are reported. Thus, this procedure is generally reserved for collecting intelligence about the use of chemical agents, rather than for immediate health and safety protection.

Laboratory Techniques

Analytical techniques in common use in the laboratory include gas chromatography (GC), gas chromatography/mass spectrometry (GC/MS), liquid chromatography/mass spectrometry (LC/MS), Nuclear Magnetic Resonance (NMR) spectrometry, and ion chromatography (IC).

Gas chromatography (GC), which separates chemical compounds in the sample by differential partitioning between the carrier gas and a stationary phase (small coated beads) in the GC column, detects an agent with a flame photometric detector (FPD), flame ionization detector (FID), or photoionization detector (PID). While the underlying technology is the same as that used in MINICAMS™, a laboratory GC has a longer, thinner column and operates at different temperatures. The laboratory instrument requires a longer amount of time for the analysis, but it provides better separation of the compounds in the sample. Since the analysis takes place in a cleaner environment than the battlefield, it is less prone to interference than the MINICAMS™.

Gas chromatographs separate the chemical compounds in a sample by differential partitioning between the carrier gas and the stationary phase in the GC column, then detect the agent with an flame photometric detector (FPD), flame ionization detector (FID), or photoionization detector (PID).

Gas chromatography/mass spectrometry (GC/MS) uses a gas chromatograph (GC) to separate the materials in a sample into relatively pure chemical compounds, then uses the mass spectrometer (MS) to identify the specific substance. The MS uses a device (typically an electron beam) to fragment and ionize the compound into all possible ions. The instrument then measures the mass-to-charge ratio of each of the fragments using an electric or magnetic field to separate the ions since heavier ions are less deflected than lighter ones. The distinctive fragmentation pattern, called a spectrum, serves as a molecular fingerprint that identifies the structure of the compound. GC and GC/MS are used for the bulk of routine analyses of samples that may contain chemical warfare agents. These techniques afford a very low rate of false positives and false negatives, and are capable of providing definitive identification of the chemical agent that is present.

GC and GC/MS are used for most routine analyses of samples that may contain chemical warfare agents. These techniques afford a very low rate of false positives as well as false negatives and are capable of providing definitive identification of the chemical agent present.

Other technologies that are used less frequently, or are currently being developed, include LC/MS, NMR, and IC. Each of these will be mentioned briefly.

Liquid chromatography/mass spectrometry (LC/MS) uses a liquid chromatograph (LC) to separate the material in a sample into relatively pure chemical compounds, then uses a mass spectrometer (MS) to uniquely identify the substance from the ions formed. It is more sensitive than GC when less volatile substances are analyzed since the material does not have to be a gas or vapor. The liquid chromatography technique is also known as high-pressure liquid chromatography (HPLC).

Nuclear Magnetic Resonance (NMR) spectrometry measures the absorption of radiofrequency (RF) energy of a particular frequency by a sample of the material being held in a strong magnetic field. NMR produces a spectrum that is a "fingerprint" of the particular chemical compound.

Ion chromatography (IC) separates ionic substances on an ion-exchange column. The column contains small beads upon which different ions adsorb at different rates. It can also be used for measuring the substances produced by environmental degradation of some chemical agents. There are experimental ion chromatographs capable of detection of chemical agents in the vapor state in the air.

EXPERIMENTAL CHEMICAL AGENT DETECTORS

There are a number of experimental systems under study for the rapid and selective detection of chemical weapons. A promising one is a sensor system based upon fluorescence colorimetry, using a color change to indicate the presence of the G nerve agents in the vapor phase. One indicator under study is naphthalene derivative with a pyridyl and a hydroxy substituent. In studies, the response time has been a few seconds, with the color changing from blue to green light under fluorescence. The sensitivity is sub-parts-per-billion (ppb) and is selective only to the G agents; however, it cannot differentiate among the agents.

Another system under study uses infrared light emitting diodes (IR-LEDs) of a wavelength at an absorbance peak of the fluorophosphonate portion of the molecule. The device uses a internal calibration sample of the agents of interest. The sensitivity is on the order of a few parts per billion, and the selectivity is excellent. It is not known if or when the device will be fielded by the U.S. military.

ANTIDOTES FOR NERVE AGENTS

The effects of nerve agent exposure can be mitigated by the use of antidotes. Military nerve agents (GA, GB, GD, and VX) are all potent inhibitors of the enzyme acetylcholinesterase. Acetylcholinesterase hydrolyzes acetylcholine, which is present at nerve synapses and transmits signals between nerve cells. When acetylcholinesterase is inhibited, acetylcholine builds up at the junction between nerve cells, effectively preventing the transmission of signals. The antidotes usually consist of an oxime with atropine, and, sometimes, followed by an anticonvulsant.

The oxime, *e.g.*, pralidoxime (pryridine-2-aldoxime methyl chloride; 2-PAM Cl), is thought to react with the inhibited acetylcholinesterase to remove the phosphonyl group from the nerve agent, and, thus, regenerate the active enzyme. However, oximes penetrate into the central nervous system (CNS) rather poorly, and thus other antidotes are required. In addition, the inhibited enzyme undergoes a further chemical reaction, known as "aging," in which the alkyl ester group hydrolyzes to give the enzyme monoester. This reaction produces a particularly stable complex that is resistant to both hydrolysis and regeneration by oximes. Aging takes place in hours for Agent GB and Agent VX but occurs in minutes for Agent GD; this makes Agent GD exposure particularly difficult to treat with oximes.

Atropine blocks the action of acetylcholine on the muscarinic receptors in the parasympathetic nervous system; thus, it counteracts many of the symptoms of nerve agent exposure. Atropine has the effect of drying the runny nose, drooling, and excessive sweating caused by exposure, and it also counteracts the respiratory depression and urinary effects of exposure.

Anticonvulsants, such as diazepam (Valium®), are also given to protect against convulsions and resulting brain damage. Soldiers and workers who handle these materials are often issued chemical agent antidote kits, which include autoinjectors containing an oxime and atropine (ComboPen®).

Individuals at risk of exposure to Agent GD can alternatively be given a carbamate pretreatment such as pyridostigmine. Carbamates reversibly inhibit acetylcholinesterase; the pyridostigmine-inhibited enzyme does not react with Agent GD. The dose taken is intended to reversibly bind up to 40% of the available acetylcholinesterase; this has minor effects on nerve function because of sufficient levels of acetylcholinesterase that remain. In the event of exposure to Agent GD, the pyridostigmine-inhibited acetylcholinesterase slowly and continuously regenerates the acetylcholinesterase with a half-life of the inhibited enzyme on the order of hours, maintaining enough of the enzyme to permit transmission of nerve signals. If the oxime antidote is given, essentially all the pyridostigmine-inhibited acetylcholinesterase is regenerated. Carbamate pretreatment is believed to provide little benefit for Agent GB or Agent VX exposure; data are not adequate to determine benefits in case of Agent GA or Agent GF exposure.

ANTIDOTES FOR VESICANTS

There is no known antidote for exposure to mustard agents (Agents H and HD). Nevertheless, prompt medical treatment of the symptoms, including frequent irrigation of affected areas and application of topical antibiotics, as well as pulmonary care, can help in speeding recovery of those who survive.

The effect of Lewisite (Agent L) can be prevented by rapid topical application of 2,3-dimercaptopropanol, known as British anti-Lewisite (BAL). BAL reacts with Lewisite to give a non-toxic arsenic derivative. However, the activity of BAL is limited because of low water solubility and some toxicity of it own. For this reason, 2,3-dimercaptopropanesulfonic acid (DMPS) and meso-2,3-dimercaptosuccinic acid (DMSA) are more commonly used in place of BAL as treatments for systemic Lewisite poisoning.

ANTIDOTES FOR OTHER CHEMICAL WARFARE AGENTS

Chlorine, phosgene, diphosgene, triphosgene, and perfluoroisobutylene all affect the lungs directly and have no specific antidotes. The only treatment is removal from the contaminated area and supportive care. Medical care includes pulmonary assistance and complete rest.

Cyanide inhibits the cellular enzyme cytochrome oxidase; this disrupts oxygen metabolism and energy generation by the cell. The effects of hydrogen cyanide and cyanogen chloride can be treated using inhaled amyl nitrite, intravenous sodium nitrite and sodium thiosulfate, and inhaled oxygen. The nitrites convert hemoglobin to methemoglobin, which removes free cyanide as cyanmethemoglobin. Thiosulfate converts cyanide to thiocyanate through the action of the enzyme rhodanase.

Finally, the only antidote for the psychotropic agent BZ is physostigmine, a acetylcholinesterase-inhibiting carbamate that crosses the blood-brain barrier.

DETOXIFICATION

Chemical warfare agents at sublethal doses can be detoxified in the body to varying degrees. For nerve agent exposure, the rate of natural regeneration of acetylcholinesterase is very slow and depends on the specific agent. Nerve agents have other effects on the nervous system that are essentially cumulative, resulting in loss of memory, ability to stand, or even respiration. For vesicants, mustard is not detoxified at all by the body. Small and repeated exposures increase sensitivity, and can lead to cumulative effects. Detoxification of Lewisite is somewhat more rapid. Choking agents (chlorine, phosgene, diphosgene, triphosgene, perfluoroisobutylene) are not detoxified, and are cumulative. Blood agents (hydrogen cyanide, cyanogen chloride) are detoxified at variable rates, depending upon a number of factors such as the amount taken in, respiratory and metabolic rates, and general physical condition. Tear gases can be detoxified by the body very rapidly. The arsenical irritants are long-lasting and cumulative.

PROTECTION AGAINST EXPOSURE TO CHEMICAL AGENTS

Gas Masks

Within two weeks of the initial use of chlorine gas in World War I, two British professors recommended that the soldiers protect themselves against the gas through use of cloths moistened with urine or earth folded in cloths, held over their nose and mouth. A month later, the troops were supplied with double layers of flannel to be dipped in a sodium hydroxide solution stored in bottles in the trenches. After another month, the British Army issued 2,500,000 "Hypo Helmets." These were bags of flannel that had been impregnated with a chemical that would react with chlorine and neutralize it. The bags had two celluloid eyepieces and were placed over the head and tucked into the collar. In the fall of 1915, the helmets were modified with a better chemical impregnant and a rubber exhaust tube.

Thus began the story of the countermeasure against chemical warfare. Virtually all lethal chemical warfare agents are toxic when inhaled, so it is crucial to remove any chemical warfare agent from breathing air. Over the years, chemical warfare agents have become even more toxic, and gas masks have become much more sophisticated. Modern U.S. gas mask models are designated M40, M40A1, M42, M42A1, XM45, M49, MCU-2/P, and JSPGM. These protective masks are equipped with replaceable cartridges that contain filters to remove aerosol particles along with beds of activated carbon and other additives to remove toxic vapors from the air stream. They also include many features in addition to the filtration system: a rubber gas-tight seal to prevent leakage around the face; a drinking tube so that wearers can drink from canteens without removing the mask; sophisticated ventilation designs to prevent moisture from fogging the eyepieces; and voice diaphragms so that the wearer's voice can be heard.

Protective Clothing

Modern chemical warfare agents are toxic by other modes in addition to inhalation. Vesicants, including sulfur mustard, the nitrogen mustards, and Lewisite, function primarily by causing skin injury. Nerve agents such as Agents GA, GB, GD, and VX can be absorbed through the skin. Thus, protective clothing is required in addition to the gas mask. Two approaches can be taken to protective clothing: impervious clothing made from rubber or plastic, or breathable clothing.

The impervious clothing is designed as a barrier to exclude any outside air from contacting the wearer. This approach has a drawback because the barrier that keeps the agent out keeps moisture in; wearers cannot lose heat by sweating and heat exhaustion from even mild physical activity is a constant danger.

Breathable clothing can be made with an impregnated chemical compound. During World War II, when mustard gas was perceived as the most severe chemical agent threat, clothing impregnated with chloramide "CC-2" was produced. With the appearance of nerve agents on the scene, activated carbon-impregnated clothing has been developed. The drawback to impregnated clothing is that use degrades the protection offered by the impregnated material.

The final parts of the body that require protection are the hands; a typical ensemble is completed with a pair of butyl rubber gloves.

The use of protective equipment against chemical agents depends upon the nature of the agent, whether nerve, blood, or choking, and whether it is vapor or liquid. It is somewhat easier to protect against vapor exposures than liquid exposures.

U.S. Military Chemical Protective Equipment

Before discussing how Chemical Protective Equipment (CPE) is used in the field, it is useful to understand the types of chemical weapons it protects against and how. Chemical warfare agents may be delivered in various forms, including gas, liquid, or aerosol. They can be non-persistent, lasting for only minutes, or persistent, remaining effective for weeks. Chemical agent clouds can cover large areas and drift into foxholes, hatches, and bunkers to cause casualties. Chemical Protective Equipment is designed to protect against both persistent and non-persistent agents.

The special filters in the protective masks absorb airborne agents and protect the lungs and eyes. The other components of CPE protect against agent contact with the skin, regardless of whether it comes in solid, liquid, or vapor form. The overboots and butyl rubber gloves are impermeable and provide a solid barrier to liquid agents. A solid barrier for the rest of the body is not practical for most combat functions because it would cause the rapid buildup of body heat and moisture. Overgarments and hoods permit some passage of air and moisture through two layers, allowing perspiration to evaporate. The polypropylene outer layer limits liquid absorption or redistributes it to reduce concentration. An inner layer filters the air and any vapor that penetrates the outer layer. This inner layer of charcoal-impregnated foam acts like the filter in the protective mask. Charcoal is highly porous and able to absorb liquid, gas, and aerosol agents. If mask filters or permeable protective garments become exposed to a chemical agent, they are discarded after wear and then replaced, in accordance with each military service's policies. For example, the U.S. Air Force chooses to air out vapor-contaminated CPE in a toxic free area, and then reuse them. Impermeable gloves and overboots can be decontaminated and recycled for use. Troops potentially exposed to high concentrations of chemical warfare agents (*e.g.*, decontamination crews) receive special impermeable overgarments.

CPE components are rated for how long they provide full protection in both chemical agent contaminated and non-contaminated environments. For example, in a contaminated environment, the Chemical Protective Overgarment (CPOG) is rated for up to six hours of protection and the Battledress Overgarment (BDO) for 24 hours. Overgarments actually exposed to chemical warfare agents are never worn again. In a non-contaminated environment, the CPOG gradually begins to lose protection after 14 days of almost full-time wear, while the BDO can last 30 days. Returning the garments to their vapor-seal bags "stops the clock" on these wear periods. The bag protects the overgarment from the degrading effects of such things as moisture, smoke, fuel solvent vapors, and sunlight. Over time, extensively worn overgarments can also become unserviceable because the charcoal migrates to the end of the sleeves and trousers, or the knees and elbows wear out, or the garment is exposed to too much mud and dirt.

The key parts of CPE include:

- *Overgarment.* The U.S. military has several models of overgarment. (If exposed to contamination, the wearer discards and replaces overgarments. They are not decontaminated or recycled. Troops normally wear the overgarment over the field uniform, but it can be worn over only underwear to reduce heat buildup.)
  - The *Battledress Overgarment* (BDO) consists of a coat and trousers in olive drab or camouflage pattern. The BDO has an outer cotton layer and an inner

layer of charcoal-impregnated polyurethane foam. It is permeable, permitting some air to filter in and out, thereby reducing heat buildup, while absorbing and trapping any chemical agents coming in contact with the BDO.

- The *Chemical Protective Overgarment (CPOG),* is similar in construction to the BDO but is an older design. It is solid olive drab with an outer layer of nylon cotton and charcoal impregnated foam inside.
- U.S. Army aircrews wear the *Aircrew Uniform Integrated Battlefield* (AUIB) instead of a normal flight suit or the BDO/CPOG. It protects against both chemical hazards and fire and includes features specialized for use in the cockpit.
- U.S. Marines have four different chemical protective suits: the *Marine Corps Standard Protective Overgarment* (OG84), the *Navy Lightweight Suit* (MK III), the *Aviator's Chemical Ensemble*, and the *British Lightweight Suit* (MK IV).
- U.S. Air Force aircrews also wear the *British Mark IV Lightweight Suit* (MK IV).

- *Chemical Protective Helmet Cover.* This cover protects against chemical and biological contamination and is made from butyl-coated nylon cloth. It has an elastic web in the hem to gather the cover and hold it on the helmet.
- *Vinyl Overboot.* Worn over combat boots, the impermeable overboot protects against chemical, radiological, and biological hazards as well as rain, mud, or snow. If contaminated, decontamination can return them to service.
- *Protective Masks.* Several models of protective masks are used by the U.S. military. All the masks protect the face and airways from airborne contamination by all known chemical or biological agents and radioactive dust. Formerly, most U.S. troops in dismounted ground operations used the M17 Series Protective Mask. The newer M40 Protective Mask is now more common. Both masks have similar basic functions and levels of protection, but the M40 is more comfortable, with improved convenience for changing filters and better voice transmission quality (voicemitter). They both include a binocular lens system, elastic head harness, voicemitters, and filters to trap nuclear, biological, and chemical (NBC) contaminants. The M17 Series is made of butyl rubber, while the M40 facepiece is made of silicone with a second layer made of butyl rubber. Masks that can be connected to vehicle air filtration systems are issued to tank crewmembers (model M25) and aircrews (model M24). The U.S. Air Force ground personnel use the M17 or M40 Series Masks or the MCU-2/P series masks. The MCU-2/P is similar to the M40 except that it has a single large eye lens instead of two.
- *Field Protective Hood.* The rubber hood attaches to and is donned with the mask. It protects the head and neck from chemical agents and other NBC hazards.
- *Chemical Protective Glove Set.* The glove set includes outer gloves made of impermeable butyl rubber and inner gloves made of thin cotton to absorb moisture. The outer gloves come in three thicknesses:
  - The 7 mil gloves are used by medical personnel, teletype operators, keyboard entry personnel, radio operators, electronic repair personnel, *etc.*, who need high touch sensitivity and who normally will not expose the gloves to harsh treatment.

  - The 14 mil gloves are used by aviators, vehicle mechanics, and weapon crews needing some touch sensitivity, but who also are unlikely to give the gloves harsh treatment.
  - The 25 mil gloves are used by troops who perform close combat tasks and other heavy labor.
- *Auxiliary Equipment.* Skin decontamination kits, antidote kits, and M8/M9 chemical agent detector paper also accompany the protective clothing as auxiliary equipment.

WORKER PROTECTION

In addition to protecting soldiers from chemical attacks on the battlefield, individuals who work with chemical warfare agents, including emergency responders, must be protected. One example is the demilitarization protective ensemble (DPE) worn by workers in U.S. facilities for the destruction of chemical weapons. Certain portions of these facilities may become contaminated with chemical agents; when workers are required to enter these areas, they wear DPE consisting primarily of an oversize plastic suit with a fitting for connection to an airline respirator and a purified air supply. A worker wearing a gas mask and a backup air supply climbs into the suit and is sealed inside with a microwave heat sealer. The worker then moves through an air lock into the contaminated area.

PERSONNEL PROTECTIVE LEVELS

Levels of protection against chemical agents are similar to general protection from occupational chemicals. There are a number of levels, based upon the threat and agent nature and concentration.

Protective Level A

Level A is defined as below the immediate danger to life and health (IDLH) concentration for nerve agents; or at or above 0.003 mg/m^3 for mustard and Lewisite. This level requires the wear of a toxicological agent protective (TAP) suit ensemble (*e.g.*, the M3), which consists of an outer butyl rubber garment; coveralls or fatigues for nerve agents or impregnated gloves, socks, and long underwear for vesicants; a butyl rubber hood; butyl rubber safety toe boots; butyl rubber gloves; and M9 or M40 mask. In certain situations, the cuffs of the sleeves and legs may be taped to the gloves and boots.

Protective Level Alternate A

This level is defined as concentration above the IDLH for nerve agents, at or above 0.003 mg/m^3 for mustard and Lewisite, or in proximity to a spilled agent in an area of known liquid contamination if a monitoring alarm is not available. It requires the wear of demilitarization protective ensemble (DPE). This is a one-time use only totally encapsulating chemical protective suit with supplied air with the respirator operating at positive pressure, plus a 10-minute emergency internal breathing system. Another way to meet this requirement is by wear of a toxicological agent protective ensemble, self-contained 1-hour (TAPES). This is a totally encapsulating positive pressure air pack suit with an integrated cooling system.

Protective Level Modified A

This level is defined as being below the immediate danger to life and health (IDLH) concentration for nerve agents; or at or above 0.003 mg/m^3 for mustard and Lewisite. It is also used where there is danger of oxygen deficiency or when a potential for agent release exists. It requires the wear of the Level A suit with modified M30 hood, pressure demand self-contained breathing apparatus (SCBA), and full-face respirator. It can also be met by wear of a self-contained toxicological environmental protective outfit-interim (STEPO-I), which is a butyl rubber total encapsulation suit with air-line tether or self-contained air supply.

Protective Level B

This is the level called for when contact with a suspect item is required, and when performing operations which may result in release of agent vapors within the work area, but without contact with a liquid agent. Level B protective gear may be worn without the butyl rubber apron for first entry monitoring of outdoor chemical weapons storage areas. It requires the wear of a butyl rubber apron extending below the top of the boots; coveralls or fatigues for nerve agents, or impregnated gloves, socks, and long underwear for vesicants; butyl rubber hood; butyl rubber safety toe boots; butyl rubber gloves; and M9-, M17- or M40-series mask worn.

Protective Level C

This is the defined level in agent areas where handling or contact with agent-filled items is involved, and when real-time monitoring is being performed. It requires the wear of a butyl rubber apron extending below the top of the boots; coveralls or fatigues; butyl rubber safety toe boots; butyl rubber gloves; and M9-, M17- or M40-series mask worn.

Protective Level D

This is the defined level for clean areas where handling or contact with agent-filled items is involved provided that real-time low level monitoring is being performed with negative results. It requires the wear of a butyl rubber apron extending below the top of the boots; coveralls or fatigues; butyl rubber safety toe boots; butyl rubber gloves; and M9-, M17- or M40-series mask carried in the ready position.

Protective Level E

This is the defined level for operating personnel who may be observing or supervising operations when contact with an item or exposure to agent would occur only in the event of an accident, and by laboratory personnel working with chemical agents. It requires the wear of a laboratory coat, coveralls, or fatigues, and M9-, M17- or M40-series mask carried in the ready position for operations; in laboratories, the mask needs only to be readily available. Gloves or safety shoes should be worn, if necessary.

Protective Level F

Street clothing and M9-, M17- or M40-series mask carried in a position where it can quickly be put on. This level is typically used by casual or transient personnel who may be required to visit chemical agent clean storage or operating areas.

MOPP

Similar to the protective levels discussed above for occupational settings are the U.S. military's levels of protection against chemical (and some biological) agents. The Mission-Oriented Protective Posture (MOPP) consists of four levels, each with different wear of the Chemical Protective Equipment (CPE).

MOPP Zero, the lowest level, is having the overgarment and helmet cover, vinyl overboots, and gloves available, and the mask and hood being carried. *MOPP One* requires that the overgarment and helmet cover be worn; the vinyl overboot avail able; and the masks, hood, and gloves being carried. MOPP Two requires the wearing of the overgarment and helmet cover and vinyl overboots; and carrying the mask, hood, and gloves. MOPP Three requires wearing the overgarment, helmet cover, vinyl overboots, mask, and hood; and carrying the gloves. MOPP Four requires the wear of all items. The commander in the field makes a determination as to the threat and then issues a MOPP order in response.

The time to put on the different levels has been estimated by the U.S. military, ending with MOPP Four, as in an actual attack. From MOPP Zero to MOPP Four would take about eight minutes. From MOPP One to MOPP Four, four minutes. From MOPP Two to MOPP Four, about one minute. From MOPP Three to MOPP Four, only a few seconds.

INITIAL MEDICAL TREATMENT

The proper medical treatment of a chemical casualty is to treat the symptoms as soon as possible following decontamination. Movement of chemical agent casualties can spread the contamination to clean areas. All casualties should be decontaminated as the situation permits, but all patients must be decontaminated before they are admitted into a clean medical facility. The admission of one contaminated patient into a hospital may contaminate the facility, thereby reducing treatment capabilities in the facility. Decontamination of many agents can be simply to remove the contaminated clothing of the victim. Decontamination of chemically contaminated patients requires the removal of their clothing and the use of a variety of decontamination kits and solutions; however, in an emergency, simple cleaning with mild soap and water, taking care not to rub the agent into the skin, will remove most of the contamination. Care must be taken with any runoff and rinse water as it may contain small amounts of the agent.

MANAGEMENT OF CHEMICAL AGENT PATIENTS

As with other agent employment, a mass casualty situation may be presented when chemical agents are employed. Additional medical personnel and equipment may have to be provided in a short period of time if an acceptable level of care is to be maintained. Treatment at early stages in the field is limited to life- or limb-saving care. Patients that can survive evacuation to the next level of medical care should not be treated in the field. This provides time for treating those patients that cannot survive the evacuation time.

Management of Exposure to Blister Agents

Typical symptoms and signs are blister formation in airways, with bloody sputum, which can result in sudden airway obstruction; nausea, vomiting, and diarrhea; apathy and lethargy; loss of white blood cells with loss of immune response; large fluid- or blood-filled blisters on the skin that heal with scarring; and conjunctivitis, sensitivity to bright light, pain, edema, and corneal perforation.

Medical management is first to decontaminate with water or a bleach/water mixture. Skin lesions are to be treated like a thermal burn. The eyes should be irrigated with copious amounts of water and treated like a corneal abrasion except do not use a tight patch. The patient may require intubation, assisted ventilation, and bronchodilators. The systemic treatments include preparing for infection, reverse isolation, and sterilization of the GI tract. The military field treatment is the same as for a burn, petrolatum gauze placed over the wound.

A number of protective measures against vesicants have been tried by various militaries: barrier creams, chlorine-containing socks and underwear, *etc.*, but most have not been successful. The best protection for a vesicant is to prevent exposure in the first place. There are a number of barrier materials that can be used to coat boots and clothing to slow the permeation of the chemical through the material. Some military equipment, such as socks, have been impregnated with chlorine-containing compounds to chemically react with vesicants; however, harm to the skin is possible. Boots can be coated with an oily or fatty material to act as a barrier to mustard agents. The U.S. military chemical protective suit is made of synthetics that prevent vesicants from passing through.

Management of Exposure to Blood Agents

The clinical effects appear very rapidly, as quick as 15 seconds after exposure with rapid breathing and gasping for breath. As soon as thirty seconds, convulsions begin. After two minutes, respiratory efforts cease and death ensues. A common sign is a cherry red color to the skin.

Medical management for most blood agents includes immediate administration of amyl nitrite, which is inhaled by patient, followed by administration of 300 mg of sodium nitrite IV over two to four minutes and 12.5 g of sodium thiosulfate IV. The military field treatment is to place a broken amyl nitrite capsule under the victim's nose and assist with breathing.

Management of Exposure to Choking Agents

The typical mechanism of toxicity is a reaction of the agent with linings of the lungs, causing small holes where water leakage into the air sacs can occur. The result is delayed pulmonary edema.

The clinical effects typically occur between twenty minutes and twenty-four hours after the exposure and include irritation of the airways and delayed pulmonary edema. The most common sign is shortness of breath due to inability to oxygenate blood.

If pulmonary edema occurs within about four hours of exposure, the individual will most likely die despite the best medical efforts. However, medical management includes admission to the hospital with strict bed rest, administration of supplemental oxygen, as needed, and preparation for intubation since conditions can worsen precipitously. The military field treatment is to assist with breathing.

Management of Exposure to Nerve Agents

Medical management is to decontaminate the patient with copious amounts of water or bleach and water mixture, and assist ventilation. An antidote for most agents is atropine—2 mg intravenously (IV) every three to five minutes—given until ventilation is eased, together with 1 gram of pralidoxime chloride (pryridine-2-aldoxime methyl chloride, 2 PAM chloride) IV over twenty minutes every hour, with a maximum of three doses. For symptoms, give 10 mg of diazepam (Valium®) IV, as needed. The military field treatment is to use the automatic ComboPen® of atropine and 2-PAM chloride, up to two doses.

Management of Exposure to Lacrimators

Medical management of these skin, eye, and nasal membrane irritants is supportive care after removal from the exposure and washing the skin and eyes with copious amounts of water. In extreme cases, administration of oxygen may be needed.

Management of Exposure to Vomiting Agents

The effects of all the vomiting agents are similar. There is severe irritation of the eyes, nose, and throat. If the agent is inhaled for a few minutes, tightness of the chest and headache are experienced. The headache develops into general nausea, which can result in vomiting in approximately three minutes. At concentrations expected to occur in combat conditions, fatalities are not expected; however, these compounds can be fatal at higher concentrations. Removal from exposure followed by supportive care, including administration of oxygen as well as electrolytes, is usually sufficient.

COMMAND STRUCTURE

Exposure to a chemical agent should be treated in a manner similar to any other emergency involving chemicals. This requires the establishment of a Command Structure with one person in charge; the Incident Commander (IC). This position and all additional duties should have been practiced routinely. There needs to be established a Hot Zone (where the release took place with subsequent contamination), a Decontamination Zone (where the agent is removed from victims), a Support Zone (where emergency treatment takes place), and, then, transfer to a medical facility.

UNIDENTIFIED CHEMICAL AGENT

Every possible attempt should be made to determine the specific chemical or mixture components before assuming the material is unidentified. However, for the public, and even many health care workers and facilities, the exact identity of a chemical agent may not be known for some period after the attack, whether in conventional warfare or terrorist activity. It is possible, even likely, that a terrorist might employ a crude mixture of agents rather than a highly-purified product due to the difficulty of manufacturing very pure agents. It is also possible that a terrorist might employ a mixture of different agents in order to confuse detection and increase anxiety. The military M8 and M9 papers and chemical agent alarms and detectors might be available to some emergency responders; if so, they should be used. While more effective medical treatment is possible when the specific identity of the agent or agents is known, there are some general procedures which may be followed in any case.

PRE-HOSPITAL MANAGEMENT

The nature of the chemical agent's route of exposure and amount of exposure are important in determining the extent of secondary contamination. Victims who have only been exposed to gas or vapor without deposition of the agent on their clothing or skin are not likely to carry significant amounts of agent beyond the immediate area where the release took place, the "Hot Zone." But victims whose skin or clothing have been soaked with a liquid agent or those having condensation of the agent on their clothing or skin may contaminate others, including medical and other response personnel. This may happen by direct contact or by off-gassing the agent vapor. If the victim has ingested a chemical agent, vomitus may be toxic and pose a hazard to others from direct contact or off-gassing.

In the Hot Zone, rescuers should be trained and appropriately attired before entering. If proper protective equipment is not available, or if rescuers are not trained in its use, they should not enter the Hot Zone. While it is difficult to watch others suffering, this may limit the number of casualties. Since the identity of the agent may not be known, worst case must be assumed. The potential for severe local effects, such as irritation and burning, and severe systemic effects, such as organ damage, must be assumed. Rescuers must wear pressure-demand, self-contained breathing apparatus (SCBA) in positive pressure mode in all response situations. They must also wear chemical protective suits, with all openings, zippers, and seams sealed with duct tape or similar sealant. As in most medical emergency situations, the rescuers, upon reaching a victim, should assess and ensure a patient airway. Observation of the victim may give a clue to the nature of the chemical agent. Most choking agents produce severe vomiting; most nerve agents cause tremors and twitching of the entire body with fluid retention and enlarged pupils of the eyes; and most blister agents produce a reddening or burning of the skin and respiratory passages. If the victim clearly displays signs of death or near death (expectant), it may be necessary in a mass casualty situation to declare that person not suitable for medical aid. If the victim can walk, lead him or her out of the Hot Zone to a Decontamination Zone. Victims unable to walk should be removed on backboards or gurneys; if these are not available, rescuers should carry or drag the victims to safety. Keep in mind that any equipment which enters the Hot Zone is now contaminated.

In the Decontamination Zone, a quick assessment is needed as to the nature and extent of the contamination. If the victims have only been exposed to gas or vapor without skin or eye irritation, they may be moved to the Support Zone. In this case, however, be sure that the agent was not mustard since effects may be delayed. In the Decontamination Zone, rescuers should wear the same protective equipment as in the Hot Zone. Victims may need supplemental oxygen; assist ventilation with a bag-valve-mask device if necessary and available. Victims who are able and cooperative may assist in their own decontamination. Remove and double-bag contaminated clothing and personal belongings. Remove wristwatches, necklaces, and rings are these may retain the agent. Flush exposed or irritated skin and hair with plain water for three to five minutes. If the agent is oily, as in the case of Agent VX, a mild soap or detergent may be used on the skin and hair. Exposed or irritated eyes should be flushed with plain water or saline solution for at least five minutes. Remove contact lenses, if present and easily removable without additional trauma to the eye. If the material appears to be corrosive or if pain or injury is evident, continue irrigation while transferring the victim to the Support Zone. If the agent has been swallowed, do not induce emesis. Conscious victims who are able to swallow should be given four to eight ounces of potable water; additional care must be immediately obtained. When decontamination is complete, move the victim to the Support Zone.

The Support Zone should only be entered by rescuers who have not been in the Hot Zone or Decontamination Zone, and only in the direction upwind of the other two zones. Victims must have been decontaminated properly, unless exposed only to gas or vapor without any evidence of eye or skin irritation prior to their entry to the Support Zone. Support Zone personnel require no special protective gear. Re-evaluate and ensure a patient airway. If trauma is suspected, immobilize the back and neck. Ensure adequate respiration and administer supplemental oxygen, if necessary. Ensure a palpable pulse. Establish intravenous access, if necessary. It may be necessary to continue irrigating exposed eyes and skin. In cases of ingestion, do not induce emesis and give four to eight ounces of potable water, if not done so earlier. In case of respiratory compromise, intubate the trachea. If the patient's condition prevents intubation, perform a cricothyroidotomy, if equipped and trained to do so. Patients with bronchospasm should be treated with aerosolizing bronchodilators. These bronchodilators and all other catecholamines must be used with caution due to the potential for cardiac dysrhythmias, especially when the nature of the agent is not known and could be addictive. Patients who are comatose, hypotensive, in seizure, or presenting cardiac dysrhythmias should be treated in accord with Advance Cardiac Life Support (ACLS) protocols, which are beyond the scope of this book. Prior to sending the patient from the Support Zone, ensure that decontamination is complete, life support has been implemented, the receiving medical facility has been notified to expect the patient, and the proper accountability official has been notified. Send with the patient a brief listing of symptoms and signs, plus information on any treatment given. All living patients must be sent to a medical facility for evaluation, regardless of condition, but should be sent out in a rational order. Prior to the patient being moved to a treatment facility, an evaluation, called triage, should be performed so as not to overburden a particular treatment facility. More seriously injured patients need the earlier treatment unless it is clear that they will not survive. Asymptomatic patients who have not had direct chemical agent exposure can be discharged from the scene after their name, address, telephone number, and other personal information has been obtained.

HOSPITAL TREATMENT

In a mass-casualty situation involving chemical agents, once the patient has been transported to a medical treatment facility, definite laboratory testing may be performed to identify the specific agent or mixture as well as the physiological condition of the patient. Even though chemical agents are involved, treatment is similar to other emergency situations involving burns, commercial chemical exposures (including pesticides), and loss of electrolytes. Medical treatment protocols for chemical agent exposures are beyond the scope of this book.

Many patients who survive nerve agent poisoning go on to essentially complete recovery; however, some will suffer life-long disability. Recovery from many blister agents is long and painful, with tissue damage so extensive that skin, eyes, and respiratory tract may never fully recover. Prompt treatment of choking agents can lead to full recovery.

GENERAL COMMENTS ON MEDICAL TREATMENT

Health service support operations in a chemical agent environment will be complex. In addition to providing care in protected environments or while dressed in protective clothing, medical personnel will have to treat chemically injured and contaminated patients in large numbers. A summary of types of injuries associated with chemical warfare follows:

- Nerve Agent Injury. Nerve agent injuries are classified as mild, moderate, or severe. Classification is based upon the signs and symptoms present in the individual. The individual may only be having minor problems or may be convulsing and exhibiting severe respiratory distress. Some individuals can return to duty after receiving a single injection of atropine and 2-PAM Chloride; others may require multiple doses of the antidote and assisted ventilation.
- Blister Agent Injury. Individuals exposed to blister agents may not know that they have been exposed to the agent for hours to days later. The first indication of exposure may be small blisters on the skin. Others will have immediate burning because of the high level of exposure. An individual with a few small blisters or reddening of the skin can receive skin ointments. An individual suffering mild injuries may require admission to a medical facility for treatment of the irritation and burns. The individual with severe injuries may have to be taken to a specialized hospital for burn care.
- Incapacitating Agent Injury. Incapacitating agents produce injury by either depressing or stimulating the central nervous system (CNS). These agents affect the CNS by disrupting the high integrative functions of memory, problem solving, attention, and comprehension. Relatively high doses produce toxic delirium, which destroys the ability to perform any task. Treat as if a psychiatric patient.
- Blood Agent Injury. Blood agents produce their effects by interfering with oxygen use at the cellular level. Inhalation is the usual route of entry. The agent prevents the oxidative process within cells. In high concentrations, there is an increase in the depth of respiration within a few seconds. The patient cannot voluntarily hold his or her breath. Violent convulsions occur after twenty to thirty seconds, with cessation of respiration within one minute. Cardiac failure usually follows within a few minutes.
- Lung-Damaging Agent Injury. Lung-damaging (choking) agents attack lung tissue, primarily causing pulmonary edema. The principle agents in this group are phosgene, diphosgene, chlorine, and chloropicrin. Treat as for pulmonary edema with assisted ventilation.

Part Four

SURVIVAL, PLANNING, AND CONTACTS

16

SUMMARY OF WEAPONS OF MASS DESTRUCTION

While a warfare or terrorist attack may seem impossible to survive, the opposite is the case much of the time. Thousands survived the horrendous atomic bomb attacks on Hiroshima and Nagasaki; a few victims even survived both explosions. Thousands lived through the destruction of the Twin Towers. While there might be many who cannot escape, a large number can live through a nuclear, biological, or chemical attack, especially if they are prepared.

Some of the guidance of being prepared for an attack from the U.S. Department of Homeland Security (DHS) summarizes what the ordinary citizen can do to enhance his or her chance of living through an attack. The information that follows is largely adapted from DHS guidance, CDC advice, WHO guidance, and general military procedures.

NUCLEAR ATTACK

A nuclear detonation is an explosion with extremely intense light and heat, a destructive over-pressure wave, and widespread radioactive material that can contaminate the air, water, and ground surfaces for miles around. While experts predict at this time that a nuclear attack is less likely than other types of warfare, terrorism by its nature is unpredictable.

If there is a nuclear detonation:

- Do not look in the direction of the blast; take cover immediately, below ground or indoors away from exterior walls if possible, although almost any shield or shelter will help protect you from the immediate effects of the blast and the pressure wave.

- Quickly assess the situation.
- Consider if you can get out of the area, or if it would be better to go or stay inside a building and follow your plan to "shelter-in-place" (discussed later in this chapter).
- In order to limit the amount of radiation you are exposed to, think about shielding, distance and time.
  - Shielding: If you have a thick shield between yourself and the radioactive materials, more of the radiation will be absorbed, and you will be exposed to less radiation (reinforced concrete and brick buildings are best).
  - Distance: The farther away you are from the blast and the fallout, the lower your exposure (this includes washing off any surface fallout).
  - Time: Minimizing the time spent exposed will reduce your risk (wash quickly and attempt to remove yourself from any sources of radiation).

Use any available information to assess the situation. If there is a significant radiation threat, health care authorities may or may not advise you to take potassium iodide tablets, depending upon a number of factors. Potassium iodide is the same chemical added to your table salt to make it iodized. It may protect your thyroid gland, which is particularly vulnerable, from radioactive iodine exposure. This is more likely in a gaseous release from a commercial nuclear reactor than from a "dirty bomb." While you might consider keeping potassium iodide in your emergency kit in the event of a reactor incident, learn what the appropriate doses are for each of your family members; remember that potassium can be toxic if taken in too large a dose. Speak with your health care provider in advance about what makes sense for your family.

RADIATION

The radiation threat from a "dirty bomb" is due to the use of common explosives to spread radioactive materials over a targeted area. It is not a nuclear detonation. The force of the explosion and radioactive contamination will be more localized. While the explosion will be immediately obvious, the presence of radiation will not be clearly defined until trained personnel with specialized equipment are on the scene. As with any radiation, you want to try to limit exposure.

RADIATION THREAT FROM FALLOUT OR A DIRTY BOMB

To limit the amount of radiation you are exposed to, remember shielding, distance, and time:

- Shielding: If you have a thick shield between yourself and the radioactive materials, more of the radiation will be absorbed, and you will be exposed to less.
- Distance: The farther away you are away from the explosion and the fallout the lower your exposure.
- Time: Minimizing time spent exposed will also reduce your risk.

As with any emergency, local authorities may not be able to immediately provide information on what is happening and what you should do. However, you should watch television, listen to the radio, or check the Internet often, if possible, for official news and information as it becomes available.

BIOLOGICAL ATTACK

A biological attack is the deliberate release of bacteria, viruses, rickettsia, or toxins, or any other biological substances that can make you sick. Many agents must be inhaled, but others can enter through a cut in the skin or be eaten to make you sick. Some biological agents, such as anthrax, do not cause contagious diseases. Others, like the smallpox virus, can result in diseases you can catch from other people.

Unlike an explosion, a biological attack may or may not be immediately obvious. While it is possible that you will see signs of a biological attack, as was the case with the anthrax mailings of 2002, it is perhaps more likely that local health care workers will report a pattern of unusual illness, or there will be a wave of sick people seeking emergency medical attention. A few locations have set up biological monitors, but even then, results will take hours to be available. You will probably learn of the danger through an emergency radio or television broadcast or some other signal used by your community. You might get a telephone call, or emergency response workers may come to your door.

In the event of a biological attack, public health officials may not immediately be able to provide information on what you should do. It will take time to determine exactly what the illness is, how it should be treated, and who is in danger. However, you should watch television, listen to the radio, or check the Internet, if possible, for official news, including the following:

- Are you in the group or in an area that authorities consider in danger?
- What are the signs and symptoms of the disease?
- Are medications or vaccines being distributed?
- Where can the medications be obtained?
- Who should get the medications?
- Where should you seek emergency medical care if you or a family member becomes sick?

If you become aware of an unusual or suspicious release of an unknown substance nearby, it doesn't hurt to protect yourself. Quickly get away. Cover your mouth and nose with layers of fabric that can filter the air, but still allow breathing. Examples include two to three layers of cotton such as a tee-shirt, handkerchief, or towel. Otherwise, several layers of tissue or paper towels may help. Wash with soap and water, and contact authorities.

At the time of a declared biological emergency, if a family member becomes sick, it is important to be suspicious. Do not automatically assume, however, that you should go to a hospital emergency room or that any illness or symptom is the result of the biological attack. Symptoms of many common illnesses may overlap. Use common sense, practice good hygiene and cleanliness to avoid spreading infection, and seek medical advice. Much of the advice for protection against a chemical agent attack below also applies to a biological attack if it is known that an attack has taken place.

CHEMICAL ATTACK

A chemical attack is the deliberate release of a toxic gas, liquid, or solid that can poison people and the environment. An unusual odor may be present, or it may not. Early symptoms include burning of the eyes, respiratory tract, and mucous membranes. The nerve agents will cause a tightness in the chest, salivation, and pulmonary edema with possible shaking, sweating, and loss of bodily functions. The vomiting agents will cause emesis. Blister agents will cause a reddening or burning of the skin and burning of the eyes and nose.

Possible Signs of Chemical Threat

- Many people suffering from watery eyes, twitching, choking, having trouble breathing or losing coordination
- Many sick or dead birds, fish, or small animals are also cause for suspicion.

If You See Signs of Chemical Attack

- Quickly try to define the impacted area or where the chemical is coming from, if possible.
- Take immediate action to get away if you can avoid the agent cloud; if not, remain indoors.
- If the chemical is inside a building where you are, get out of the building without passing through the contaminated area, if possible, and move upwind.
- Otherwise, it may be better to move as far away from where you suspect the chemical release is, and "shelter-in-place" (discussed later in this chapter).
- If you are outside, quickly decide what is the fastest and safest escape route from the chemical threat. Consider if you can get out of the area, or if you should follow plans to "shelter-in-place."

HIGH RISE BUILDINGS

If you find yourself in a tall building when an emergency occurs:

- Note where the closest emergency exit is.
- Be sure you know another way out in case your first choice is blocked.
- Take cover against a desk or table if things are falling.
- Move away from file cabinets, bookshelves or other things that might fall.
- Face away from windows and glass.
- Move away from exterior walls.
- Determine if you should stay put by "shelter-in-place," or get away.
- Listen for and follow instructions.
- Take your emergency supply kit, unless there is reason to believe it has been contaminated.
- Do not use elevators.
- Stay to the right while going down stairwells to allow emergency workers to come up.

If there is an explosion:

- Take shelter against your desk or a sturdy table.
- Exit the building as soon as safe to do so.
- Do not use elevators.
- Check for fire and other hazards.
- Take your emergency supply kit, if time allows.

If there is a fire:

- Exit the building as soon as safe to do so.
- Crawl low if there is smoke.
- Use a wet cloth, if possible, to cover your nose and mouth.
- Use the back of your hand to feel the upper, lower, and middle parts of closed doors.
- If the door does not feel hot, brace yourself against it and open slowly, being prepared to quickly close it.
- If the door is hot, do not open it; look for another way out.
- Do not use elevators.
- If you catch fire, do not run; stop-drop-and-roll to put out the fire.
- If you are at home, go to a previously designated meeting place.
- Account for your family members, and carefully supervise small children.
- Never go back into a burning building.

If you are trapped in debris:

- If possible, use a flashlight to signal your location to rescuers; do not use a cigarette lighter since explosive gases may be present.
- Avoid unnecessary movement so that you don't kick up dust or further injury yourself.
- Cover your nose and mouth with anything you have on hand; dense-weave cotton material can act as a good filter; try to breathe through the material.
- Tap on a pipe or wall so that rescuers can hear where you are.
- If possible, use a whistle to signal rescuers.
- Shout only as a last resort or if you can hear rescue workers; shouting can cause you to inhale dangerous amounts of potentially toxic dust.

PRE-ATTACK PLANNING

When preparing for any possible emergency situation, it's best to think first about the basics of survival: fresh water, food, clean air, and protection from the elements. Store and prepare enough food for at least three days of survival. Store enough clean water for both drinking and cleaning. Learn how to improvise with what you have on hand to protect your mouth, nose, eyes, and cuts in your skin. Have blankets, bedding, and spare clothing on hand. Have a First Aid kit, and know how to treat minor injuries. Plastic sheeting and duct tape may be useful to keep out harmful contaminants. If you have these basic supplies you are better prepared to help your loved ones when they are hurt.

A checklist of all needed items should be prepared in advance and regularly checked to ensure all the items are present and serviceable. Assemble clothing, bedding, tools, and other basic supplies, including spare electrical batteries. Be sure to include any special needs items for babies, adults, seniors, and people with disabilities. Ensure there is sufficient supply of needed medications on hand.

You should plan in advance and practice what you will do in an emergency. Be prepared to assess the situation, and use common sense and whatever you have on hand to take care of yourself and your loved ones. Think about the places where your family spends time: school, work, ball fields, and other places you frequent. Ask about their emergency plans. Find out how they will communicate with families during an emergency. If they do not have an emergency plan, consider helping develop one.

CREATE A FAMILY PLAN

Your family may not be together when any disaster strikes, so plan how you will contact one another, and review what you will do in different situations.

- It may be easier to make a long-distance phone call than to call across town, so an out-of-state contact may be in a better position to communicate among separated family members.
- Be sure every member of your family knows the phone number and has coins or a prepaid phone card to call the emergency contact.
- You may have trouble getting through, or the telephone system may be down altogether, but be patient.

EMERGENCY INFORMATION

Find out what kinds of disasters, natural or man-made, are most likely to occur in your area and how you will be notified. Methods of getting your attention vary from community to community. One common method is to broadcast via emergency radio and television broadcasts. You might also hear a special siren, or get a telephone call, or emergency workers may go door-to-door. Information is available from the closest chapter of the American Red Cross for emergency systems that apply to your community.

EMERGENCY PLANS

You may also want to inquire about emergency plans at places where your family spends time: at work, athletic fields, or daycare and school. If no plans exist, consider volunteering to help create one. Talk to your neighbors about how you can work together in the event of an emergency. You will be better prepared to safely reunite your family and loved ones during an emergency if you think ahead and communicate with others in advance.

DECIDING TO STAY OR GO

Depending on your circumstances and the nature of the attack, the first important decision is whether you stay put or try to get away. You should understand and plan for both possibilities. Use common sense and available information, including what you are learning here, to determine if there is immediate danger.

In any emergency, local authorities may or may not immediately be able to provide information on what is happening and what you should do. However, you should monitor television or radio news reports for information or official instructions as they become available. If you're specifically told to evacuate or seek medical treatment, do so immediately.

STAYING PUT

Whether you are at home, work or elsewhere, there may be situations when it's simply best to stay where you are and avoid any uncertainty outside. Heavy traffic may clog escape routes to the point where no one can move.

There are other circumstances when staying put and creating a barrier between yourself and potentially contaminated air outside, a process known as "shelter-in-place," is a matter of survival. Use available information and judgement to assess the situation. If you see large amounts of debris in the air, or if local authorities say the air is badly contaminated, you may want to take this kind of action.

To "Shelter-in-Place"

People should choose a room in their house or apartment for their shelter. The best room to use for the shelter is a room with as few windows and doors as possible. A large room, preferably with a water supply, is desirable, something like a master bedroom that is connected to a bathroom. For chemical incidents, this room should be as high in the structure as possible to avoid vapors or gases that are heavier than air and will sink. Note that this guideline is different from the sheltering-in-place technique used in tornadoes and other severe weather, when the shelter should be low in the home.

People might not be at home if the need to shelter in place ever arises, but if they are, it is good to have the following items on hand. Ideally, all of these items should be kept in that room to save time:

- First aid kit
- Food and bottled water. One gallon of water per person in plastic bottles as well as ready-to-eat foods that will keep without refrigeration should be stored at the shelter-in-place location. If bottled water no longer is available, water in a toilet tank (the one at the back of the toilet, *not* the toilet bowl) is suitable for drinking, unless you have been notified that the water supply has been previously contaminated.
- Flashlight, battery-powered radio or television, and extra batteries for both
- Duct tape and scissors or knife
- Towels and plastic sheeting
- A working telephone

How will you know if you need to shelter in place?

- People will hear from the local police, emergency coordinators, or government on radio and television if they need to shelter in place.
- If there is a "code red" or "severe" terror alert, people should pay attention to radio and television broadcasts to know right away whether a shelter-in-place alert is announced for their area.

- If people are away from their shelter-in-place location when a chemical incident occurs, they should follow the instructions of emergency coordinators to find the nearest shelter. If children are at school, they will be sheltered there. Unless instructed to do so, parents should not try to get to the school to bring their children home.

WHAT TO DO IN A CHEMICAL AGENT INCIDENT

People should act quickly and follow the instructions of their local emergency coordinators. Every situation can be different, so local emergency coordinators might have special instructions to follow. In general, do the following:

- Go inside as quickly as possible.
- *If there is time*, shut and lock all outside doors and windows. Locking them may provide a tighter seal against the chemical. Turn off the air conditioner or heater. Turn off all fans, too. Close the fireplace damper and any other place that air can come in from the outside.
- Go in the shelter-in-place room and shut the door.
- Tape plastic over any windows in the room. Use duct tape around the windows and doors and make an unbroken seal. Use the tape over any vents into the room, and seal any electrical outlets or other openings. Sink and toilet drain traps should have water in them (you can use the sink and toilet as you normally would, unless notified that the water supply is contaminated). Push a wet towel up against the crack between the door and the floor to seal it. If it is necessary to drink water, drink the stored water, not water from the tap.
- Turn on the radio. Keep a telephone close at hand, but don't use it unless there is a serious emergency.

Sheltering in this way should keep people safer than if they are outdoors. They will most likely not be in the shelter for more than a few hours. People should listen to the radio for an announcement indicating that it is safe to leave the shelter.

In summary, remember:

- Bring your family and pets inside.
- Lock doors; close windows, air vents, and fireplace dampers.
- Turn off fans, air conditioning and forced air heating systems.
- Take your emergency supply kit, unless you have reason to believe it has been contaminated.
- Go into an interior room with few windows, if possible.
- Seal all windows, doors, and air vents with plastic sheeting and duct tape; consider measuring and cutting the sheeting in advance to save time.
- Be prepared to improvise and use what you have on hand to seal gaps so that you create a barrier between yourself and any contamination.

- Local authorities may not immediately be able to provide information on what is happening and what you should do; however, you should watch television, listen to the radio, or check the Internet often, if possible, for official news and instructions as they become available.

GETTING AWAY

There may be conditions under which you will decide to get away, or there may be situations when you are ordered to leave. Plan how you will assemble your family, and anticipate where you will go. Choose several destinations in different directions so you have options in an emergency. Follow all instructions exactly since most localities have preplanned escape routes and procedures. Going the wrong way might either be dangerous or harmful to others.

Create and execute an evacuation plan:

- Plan places where your family will meet, both within and outside of your immediate neighborhood.
- If you have a car, keep at least a half tank of gas in it at all times in case you need to evacuate.
- Become familiar with alternate routes and other means of transportation out of your area.
- If you do not have a car, plan how you will leave if you have to.
- Take your emergency supply kit, unless you have reason to believe it has been contaminated.
- Lock the door behind you.
- Take your pets with you, but understand that only service animals, like seeing-eye dogs, may be permitted in public shelters; plan how you will care for your pets in an emergency; and have extra food, water and supplies for your pet.

If time allows:

- Call or e-mail your out-of-state contact specified in your family communications plan.
- Tell him or her where you are going.
- If there is damage to your home and you are instructed to do so, shut off water, gas, and electricity before leaving.
- Leave a note telling others when you left and where you are going.
- Check with neighbors who may need a ride.

Learn how and when to turn off utilities:

If there is damage to your home or you are instructed to turn off your utilities:

- Locate the electric, gas, and water shut-off valves.
- Keep necessary tools near gas and water shut-off valves.
- Teach family members how to turn off utilities.
- If you do turn the gas off, a professional must turn it back on; do not attempt to do this yourself.

AT WORK OR AT SCHOOL

Schools, daycare providers, workplaces, neighborhoods, and apartment buildings should all have site-specific emergency plans. Ask about plans at the places where your family spends the most time: work, school, and other places you frequent. If none exist, consider volunteering to help develop one. You will be better prepared to safely reunite your family and loved ones during an emergency if you think ahead and communicate with others in advance.

Employers:

If you are an employer, make sure your workplace has a building evacuation plan that is regularly practiced.

- Take a critical look at your heating, ventilation, and air conditioning (HVAC) system to determine if it is secure, or if it could feasibly be upgraded to better filter potential contaminants, and be sure you know how to turn it off, if you need to.
- Think about what to do if your employees can't go home.
- Make sure you have appropriate supplies on hand.
- Have a business crisis management plan to ensure continued operation of your business.

U.S. Occupational Safety and Health Administration (OSHA) Evacuation Matrix:

OSHA has produced a matrix for evacuation in the event of an attack. OSHA offers the guidance to assist employers and workers who are interested in implementing plans and procedures that may reduce the likelihood of a terrorist incident and reduce the effect of a terrorist release, should a terrorist incident occur at a workplace. However, the guidance does not create legal obligations for employers or create rights for third parties. Legal obligations under the OSH Act are created by statute, regulations, and standards.

Recent terrorist events in the United States underscore the importance of workplace evacuation planning. Consequently, OSHA developed the Evacuation Planning Matrix to provide employers with planning considerations and on-line resources that may help employers reduce their vulnerability to a terrorist act or the impact of a terrorist release. Terrorist incidents are not emergencies that OSHA expects an employer to reasonably anticipate. However, if a terrorist release does occur in or near your workplace, an effective evacuation plan increases the likelihood that your employees will reach shelter safely.

Since terrorism can impact employers and workers, OSHA is committed to strengthening workplace planning and preparedness so that employers and workers may better protect themselves and reduce the likelihood that they may be harmed in the event of a terrorist incident. OSHA continues to work with other Federal response agencies, including the Federal Emergency Management Agency (FEMA), the Environmental Protection Agency (EPA), the U.S. Soldier Biological and Chemical Command (SBCCOM), the Centers for Disease Control and Prevention (CDC) and, within CDC, the National Institute for Occupational Safety and Health (NIOSH), to provide accurate, current information in this rapidly developing area of occupational safety and health.

In assessing the risk of a terrorist release, OSHA drew on the FBI definition of terrorism and defined terrorist release as the release of a chemical, biological, radiological or nuclear material (commonly identified as a Weapon of Mass Destruction (WMD), or of another hazardous substance, performed as a violent act dangerous to human life and intended to further political or social objectives.

In order to use this evacuation guidance effectively, an employer must first assess the risk of a terrorist release in the workplace. The level of risk is a combination of workplace vulnerabilities, recognized threat, and anticipated consequences of the event. This kind of assessment is not a typical safety and health evaluation. However, guidance on conducting such an assessment is becoming more widely available. For many employers, *Best Practices in Workplace Security*, a homeland security guide developed by the State of South Carolina and available online at http://www.llr.state.sc.us/workplace/Full%20Report.pdf, in PDF format, can offer valuable assistance. Its Worksite Risk Assessment List helps employers assess risk based on the following terrorism risk factors if the facility:

- uses, handles, stores or transports hazardous materials.
- provides essential services, *e.g.*, sewer treatment, electricity, fuels, telephone, *etc.*
- has a high volume of pedestrian traffic.
- has limited means of egress, such as a high rise complex or underground operations.
- is considered a high profile site, such as a water dam, military installation, or classified site; or is part of the transportation system, such as shipyard, bus line, trucking, airline.

If these risk factors apply to your work site and cannot be eliminated, you may face greater vulnerability to a terrorist release than other workplaces. To assess the potential threat and consequences of a terrorist release at or near your workplace, consult local law enforcement, the local FBI, and/or the local emergency planning committee (LEPC) (the EPA's database for LEPCs is available online at http://www.epa.gov/ceppo/lepclist.htm). You will need information provided by these agencies to complete your overall risk assessment and to determine which of the three risk zones noted below best characterizes your workplace.

Chemical facilities can use the U.S. Department of Justice (DoJ) Chemical Facility Vulnerability Assessment Methodology, online at http://www.ojp.usdoj.gov/nij/pubs-sum/195171.htm, to assess workplace vulnerabilities. Although this document also discusses threat and consequence assessment, you still will need input from local law enforcement, local FBI, and/or your local LEPC to complete your evaluation.

Using OSHA's Evacuation Planning Matrix The Matrix is not a compliance tool for conducting a comprehensive compliance evaluation of an emergency plan developed to comply with the Emergency Action Plan Standard (29 CFR 1910.38) or the Hazardous Waste Operations and Emergency Response Standard (29 CFR 1910.120(q)). Rather, this document covers the general aspects of emergency planning and includes broad questions to help employers review their plan in light of an indoor or outdoor terrorist release. The document also offers basic planning and preparedness measures for workplaces in each of three risk zones and on-line resources for assistance. After you complete the terrorism risk assessment, review the description of each

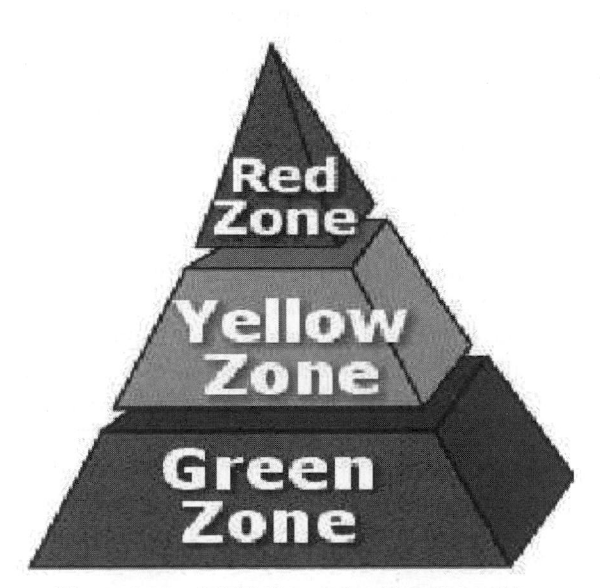

Figure 16.1 OSHA Evacuation Zone Pyramid

risk zone to see where your workplace fits best then examine the planning considerations for that zone.

OSHA Terrorist Release Risk Categories OSHA shows the zones in the shape of a pyramid to represent how the nation's workplaces appear to be distributed within the zones. Based on information currently available, the vast majority of American workplaces are at low risk for a terrorist release, *i.e.* are in the Green Zone. The questions, recommendations, and on-line resources in each risk zone build on those in the zone below it. For example, the Yellow Zone includes both the information in the Green Zone and additional information for Yellow Zone workplaces.

- Green Zone: Workplaces that are not likely to be a target for a terrorist release because they are characterized by limited vulnerability, limited threat, and limited potential for significant impact (consequence). Note: If the workplaces near you seem to be in a higher zone, you may wish to review and implement the planning/preparedness considerations in the Yellow Zone.
- Yellow Zone: Workplaces that may be targets because they are characterized by high vulnerability or high threat or a potentially significant impact (consequence) but not more than one of these. Note: If the workplaces near you seem to be in a higher zone, you may wish to review and implement the planning/preparedness considerations in the Red Zone.
- Red Zone: Workplaces that are most likely to be targets because they are characterized by two or more of the following: high vulnerability, high threat, and potentially catastrophic impact (consequence). Such workplaces need to consider sheltering employees in place as well as evacuation, and may consider assigning some terrorist incident response roles to their own employees.

Note: The color-coded risk levels in this OSHA Matrix do not equate to the Threat Levels in the Homeland Security Advisory System developed by the Department of Homeland Security (DHS). However, employers that place themselves in the Yellow or Red risk levels may consider implementing sequential preparedness measures consistent with those listed in the Homeland Security Presidential Directive-3 (which describes Threat Levels) for federal agencies.

Because of the vast number and types of workplaces in the U.S., this Matrix provides broad information applicable to most workplaces. If you want to modify your plan to address specific considerations, you can get additional information from online resources identified. For additional information about workplace emergency planning, see OSHA's Emergency Response Technical Links webpage.

As a nation, our understanding of the risk of terrorist releases and the agents involved continues to evolve. It is likely that OSHA's recommendations for preparedness, training, and equipment also will evolve. OSHA remains committed to helping employers and workers protect themselves from the risk of terrorism in the workplace and is working closely with other Federal agencies to provide employers with current information and guidance.

Neighborhoods and Apartment Buildings:

A community working together during an emergency makes sense. Sharing plans and communicating in advance is a good strategy.

- Talk to your neighbors about how you can work together during an emergency.
- Find out if anyone has specialized equipment such as a power generator or radiac instrument, or expertise such as medical knowledge, that might help in a crisis.
- Decide who will check on elderly or disabled neighbors.
- Make back-up plans for children in case you can't get home in an emergency.

Schools and Daycare Facilities:

If you are a parent, or the guardian of an elderly or disabled adult, make sure schools and daycare providers have emergency response plans.

- Ask how they will communicate with families during a crisis.
- Ask if they store adequate food, water and other basic supplies.
- Find out if they are prepared to "shelter-in-place," if need be, and where they plan to go if they must get away.

In a moving vehicle:

- If there is an explosion or other factor that makes it difficult to control the vehicle, pull over, stop the car, and set the parking brake.
- If the emergency could impact the physical stability of the roadway, avoid overpasses, bridges, power lines, signs and other hazards.
- If a power line falls on your car, you are at risk of electrical shock; stay inside until a trained person removes the wire.
- Listen to the vehicle radio for information and instructions as they become available.

IF YOU THINK YOU HAVE BEEN EXPOSED TO A CHEMICAL AGENT

If your eyes are watering, your skin is stinging, you are having trouble breathing, or you have a tightness in the chest, you may have been exposed to a chemical agent. If you think you may have been exposed to a chemical agent:

- Strip off clothing immediately, especially if a liquid agent is present.
- Look for a hose, fountain, or any other source of water, and wash with soap, if possible, being sure not to scrub the chemical into your skin.
- Seek emergency medical attention.

CONTACTS AND SOURCES OF INFORMATION

U.S. Department of Homeland Security (DHS):

The agencies which are part of the Department of Homeland Security are housed in one of four major directorates: Border and Transportation Security, Emergency Preparedness and Response, Science and Technology, and Information Analysis and Infrastructure Protection.

The Border and Transportation Security directorate brings the major border security and transportation operations under one roof, including:

- The U.S. Customs Service (Treasury)
- The Immigration and Naturalization Service (some parts) (Justice)
- The Federal Protective Service (GSA)
- The Transportation Security Administration (Transportation)
- Federal Law Enforcement Training Center (Treasury)
- Animal and Plant Health Inspection Service (some parts)(Agriculture)
- Office for Domestic Preparedness (Justice)

The Emergency Preparedness and Response directorate oversees domestic disaster preparedness training and coordinates government disaster response. It brings together:

- The Federal Emergency Management Agency (FEMA)
- Strategic National Stockpile and the National Disaster Medical System (Department of Health and Human Services)
- Nuclear Incident Response Team (Department of Energy)
- Domestic Emergency Support Teams (Department of Justice)
- National Domestic Preparedness Office (Federal Bureau of Investigation)

The Science and Technology directorate seeks to utilize all scientific and technological advantages when securing the homeland. The following assets are part of this effort:

- CBRN Countermeasures Programs (Department of Energy)
- Environmental Measurements Laboratory (Department of Energy)
- National BW Defense Analysis Center (Department of Defense)
- Plum Island Animal Disease Center (Department of Agriculture)

The Information Analysis and Infrastructure Protection directorate analyzes intelligence and information from other agencies (including the CIA, FBI, DIA and

NSA) involving threats to homeland security and evaluate vulnerabilities in the nation's infrastructure. It brings together:

- Critical Infrastructure Assurance Office (Department of Commerce)
- Federal Computer Incident Response Center (General Services Administration)
- National Communications System (Department of Defense)
- National Infrastructure Protection Center (Federal Bureau of Investigation)
- Energy Security and Assurance Program (Department of Energy)

The U.S. Secret Service and the U.S. Coast Guard are also located in the Department of Homeland Security, remaining intact and reporting directly to the Secretary. In addition, the Immigration and Naturalization Service (INS) adjudications and benefits programs reports directly to the Deputy Secretary as the Bureau of Citizenship and Immigration Services.

U.S. Federal Emergency Management Agency (FEMA) State and Territory Offices

Alabama Emergency Management Agency
5898 County Road 41
P.O. Drawer 2160
Clanton, Alabama 35046-2160
(205) 280-2200
(205) 280-2495 Fax
http://www.aema.state.al.us

Alaska Division of Emergency Services
P.O. Box 5750
Fort Richardson, Alaska 99505-5750
(907) 428-7000
(907) 428-7009 Fax
http://www.ak-prepared.com

American Samoa Territorial Emergency Management Coordination (TEMCO)
American Samoa Government
P.O. Box 1086
Pago Pago, American Samoa 96799
(011)(684) 699-6415
(011)(684) 699-6414 Fax

Arizona Division of Emergency Management
5636 E. McDowell Rd
Phoenix, Arizona 85008
(602) 244-0504 or 1-800-411-2336
http://www.dem.state.az.us

Arkansas Department of Emergency Management
P.O. Box 758
Conway, Arkansas 72033
(501) 730-9750
(501) 730-9754 Fax
http://www.adem.state.ar.us/

California Governor's Office of Emergency Services
P.O. Box 419047
Rancho Cordova, California 95741-9047
(916) 845-8510
(916) 845-8511 Fax
http://www.oes.ca.gov/

Colorado Office of Emergency Management
Division of Local Government
Department of Local Affairs
15075 South Golden Road
Golden, Colorado 80401-3979
(303) 273-1622
(303) 273-1795 Fax
www.dola.state.co.us/oem/oemindex.htm

Connecticut Office of Emergency Management
Military Department
360 Broad Street
Hartford, Connecticut 06105
(860) 566-3180
(860) 247-0664 Fax
http://www.mil.state.ct.us/OEM.htm

Delaware Emergency Management Agency
165 Brick Store Landing Road
Smyrna, Delaware 19977

(302) 659-3362
(302) 659-6855 Fax
http://www.state.de.us/dema/index.htm

District of Columbia Emergency Management Agency
2000 14th Street, NW, 8th Floor
Washington, D.C. 20009
(202) 727-6161
(202) 673-2290 Fax
http://www.dcema.dc.gov

Florida Division of Emergency Management
2555 Shumard Oak Blvd.
Tallahassee, Florida 32399-2100
(850) 413-9969
(850) 488-1016 Fax
www.floridadisaster.org

Georgia Emergency Management Agency
P.O. Box 18055
Atlanta, Georgia 30316-0055
(404) 635-7000
(404) 635-7205 Fax
http://www.State.Ga.US/GEMA/

Government of Guam
Office of Civil Defense
P.O. Box 2877
Hagatna, Guam 96932
(011)(671) 475-9600
(011)(671) 477-3727 Fax
http://ns.gov.gu/

Hawaii State Civil Defense
3949 Diamond Head Road
Honolulu, Hawaii 96816-4495
(808) 733-4300
(808) 733-4287 Fax
http://www.scd.state.hi.us

Idaho Bureau of Disaster Services
4040 Guard Street, Bldg. 600
Boise, Idaho 83705-5004
(208) 334-3460
(208) 334-2322 Fax
http://www.state.id.us/bds/bds.html

Illinois Emergency Management Agency
110 East Adams Street
Springfield, Illinois 62701
(217) 782-2700
(217) 524-7967 Fax
http://www.state.il.us/iema

Indiana State Emergency Management Agency
302 West Washington Street
Room E-208 A
Indianapolis, Indiana 46204-2767
(317) 232-3986
(317) 232-3895 Fax
http://www.ai.org/sema/index.html

Iowa Division of Emergency Management
Department of Public Defense
Hoover Office Building
Des Moines, Iowa 50319
(641) 281-3231
(641) 281-7539 Fax
http://www.state.ia.us/government/dpd/emd/index.htm

Kansas Division of Emergency Management
2800 SW Topeka Boulevard
Topeka, Kansas 66611-1287
(785) 274-1401
(785) 274-1426 Fax
http://www.ink.org/public/kdem/

Kentucky Emergency Management
Emergency Operations Center (EOC) Building
100 Minuteman Parkway Bldg. 100
Frankfort, Kentucky 40601-6168
(502) 607-1682
(502) 607-1614 Fax
http://kyem.dma.state.ky.us

Louisiana Office of Emergency Preparedness
7667 Independence Blvd.
Baton Rouge, Louisiana 70806
(225) 925-7500
(225) 925-7501 Fax
http://www.loep.state.la.us

Maine Emergency Management Agency
State Office Building, Station 72
Augusta, Maine 04333
(207) 626-4503
(207) 626-4499 Fax
http://www.state.me.us/mema/memahome.htm

Commonwealth of the Northern Mariana Islands
CNMI Emergency Management Office
Office of the Governor
P.O. Box 10007
Saipan, Mariana Islands 96950
(670) 322-9529
(670) 322-7743 Fax
http://www.cnmiemo.org/

Marshall Islands National Disaster Management Office
Office of the Chief Secretary
P.O. Box 15
Majuro, Republic of the Marshall Islands 96960-0015
(011)(692) 625-5181
(011)(692) 625-6896 Fax

Maryland Emergency Management Agency
Camp Fretterd Military Reservation
5401 Rue Saint Lo Drive
Reistertown, Maryland 21136
(410) 517-3600
(877) 636-2872 Toll-Free
(410) 517-3610 Fax
http://www.mema.state.md.us/

Massachusetts Emergency Management Agency
400 Worcester Road
Framingham, Massachusetts 01702-5399
(508) 820-2000
(508) 820-2030 Fax
http://www.state.ma.us/mema

Michigan Division of Emergency Management
4000 Collins Road
P.O. Box 30636
Lansing, Michigan 48909-8136
(517) 333-5042
(517) 333-4987 Fax
http://www.msp.state.mi.us/division/emd/emdweb1.htm

Federated States of Micronesia
National Disaster Control Officer
P.O. Box PS-53
Kolonia, Pohnpei-Micronesia 96941
(011)(691) 320-8815
(001)(691) 320-2785 Fax

Minnesota Division of Emergency Management
Department of Public Safety
Suite 223
444 Cedar Street
St. Paul, Minnesota 55101-6223
(651) 296-2233
(651) 296-0459 Fax
http://www.dps.state.mn.us/emermgt/

Mississippi Emergency Management Agency
P.O. Box 4501-Fondren Station
Jackson, Mississippi 39296-4501
(601) 352-9100
(800) 442-6362 Toll Free
(601) 352-8314 Fax
http://www.mema.state.ms.us

Missouri Emergency Management Agency
P.O. Box 16
2302 Militia Drive
Jefferson City, Missouri 65102
(573) 526-9100
(573) 634-7966 Fax
http://www.sema.state.mo.us/semapage.htm

Montana Division of Disaster & Emergency Services
1100 North Main
P.O. Box 4789
Helena, Montana 59604-4789
(406) 841-3911
(406) 444-3965 Fax
http://www.state.mt.us/dma/des/index.shtml

Nebraska Emergency Management Agency
1300 Military Road
Lincoln, Nebraska 68508-1090
(402) 471-7410
(402) 471-7433 Fax
http://www.nebema.org

Nevada Division of Emergency Management
2525 South Carson Street
Carson City, Nevada 89711
(775) 687-4240
(775) 687-6788 Fax
http://dem.state.nv.us/

New Hampshire Governor's Office of Emergency Management
State Office Park South
107 Pleasant Street
Concord, New Hampshire 03301
(603) 271-2231
(603) 225-7341 Fax

New Jersey Office of Emergency Management
Emergency Management Bureau
P.O. Box 7068
West Trenton, New Jersey 08628-0068
(609) 538-6050 Monday–Friday
(609) 882-2000 Ext 6311 (24 hours, 7 days per week)
(609) 538-0345 Fax
http://www.state.nj.us/oem/county/

New Mexico Department of Public Safety
Office of Emergency Services & Security
P.O. Box 1628
13 Bataan Boulevard
Santa Fe, New Mexico 87505
(505) 476-9600
(505) 476-9695 Fax
http://www.dps.nm.org/emergency/index.htm
http://www.dps.nm.org/emc.htm

New York State Emergency Management Office
1220 Washington Avenue
Building 22, Suite 101
Albany, New York 12226-2251
(518) 457-2222
(518) 457-9995 Fax
http://www.nysemo.state.ny.us/

North Carolina Division of Emergency Management
116 West Jones Street
Raleigh, North Carolina 27603
(919) 733-3867
(919) 733-5406 Fax
http://www.dem.dcc.state.nc.us/

North Dakota Division of Emergency Management
P.O. Box 5511
Bismarck, North Dakota 58506-5511
(701) 328-8100
(701) 328-8181 Fax
http://www.state.nd.us/dem

Ohio Emergency Management Agency
2855 W. Dublin Granville Road
Columbus, Ohio 43235-2206
(614) 889-7150
(614) 889-7183 Fax
http://www.state.oh.us/odps/division/ema/

Oklahoma Office of Civil Emergency Management
Will Rogers Sequoia Tunnel 2401 N. Lincoln
Oklahoma City, Oklahoma 73152
(405) 521-2481
(405) 521-4053 Fax
http://www.odcem.state.ok.us/

Oregon Emergency Management
Department of State Police
595 Cottage Street, NE
Salem, Oregon 97310
(503) 378-2911 ext. 225
(503) 588-1378
http://www.osp.state.or.us/oem/oem.htm

Palau NEMO Coordinator
Office of the President
P.O. Box 100
Koror, Republic of Palau 96940
(011)(680) 488-2422
(011)(680) 488-3312

Pennsylvania Emergency Management Agency
P.O. Box 3321
Harrisburg, Pennsylvania 17105-3321
(717) 651-2001

(717) 651-2040 Fax
http://www.pema.state.pa.us/

Puerto Rico Emergency Management Agency
P.O. Box 966597
San Juan, Puerto Rico 00906-6597
(787) 724-0124
(787) 725-4244 Fax

Rhode Island Emergency Management Agency
645 New London Ave
Cranston, Rhode Island 02920-3003
(401) 946-9996
(401) 944-1891 Fax
http://www.state.ri.us/riema/riemaaa.html

South Carolina Emergency Management Division
1100 Fish Hatchery Road
West Columbia South Carolina 29172
(803) 737-8500
(803) 737-8570 Fax
http://www.state.sc.us/epd/

South Dakota Division of Emergency Management
500 East Capitol
Pierre, South Dakota 57501-5070
(605) 773-6426
(605) 773-3580 Fax
http://www.state.sd.us/state/executive/military/sddem.htm

Tennessee Emergency Management Agency
3041 Sidco Drive
Nashville, Tennessee 37204-1502
(615) 741-4332
(615) 242-9635 Fax
http://www.tnema.org

Texas Division of Emergency Management
5805 N. Lamar
Austin, Texas 78752

(512) 424-2138
(512) 424-2444 or 7160 Fax
http://www.txdps.state.tx.us/dem/

Utah Division of Emergency Services and Homeland Security
1110 State Office Building
P.O. Box 141710
Salt Lake City, Utah 84114-1710
(801) 538-3400
(801) 538-3770 Fax
http://www.des.utah.gov

Vermont Emergency Management Agency
Department of Public Safety
Waterbury State Complex
103 South Main Street
Waterbury, Vermont 05671-2101
(802) 244-8721
(802) 244-8655 Fax
http://www.dps.state.vt.us/

Virgin Islands Territorial Emergency Management—VITEMA
2-C Contant, A-Q Building
Virgin Islands 00820
(340) 774-2244
(340) 774-1491

Virginia Department of Emergency Management
10501 Trade Court
Richmond, Virginia 23236-3713
(804) 897-6502
(804) 897-6506 Fax
http://www.vdem.state.va.us

State of Washington Emergency Management Division
Building 20, M/S: TA-20
Camp Murray, Washington 98430-5122
(253) 512-7000
(253) 512-7200 Fax
http://www.wa.gov/wsem/

West Virginia Office of Emergency Services
Building 1, Room EB-80 1900 Kanawha Boulevard, East
Charleston, West Virginia 25305-0360
(304) 558-5380
(304) 344-4538 Fax
http://www.state.wv.us/wvoes

Wisconsin Emergency Management
2400 Wright Street
P.O. Box 7865
Madison, Wisconsin 53707-7865
(608) 242-3232
(608) 242-3247 Fax
http://badger.state.wi.us/agencies/dma/wem/index.htm

Wyoming Emergency Management Agency
5500 Bishop Blvd
Cheyenne, Wyoming 82009-3320
(307) 777-4920
(307) 635-6017 Fax
http://wema.state.wy.us

U.S. Army Medical Research Institute for Infectious Diseases (USAMRIID)

Commander
USAMRIID
Attn: MCMR-UIZ-R
1425 Porter St.
Fort Detrick
Frederick, MD 21702-5011
USAMRIIDweb@amedd.army.mil

U.S. Defense Threat Reduction Agency (DTRA)

For general information, contact:
(703) 767-5870 (voice)
1-800-701-5096 (voice)
(703) 767-4450 (Fax)
dtra.publicaffairs@dtra.mil

For detailed information on specific topics, contact any of the following offices on the Internet:

Chemical-Biological Defense: cb@dtra.mil
Cooperative Threat Reduction Program: ct@dtra.mil
Freedom of Information Act (FOIA) Program: foia@dtra.mil
Technology Development: dtra.publicaffairs@dtra.mil

DTRA has developed the Hazard Prediction and Assessment Capability (HPAC) automated software system to provide the means to accurately predict the effects of hazardous material releases into the atmosphere and its impact on civilian and military populations. The system uses integrated source terms, high-resolution weather forecasts and particulate transport analyses to model hazard areas produced by military or terrorist incidents and industrial accidents.

The Hazard Prediction and Assessment Capability system is a collateral assessment tool available by license for government, government-related, or academic use. It provides the means to accurately predict the effects of hazardous material releases into the atmosphere and its impact on civilian and military populations. It models nuclear, biological, chemical, radiological and high explosive collateral effects resulting from conventional weapon strikes against enemy weapons of mass destruction production and storage facilities. The HPAC system also predicts downwind hazard areas resulting from a nuclear weapon strike or reactor accident and has the capability to model nuclear, chemical and biological weapon strikes or accidental releases.

For more information about the Hazard Prediction and Assessment Capability system contact:

Defense Threat Reduction Agency
Consequences Assessment Branch (TDOC)
6801 Telegraph Road
Alexandria, VA 22310-3398
ACEhelp@dtic.mil
(703) 325-6106
(703) 325-0398 (Fax)

If you meet the qualifications, direct software requests to https://register.dtic.mil/dtra (register online) or visit the users internet site at https://acecenter.dtic.mil (password protected). Note that downloading the HPAC program requires browsers with 128-bit SSL capabilities.

U.S. Centers for Disease Control and Prevention (CDC)

Centers for Disease Control and Prevention
1600 Clifton Rd.

Atlanta, GA 30333

(404) 639-3311

Public inquiries (Mon–Fri 8 AM–11PM Eastern Time; Sat–Sun 10 AM–8PM Eastern Time):

English (888) 246-2675

Español (888) 246-2857

TTY (866) 874-2646

Contacts for use by State and Local Health Officials and healthcare providers:

CDC Emergency Response Hotline (24 hours) 770-488-7100

Program questions 404-639-0385

CDC Bioterrorism Preparedness and Response Program:

Program questions 404-639-0385

U.S. Occupational Safety and Health Administration (OSHA) Regional Offices

Region 10 (WA, AK, ID, OR)

Regional Office

1111 Third Avenue, Suite 715

Seattle, WA 98101-3212

(206) 553-5930

(206) 553-6499 Fax

Region 9 (CA, NV, AZ, HI, American Samoa, Guam)

Regional Office

71 Stevenson Street, Room 420

San Francisco, CA 94105

(415) 975-4310 (Main Public 8:00 AM–4:30 PM Pacific)

(800) 475-4019 (For Technical Assistance)

(800) 475-4020 (For Complaints—Accidents/Fatalities)

(800) 475-4022 (For Publication Requests)

(415) 975-4319 Fax

In case of emergency, call 1-800-321-OSHA

Region 8 (ND, SD, MT, UT, CO, WY)
Regional Office
1999 Broadway, Suite 1690
P.O. Box 46550
Denver, CO 80201-6550
(303) 844-1600
(303) 844-1616 Fax

Region 7 (MO, NE, IA, KS)
Regional Office
City Center Square
1100 Main Street, Suite 800
Kansas City, MO 64105
(816) 426-5861
(816) 426-2750 Fax

Region 6 (AR, LA, TX, NM, OK)
Regional Office
525 Griffin Street, Room 602
Dallas, TX 75202
(214) 767-4731
(214) 767-4693 Fax

Region 5 (OH, IN, WI, MN, MI, IL)
Regional Office
230 South Dearborn Street, Room 3244
Chicago, IL 60604
(312) 353-2220
(312) 353-7774 Fax

Region 4 (NC, SC, GA, FL, TN, KY, AL, MS)
Regional Office
61 Forsyth Street, SW
Atlanta, GA 30303
(404) 562-2300
(404) 562-2295 Fax

Region 3 (PA, MD, WV, DE, VA, DC)
Regional Office
U.S. Department of Labor/OSHA
The Curtis Center-Suite 740 West
170 S. Independence Mall West
Philadelphia, PA 19106-3309
(215) 861-4900
(215) 861-4904 Fax

Region 2 (NJ. NY, VI, PR)
Regional Office
201 Varick Street, Room 670
New York, NY 10014
(212) 337-2378
(212) 337-2371 Fax

Region 1 (ME, NH, VT, RI, MA)
Regional Office
JFK Federal Building, Room E340
Boston, Massachusetts 02203
(617) 565-9860
(617) 565-9827 Fax

17

REFERENCES

GENERAL COMMENTS

In order for this list to be as complete as possible, a number of these references are from activist organizations, including former Soviet clients, as well as from news magazines and political entities; therefore, article content should be approached with caution. Anonymous articles appear first. Inclusion does not imply agreement with the material presented.

REFERENCES

Anonymous

"A Peace Initiative: A Step towards Abolishing Chemical Weapons." *World Marxist Review* 28 (November 1985): 111–114.

"Briefing: At Hand to Deal with an Underhand Attack" *Jane's Defence Weekly* 26 (August 14, 1996): 17.

"Briefing: Counter-Terrorism Technology" *Jane's Defence Weekly* 26 (August 14, 1996): 18.

"Chemical and Biological Warfare Protection for the XM1 Tank." *Military Review* 60 (October 1980): 82.

"Chemical and Biological Warfare: Can a Lid Be Kept On?" *Economist* 284 (July 24, 1982): 74–76.

"Chemical and Biological Weapons." *Arms Control Today* 16 (September 1986): 8–22.

"Chemical Arms Control after the Paris Conference." *Arms Control Today* 19 (January/February 1989): 3–6.

"Chemical Warfare in Southeast Asia and Afghanistan: An Update." *Department of State Bulletin* 82 (December 1982): 44–53.

"Chemical Warfare: Ban the World's Machineguns." *Economist* 307 (June 4, 1988): 19–20.

"Chemical Warfare: The New Face of War." *World Press Review* 36 (March 1989): 11–14.

"Chemical Weapons: A Cloud of Talk." *Economist* 310 (January 14, 1989): 38–39.

"Coming: Nerve Gas That Is Safer." *U.S. News & World Report* 95 (September 26, 1983): 17.

"Conference on Disarmament Continues Work on Chemical Weapons." *U.N. Chronicle* 27 (September 1990): 34.

"Conference on Disarmament Focuses on Chemical Weapons Ban, Nuclear Tests." *U.N. Chronicle* 27 (December 1990): 48–49.

"Conference on Disarmament Urged to Speed Work on Chemical Weapons Ban." *U.N. Chronicle* 27 (June 1990): 30–31.

"Curbing Chemical Warfare." *World Press Review* 31 (June 1984): 35–40.

"Dying in the Rain." *National Review* 36 (April 6, 1984): 19–20.

"Gas Warfare." *U.S. News & World Report* 88 (June 16, 1980): 37.

"GEN Bernard Rogers: Chemical Deterrence Is Imperative." (Interview) *National Guard* 39 (May 1985): 27–28.

"Kaddafi Builds a Poison-Gas Factory." *Newsweek* 112 (November 7, 1988): 72.

"Libya's Chemical Weapons Plant." (U.S. State Department statement, January 1, 1989). *Department of State Bulletin* 89 (March 1989): 71.

"NATO Report Warns of Threat from Weapons of Mass Destruction, 3124465" Armed Forces Newswire Service (March 29, 1996): 221 words.

"Nerve Gas Sparks a Nervous Debate." *U.S. News & World Report* 95 (July 25, 1983): 10.

"New Generation of Nerve-Gas Weapons." *U.S. News & World Report* 95 (September 5, 1983): 65.

"Paris Conference Calls for Complete Ban on Chemical Weapons." *U.N. Chronicle* 26 (June 1989): 38–39.

"Portillo: Conservative on the Defence." *Defense News* (October 30, 1995/November 5, 1995): 19.

"Progress Continues on Text for Chemical Weapons Ban." *U.N. Chronicle* 25 (June 1988): 24–26.

"Prohibition of Chemical Weapons Conference Held in Paris." *Department of State Bulletin* 89 (March 1989): 4–10.

"Review Conference Held on Biological and Toxin Weapons Convention." *Department of State Bulletin* 86 (December 1986): 40–47.

"Security Council Members Condemn Use of Chemical Weapons in Iran-Iraq Conflict; Demand Observance of Geneva Protocol." *U.N. Chronicle* 24 (August 1987): 33–34.

"Should U.S. Gear Up for Gas Warfare?" *U.S. News & World Report* 89 (November 3, 1980): 45–46.

"Stop the Gas: A Ban on Chemical Warfare Doesn't Have to Wait for Nuclear Disarmament." *Economist* 282 (March 6, 1982): 1–15.

"U.S. Chemical Weapons Production: Poisoning the Atmosphere." *Defense Monitor* 18 no. 3 (1989): entire issue.

"U.S. Plan Lets Troops Fight Terrorists." Periscope Daily Defense News Capsules (July 6, 1995). Available from: LEXIS/NEXIS Library: News File: ALLNWS.

"Washington's Yellow Science." *Progressive* 48 (March 1984): 10–11.

World Health Organization (WHO). Vaccination against Argentine Hemorrhagic Fever. *Wkly Epid. Rec.* 68 (1993): 233–234.

Authored

Adams, R.W. "The Threat of Chemical Warfare." *Marine Corps Gazette* 69 (March 1985): 52–59.

Adams, Valerie. "Retaliatory Chemical Warfare Capability—Some Problems for NATO." *Journal of the Royal United Services Institute for Defence Studies* 130 (December 1985): 15–19.

Adelman, Kenneth. "Chemical Weapons: Restoring the Taboo." *Orbis* 30 (Fall 1986): 443–455.

Alter, Jonathan. "Nerve Gas: A New Arms Race?" *Newsweek* 102 (July 25, 1983): 17–18.

Anderson, Andy. "Chemical Warfare Threat in Battle—The Effects of Degradation." *NATO's Sixteen Nations* 34 (December 1989): 57–60.

Anderson, Jack, and Dale Van Atta. "Poison and Plague: Russia's Secret Terror Weapons." *Reader's Digest* 125 (September 1984): 54–58.

Appel, John G., and Charles G. Shaw. "Fighting and Winning when the Enemy Turns to NBC on Battlefield." *Army* 38 (August 1988): 42–46.

Astafyev, Aleksandr. "Storming the Chemical Arsenals." *Soviet Military Review* no. 7 (July 1989): 52–55.

Babiasz, Frank E. "Threat: Chemical Warfare." *United States Army Aviation Digest* 26 (June 1980): 17–18.

Bachman, Denise M., and Joseph Cartelli. "Focal Point—An Alternate to the Program Executive Office in Chemical Nuclear Matters." *Program Manager* 19 (July/August 1990): 36–37.

Bailey, Kathleen C. "Chemical Weapons Proliferation." (Address, July 16, 1988) *Vital Speeches of the Day* 54 (October 1, 1988): 749–751.

Bambini, Adrian P., Jr. "Chemical Warfare and the NATO Alliance." *Military Review* 61 (April 1981): 27–33.

Bartley, Robert L., and William P. Kucewicz. "Yellow Rain and the Future of Arms Agreements." *Foreign Affairs* 61 (Spring 1983): 805–826.

Bartley, Robert L., and William P. Kucewicz. "Yellow Rain." *Foreign Affairs* 61 (Summer 1983): 1185–1191.

Bass, Alison. "One Scientist's Crusade." *Technology Review* 89 (April 1986): 42–54.

Bay, Charles H. "An Update on the Other Gas Crisis: Chemical Weapons." *Parameters* 10 (December 1980): 27–35.

Beal, Clifford. "Invisible Enemy (Biological Weapons) Genie." *International Defense Review* 28 no. 3 (March 1995): 36–41.

Beardsley, Tim. "Easier Said than Done." *Scientific American* 263 (September 1990): 48.

Bennett, Ralph Kinney. "The Growing Menace of Chemical Weapons." *Reader's Digest* 135 (July 1989): 82–87.

Benz, K. G. "NBC Defense—An Overview." *International Defense Review* Part. 1, "Protection Equipment." 16 no. 12 (1983): 1783–1790; Part. 2, "Detection and Decontamination." 17 no. 2 (1984): 159–164.

Beres, Louis Rene. "Israeli Security in a Changing World." *Strategic Review* 18 (Fall 1990): 10–22.

Bergmeister, Francis X. "Chemical Warfare Logistics: Neglect in Need of Reform." *Marine Corps Gazette* 72 (December 1988): 25–26.

Bernstein, Barton J. "Churchill's Secret Biological Weapons." *Bulletin of the Atomic Scientists* 43 (January/February 1987): 46–50.

Bernstein, Barton J. "The Birth of the U.S. Biological Warfare Program." *Scientific American* 256 (June 1987): 116–121.

Bernstein, Barton J. "Why We Didn't Use Poison Gas in World War II." *American Heritage* 36 (August-September 1985): 40–45.

Beyer, Lisa. "Coping with Chemicals." *Time* 137 (February 15, 1991): 47–48.

Biersner, Robert J., and Paul O. Davis. "Needed: Chemical Warfare Defense Doctrine." *Proceedings of the United States Naval Institute* 112 (November 1986): 116–120.

Bodansky, Yossef. "Iranian and Bosnian Leaders Embark on a New Major Escalation of Terrorism against the West." Defense & Foreign Affairs' Strategic Policy 21 no. 8 (August 31, 1993): 6–9.

Bodansky, Yossef. "Saudi Arabia's Leadership Takes Significant Steps to Reduce the Pressure from Radicals." *Defense & Foreign Affairs*' Strategic Policy 23 no. 5 (May 31, 1995): 4–6.

Booth, Diane. "NBC Detection Research: So that Troops Can Survive and Fight On." *Army* 37 (December 1987): 40–44.

Borisov, K. "For Strict Observance of the Bacteriological Weapons Ban." *International Affairs* (Moscow) (August 1983): 95–98.

Bornmann, Karl Gerhard. "Modern Weapons and Equipment Increase the Striking Power of Counter-Terrorist Groups." *Military Technology* 6 (August 1982): 155–158.

Boyle, Dan. "End to Chemical Weapons—What Are the Chances?" *International Defense Review* 21 (September 1988): 1087–1089.

Brennan, R.J., J.F. Waeckerle, T.W. Sharp, and S.R. Lillibridge. Chemical Warfare Agents: Emergency Medical and Emergency Public Health Issues. *Ann. Emerg. Med.* 1999 Aug; 34(2): 191–204.

Brennan, Frank J., Jr. "Gas! How an NBC Defense Company Reacts." *National Guard* 38 (March 1984): 15–17.

Budavari, S.; M.J. O'Niel; A. Smith; P.E. Heckelmann, eds. *The Merck Index*, 11th ed., Merck & Co.: Rahway, NJ. 1996.

Budiansky, Stephen. "Qualified Approval for Binary Chemical Weapons." *Science* 234 (November 21, 1986): 930–932.

Burgess, J.L., M.C. Keifer, S. Barnhart, *et al.* Hazardous Materials Exposure Information Service: Development, Analysis, and Medical Implications. *Ann. Emerg. Med.* 1997 Feb; 29(2): 248–254.

Burt, Richard R. "Use of Chemical Weapons in Asia." *Department of State Bulletin* 82 (January 1982): 52–54.

Burton, John C., III. "CB Winds of Change." *Defense & Foreign Affairs* 8 (No. 7, 1980): 12–15; *Military Review* 60 (December 1980): 22–30.

Bush, George. "U.S. Proposes Banning Chemical Weapons." *Department of State Bulletin* 84 (June 1984): 40–43.

Caen, Deborah. "Chemical Warfare in the Gulf War: Kurdish Victims Win Sympathy and Attention in Western Europe." *Asian Defence Journal* No.2 (February 1989): 50–52.

Campbell, Duncan. "Thatcher Goes for Nerve Gas." *New Statesman* 109 (January 11, 1985): 8–10.

Canine, Craig. "Is Baghdad Using Poison Gas?" *Newsweek* 103 (March 19, 1984): 39–40.

Cannistraro, Vincent M. "Covert Violence Attracts Small Nation Sponsorship." *Signal* 46 no. 4 (December 1991): 32–36.

Carreon, Rodolfo. "NBC Defense in Aviation Operations." *United States Army Aviation Digest* (October 1987): 38–45.

Casarett, Alison P. *Radiation Biology*. Prentice-Hall, Inc.; Englewood Cliffs, NJ. 1968.

Cember, Herman. *Introduction to Health Physics*. 2nd ed. Pergamon Press; New York, NY. 1983.

Champlin, Danny W. "NBC Defense—Task or Condition?" *Marine Corps Gazette* 74 (August 1990): 17–18.

Chapman, Betty. "Chemical Warfare: The Dirty Weapon." *National Defense* 64 (June 1980): 33–37.

Chapman, Betty. "Navy CW Defense." *National Defense* 67 (April 1983): 54–57.

Chapman, Suzann. "U.S. Sees Expanded Troop Threat" *Air Force Magazine* (September, 1996): 16.

Chase, Eric L. "Will We End the Era of Terrorism?" *Marine Corps Gazette* 76 no. 4 (April 1992): 67–68.

Child, P.L., R.B. MacKenzie, L.R. Valverde, and K.M. Johnson. Bolivian Hemorrhagic Fever: A Pathologic Description. *Arch. Pathol. Lab. Med.* 83 (1967): 434–445.

Cohen, Raymond, and Robin Ranger. "Enforcing Chemical Weapons Ban." *International Perspectives* 18 (July/August 1989): 9–12.

Cole, Leonard A. The Specter of Biological Weapons. *Scientific American* 1996:60–65.

Cole, Leonard A. "Operation Bacterium." *Washington Monthly* 17 (July/August 1985): 38–42.

Cole, Leonard A. "Yellow Rain or Yellow Journalism?" *Bulletin of the Atomic Scientists* 40 (August/September 1984): 36–38.

Compton, J. A. F., *Military Chemical and Biological Agents*. Telford Press: Caldwell, NJ, 1988.

Conant, Carleton A. "Libya's CW Gamble: Washington Has Made Allegations of a Major Build-up of Chemical Weapon Manufacturing Capability." *Defense & Foreign Affairs* 17 (January 1989): 30–32.

Copley, Gregory. "Feature Reports: Not to be Forgotten." *Defense & Foreign Affairs* (February, 1989): 33.

Copley, Gregory. "Strategic Trends: Crisis Mismanagement." *Defense & Foreign Affairs*' Strategic Policy (June, 1995): 4.

Cordesman, Anthony H. "Creating Weapons of Mass Destruction." *Armed Forces Journal International* 126 (February 1989): 54.

Cox, R.D. Decontamination and Management of Hazardous Materials Exposure Victims in the Emergency Department. *Ann. Emerg. Med.* 1994 Apr; 23(4): 761–70.

Cox, Frank. "Training with the Real Thing." *Soldiers* 43 (August 1988): 6–9.

Crenshaw, Martha. "Current Research on Terrorism: The Academic Perspective." *Studies in Conflict and Terrorism* 15 no.1 (January-March 1992): 1–11.

Croddy, Eric. "Urban Terrorism–Chemical Warfare in Japan." *Jane's Intelligence Review* 7 (November 1, 1995): 520.

Dando, M.R., and T. Phillips. "Global Security." *Biologist* 43 no. 4 (September 1 1996): 158.

Dashiell, Thomas. "A Realistic Look at Chemical Warfare." *Defense* (January 1981): 16–20.

Daskal, Steven E. "Developments in U.S. Chemical Warfare Defense." *National Defense* 69 (September 1985): 28–32.

Daskal, Steven E. "NBC: Requirements for Defense." *Journal of Defense & Diplomacy* 4 (December 1986): 41–46.

Davidson, C. J. "Situation Report on Chemical Warfare." *Royal United Services Institute for Defence Studies Journal* 125 (June 1980): 63–65.

Dick, C.J. "Soviet Chemical Warfare Capabilities." *International Defense Review* 14 No. 1 (1981): 31–38.

Dick, Charles. "The Soviet Chemical and Biological Warfare Threat." *Royal United Services Institute for Defence Studies Journal* 126 (March 1981): 45–51.

Dickson, David. "Approval Seen for New U.S. Chemical Weapons." *Science* 232 (May 2, 1986): 567–568.

Ditzian, Jan L. "Designing for the NBC Environment." *National Defense* 68 (March 1984): 34–38.

Dodds, Henry. "New Soviet Respirator." *Jane's Soviet Intelligence Review* 2 (October 1990): 468.

Donnelly, C.N. "Winning the NBC War: Soviet Army Theory and Practice." *International Defense Review* 14 No. 8 (1981): 989–996.

Douglass, Joseph D., Jr. "Beyond Nuclear War." *Journal of Social, Political, and Economic Studies* 15 (Summer 1990): 141–156.

Douglass, Joseph D., Jr. "BioChem Warfare: New Dimensions and Implications." *Defense & Foreign Affairs* 15 (April 1987): 41–45.

Douglass, Joseph D., Jr. "Biochemical Warfare: A Warning." *Defense Science* 3 (October 1984): 66–69.

Douglass, Joseph D., Jr. "Chemical Weapons: An Imbalance of Terror." *Strategic Review* 10 (Summer 1982): 36–47.

Douglass, Joseph D., Jr. "Expanding Threat of Chemical-Biological Warfare: A Case of U.S. Tunnel-Vision." *Strategic Review* 14 (Fall 1986): 37–46.

Douglass, Joseph D., Jr. "Soviets Surge in Biochemical Warfare: West Remains Drugged with Apathy." *Armed Forces Journal International* 126 (August 1988): 54.

Douglass, Joseph D., Jr., and H. Richard Lukens. "The Expanding Arena of Chemical-Biological Warfare." *Strategic Review* 12 (Fall 1984): 71–80.

Dunn, Lewis A. "Chemical Weapons Arms Control—Hard Choices for the Bush Administration." *Survival* 31 (May/June 1989): 209–224.

Eagan, Sean P. "From Spikes to Bombs: The Rise of Eco-Terrorism." *Studies in Conflict and Terrorism* 19 no. 1 (January/March 1996): 1–18.

Eagleburger, Lawrence S. "Yellow Rain: The Arms Control Implications." *Department of State Bulletin* 83 (April 1983): 77–78.

Eaton, D.L. and J.D. Groopman. *The Toxicology of Aflatoxins*. Academic Press, New York. NY; 1994.

Eickhoff, Theodore C. "Airborne Disease: Including Chemical and Biological Warfare." *Amer. J. Epidemiol.* 144 no. 8 Supp. (October 15, 1996): S39.

Elbe, Frank. "Banning Chemical Weapons." *International Perspectives* (January/ February 1985): 16–18.

Enria, D.A., Franco S. Garcia, A. Ambrosio, D. Vallejos, S. Levis, and J. Maiztegui. Current Status of the Treatment of Argentine Hemorrhagic Fever. *Med. Microbiol. Immunol.* 175 (1986): 173–176.

Enria, D.A., and J.U. Maiztegui. Antiviral Treatment of Argentine Hemorrhagic Fever. *Antiviral Res.* 23 (1994): 23–31.

Eshel, Tamir. "Desert Operations Demand Lighter NBC Protective Gear." *Armed Forces Journal International* 128 (September 1990): 28.

Evancoe, Paul R. "Environmental Terrorism Lurks on Threat Horizon." *National Defense* 78 (March 1994): 22–23.

Evancoe, Paul R. and Knight Campbell. "Chemical Weapons Lurk at Terrorists' 'Surprise'." *National Defense* 79 (January 1995): 24–25.

Ewin, J. V. Haase. "Chemical and Biological Warfare: The New Frontier." *Asian Defence Journal* no. 6 (June 1986): 84–86.

Ewin, J. V. Haase. "NBC: Combat at the Cellular Level." *Asian Defence Journal* no. 8 (August 1986): 105–106.

Ewin, J. V. Haase. "New French NBC Protection." *Asian Defence Journal* no. 12 (December 1988): 115–116.

Ewin, J. V. Haase. "Planning Medical Defence Against Chemical Weapons." *Asian Defence Journal* no. 8 (August 1987): 88.

Fadiman, Anne. "Yellow Rain." *Life* 7 (August 1984): 23–24.

Fair, Stanley D. "Mussolini's Chemical War." *Army* 35 (January 1985): 44–48.

Falkenrath, Richard, Robert Newman, and Bradley Thayer. *America's Achilles' Heel: Nuclear, Biological, and Chemical Terrorism and Covert Attack.* MIT. 2000.

Feeney, Joseph J. "Program Manager's Guide to Producing Survivable Systems." *Program Manager* 18 (March/April 1989): 16–23.

Feith, Douglas J. "Separating Realism from Rhetoric in Chemical Warfare Negotiations." *Defense* (October 1985): 8–14.

Filippov, Pavel. "Diverting Attention." *New Times* no. 49 (December 1982): 22–24.

Finder, Joseph. "Biological Warfare, Genetic Engineering, and the Treaty that Failed." *Washington Quarterly* 9 (Spring 1986): 5–14.

Findlay, Trevor. "Chemical Disarmament and the Environment." *Arms Control Today* 20 (September 1990): 12–16.

Fletcher, Thomas H. "Chemical Defense Equipment—A Perception." *Army Logistician* 14 (May/June 1982): 26–27.

Flowerree, Charles C. "Elimination of Chemical Weapons: Is Agreement in Sight?" *Arms Control Today* 18 (April 1988): 7–10.

Flowerree, Charles C. "The Politics of Arms Control Treaties: A Case Study." *Journal of International Affairs* 37 (Winter 1984): 269–282.

Fox, Jeffrey L. "Shiga Toxin: No Smoking Gun." *Science* 223 (February 24, 1984): 799.

Fritz, B. "A New NBC Clothing Concept from France." *International Defense Review* 15 no. 11 (1982): 1587–1588.

Fritz, B. "The AMF80 Modular NBC Shelter." *International Defense Review* 15 no. 11 (1982): 1590–1592.

Gander, T. J. "Iraq—Chemical Warfare Potential." *Jane's Soviet Intelligence Review* 2 (October 1990): 441.

Gander, T. J. "Soviet Air-Launched Chemical Munitions." *Jane's Soviet Intelligence Review* 1 (June 1989): 256–257.

Garelik, Glenn. "Toward a Nerve-Gas Arms Race." *Time* 131 (January 11, 1988): 28.

Garofala, Janet. "The Russian Connection—Controlled Substances that Can Kill You." *United States Army Aviation Digest* 29 (November 1983): 37–39.

Geisenheyner, Stefan. "Chemical Warfare." *Asian Defence Journal* (October 1984): 38–40.

Geraghy, Tony. "Targets of Terror: Getting Mad and Getting Even." *Journal of Defense & Diplomacy* 9 no. 1-2 (January/February 1991): 12–18.

Gertz, Bill. "Horror Weapons." *Air Force Magazine* 79 no.1 (January 1996): 44–48.

Glickman, Leonard. "Biochemical Warfare Must Be Examined, Controlled." *Journal of Defense & Diplomacy* 6 (January 1988): 56–58.

Gold, Theodore S. "U.S. Chemical Warfare Policy and Program." *NATO's Sixteen Nations* 28 (February/March 1983): 66–70.

Goldbatt, L..A. *Aflatoxin*. Academic Press, New York, NY; 1969.

Goldblat, Jozef. "Chemical Weapons Verification." *Bulletin of the Atomic Scientists* 41 (May 1985): 19.

Goldblat, Jozef. and Thomas Bernauer. "The U.S.–Soviet Chemical Weapons Agreement of June 1990: Its Advantages and Shortcomings." *Bulletin of Peace Proposals* 21 (December 1990): 355–362.

Graveley, A. F. "Defence or Deterrence? The Case for Chemical Weapons." *Royal United Services Institute for Defence Studies Journal* 126 (December 1981): 13–20.

Gudkov, Yuri. "Toxic Death Plan for Europe." *New Times* no. 5 (January 1982): 5–7.

Haar, Barendter, *et al.* "Verification of Non-Production of Chemical Weapons—An Adequate System Is Feasible." *NATO's Sixteen Nations* 32 (August 1987): 46–48.

Hamm, Manfred R. "Deterrence, Chemical Warfare, and Arms Control." *Orbis* 29 (Spring 1985): 119–163.

Harris, Elisa D. "Sverdlovsk and Yellow Rain: Two Cases of Soviet Noncompliance?" *International Security* 11 (Spring 1987): 41–95.

Harris, Paul. "British Preparations for Offensive Chemical Warfare." *Royal United Services Institute for Defence Studies Journal* 125 (June 1980): 56–62.

Heathcote, J.G. and J.R. Hibbert. *Aflatoxins: Chemical and Biological Aspect.* Elsevier, New York, NY; 1978.

Henderson, D.A., T.V. Inglesby, J.G. Bartlett, *et al.* Smallpox as a Biological Weapon: Medical and Public Health Management: Consensus Statement. *JAMA* 281 (1999): 2127–2137.

Hitchens, Theresa. "Legislation Would Reorder U.S. Chemical Warfare Priorities." *Defense News* (April 15, 1996/April 21, 1996): 21.

Hitchens, Theresa. "Wargame Finds U.S. Falls Short In Bio War." *Defense News* (August 28, 1995/September 3, 1995): 1.

Hoffman, Bruce. "American Right-Wing Extremism." *Jane's Intelligence Review* 7 no.7 (July 1995): 329–330.

Hoffman, Bruce. Current Research on Terrorism and Low-Intensity Conflict." *Studies in Conflict and Terrorism* 15 no. 1 (January/March 1992): 25–37.

Hoffman, Bruce. "Future Trends in Terrorist Targeting and Tactics." *Special Warfare* 6 no. 3 (July 1993): 30–35.

Hoffman, Bruce. "'Holy Terror': The Implications of Terrorism Motivated by a Religious Imperative." *Studies in Conflict and Terrorism* 18 no. 4 (October/December 1995): 271–284.

Huff, J.S. Lessons Learned from Hazardous Materials Incidents. *Emerg. Care Quart.* 1991; 7: 17–22.

Hughes, David. "USAF May Speed Production of New Suits to Protect Crews from Chemical Weapons." *Aviation Week & Space Technology* 111 (August 10, 1990): 27–28.

Hurwitz, Elliott. "Terrorists and Chemical/Biological Weapons." *Naval War College Review* 35 (May/June 1982): 36–40.

Inglesby, T.V., D.A. Henderson, J.G. Bartlett, *et al.* Anthrax as a Biological Weapon: Medical and Public Health Management: Consensus Statement. *JAMA* 281 (1999): 1735–1745.

Isaacs, John. "20-year Battle on Chemical Weapons Is Over." *Bulletin of the Atomic Scientists* 46 (July/August 1990): 3–4.

Isaacs, John. "Nervous about Nerve Gas." *Bulletin of the Atomic Scientists* 39 (December 1983): 7–8.

Jackson, James Heitz. "When Terrorists Turn to Chemical Weapons." *Jane's Intelligence Review* 4 (November 1, 1992): 520.

Jackson, James O. "Anger and Recrimination." *Time* 133 (January 30, 1989): 34.

Jamieson, Alison. "Recent Narcotics and Mafia Research." *Studies in Conflict and Terrorism* 15 no.1 (January/March 1992): 39–51.

Janke, Peter. "Terrorism: Trends and Growth?" Based on a presentation to the Royal United Services Institute, 11 May 1993. *Royal United Services Institute for Defence Studies Jou*rnal 138 no. 4 (August 1993): 24–28.

Jarman, Robert, and Philip Jarman. "Soviet Chemical Warfare: A Present Danger." *Defense & Foreign Affairs* 11 (June 1983): 30–31.

JCAHO: 1996 *Comprehensive Accreditation Manual for Hospitals*. Oakbrook Terrace, IL: Joint Commission on Accreditation of Healthcare Organizations; 1996.

Jenne, Michael. "Proposal—A Chemical Deterrence Force." *National Defense* 73 (September 1988): 27–30.

Johnson, David E. "The Medic on the Chemical Battlefield." *Infantry* 73 (March/April 1983): 24–26.

Johnson, Gregory. "Flying the Poisoned Skies." *Proceedings of the United States Naval Institute* 113 (June 1987): 72–75.

Johnson, Gregory. "Helicopters in a Chemical Environment." *Marine Corps Gazette* 70 (May 1986): 52–53.

Johnson, K.M., M.L. Kuns, R.B. MacKenzie, P.A. Webb, and C.E. Yunker. Isolation of Machupo Virus from Wild Rodent, *Calomys callosus. Am. J. Trop. Med. Hyg.* 15 (1966): 103–106.

Johnson, K.M. Epidemiology of Machupo Virus Infection: III. Significance of Virological Observations in Man and Animals. *Am. J. Trop. Med. Hyg.* 14 (1965): 816–818.

Johnson, Loch. *Bombs, Bugs, Drugs and Thugs: Intelligence and America's Quest for Security*. New York University Press. 2000.

Jones, David T. "Eliminating Chemical Weapons: Less than Meets the Eye." *Washington Quarterly* 12 (Spring 1989): 83–92.

Joseph, Robert G. "Regional Implications of NBC Proliferation." *Joint Forces Quarterly* 9 (Autumn 1995): 64–69.

Judge, John F. "Congressional Perceptions of Soviet Military Power." *Defense Electronics* 17 (September 1985): 146–148.

Kelly, Orr. "Why Reagan Seeks Buildup of Chemical Arms." *U.S. News & World Report* 92 (February 22, 1982): 28.

Kirk, M.A., J. Cisek,, and S.R. Rose. Emergency Department Response to Hazardous Materials Incidents. *Emerg. Med. Clin. North Am.* 1994 May; 12(2): 461–81.

Kitfield, James. "Age of Superterrorism" *Government Executive* 27 no 7 (July 1995): 46–48.

Kitfield, James. "Perils of Proliferation." *Government Executive* 22 (October 1990): 34–35.

Knoll Glenn F. *Radiation Detection and Measurement*. John Wiley and Sons; New York, NY. 1979.

Koller, Duncan G. "NATO in a Chemical Warfare Environment: Defense Is Not Enough." *Joint Perspectives* 3 (Summer 1982): 48–57.

Koslow, Evan E. "Would You Go to War Wearing this Equipment?" *Armed Forces Journal International* 127 (May 1990): 55–56.

Kozyrev, N., and D. Pogorzhelsky. "Inspection or Snooping?" New *Times* no. 19 (May 1984): 14–15.

Krause, Joachim. "Proliferation Risks and Their Strategic Relevance: What Role For NATO?" *Survival* 37 no. 2 (Summer 1995): 135–148.

Kroesen, Fredrick J. "Chemical War—Deadly for Our Side." *National Guard* 39 (May 1985): 22–26.

Kruze, Douglas R. "Denial Systems Deflect Terrorists from Mischief." *National Defense* 79 no. 508 (May/June 1995): 55–56.

Kuntsevich, Anatoly. "Devils' Brew for Europe." New *Times* (July 7, 1986): 18–20.

Kuntsevich, Anatoly. "What Lies Behind the Binary Programme?" *International Affairs* (Moscow) (November 1986): 32–39.

Kupperman, Robert H. "United States Becoming Target for Terror Forays." *National Defense* 79 no. 504 (January 1995): 22–23.

Kyle, Deborah, and Benjamin F. Schemmer. "*Army* Secretary (John O. Marsh, Jr.), Defense Logistics Agency Pressured from Fixing Chemical Warfare Mess." *Armed Forces Journal International* 118 (April 1981): 27–29.

Kyle, Deborah. "Chemical Warfare." *Armed Forces Journal International* 119 (November 1981): 56–57.

Lami, Lucio. "Yellow Rain: The Conspiracy of Closed Mouths." *Commentary* 76 (October 1983): 60–61.

Latter, Richard. "Increased Danger of Biological Weapons Proliferation." *Jane's Intelligence Review* 6 no. 2 (February 1994): 93–95.

Ledeen, Michael. "The Curious Case of Chemical Warfare." *Commentary* 88 (July 1989): 37–41.

Levinson, Macha. "Chemical Deterrence—Will It Work?" *International Defense Review* 19 no. 6 (1986): 731–732.

Levinson, Macha. "Custom-made Biological Weapons." *International Defense Review* 19 no. 11 (1986): 1611–1612.

Levite, Ariel. "Israel Intensifying Preparations to Counter Chemical Attack." *Armed Forces Journal International* 127 (May 1990): 60.

Levitin, H.W., and H.J. Siegelson. Hazardous Materials. Disaster Medical Planning and Response. *Emerg. Med. Clin. North Am.* 1996 May; 14(2): 327–48.

Litallen, Dennis J. "NBC Training." *Marine Corps Gazette* 72 (December 1988): 23–24.

Locher, James R. "Combating the Terrorists' War: The Front Is Everywhere." Remarks to DoD Worldwide Anti-Terrorism Conference, Ft. Leavenworth, 23 June 1992. Defense Issues 7 no. 41 (1992): 1–4.

Lord, Carnes. "Rethinking On-Site Inspection in the U.S. Arms Control Policy." *Strategic Review* 12 (Spring 1985): 45–51.

Lowitz, Donald S. "U.S. Activities in the Conference on Disarmament." *Department of State Bulletin* 85 (November 1985): 29–31.

Mackenzie, R.B., M.L. Kuns, and P.A. Webb. Possibilities for Control of Hemorrhagic Fevers in Latin America. Pan American Health Organization; Scientific Publication No.147:260-265. First International Conference on Vaccines against Viral and Rickettsial Diseases of Man, 1966, Washington, D.C.

Manegold, C. S. "In Pursuit of Poison: Libya's Chemical-War Plant Is in Production." *Newsweek* 115 (March 19, 1990): 33.

Marbach, William D. "The New Arms Race?" *Newsweek* 106 (July 1, 1985): 42.

Marshall, Eliot. "Bugs in the Yellow Rain Theory." *Science* 220 (June 24, 1983): 1356–1358.

Marshall, Eliot. "Chemical Genocide in Iraq?" *Science* 241 (September 30, 1988): 1752.

Marshall, Eliot. "Chemical Weapons: A Plan for Europe." *Science* 234 (December 5, 1986): 1194.

Marshall, Eliot. "Iraq's Chemical Warfare: Case Proved." *Science* 224 (April 13, 1984): 130–132.

Masland, Tom. "Are We Ready for Chemical War?" *Newsweek* 117 (March 4, 1991): 29.

Matteson, Robert. "Chemical Decontamination—A First for Many." *Army Logistician* 15 (January/February 1983): 16–17.

McCormack, Michael S., and E. E. Whitehead. "APACHE Aviation Performance Assessment in a Chemical Environment." *United States Army Aviation Digest* 29 (November 1983): 40–43.

McEwen, Michael T. "Soviet Armor—CW Threat." *Infantry* 71 (November/December 1981): 43–44.

McGeorge, Harvey J. "Bugs, Gas and Missiles." *Defense & Foreign Affairs* 17 (May/June 1990): 14–19.

McGeorge, Harvey J. "Chemical Addition." *Defense & Foreign Affairs* 17 (April 1989): 16–19.

McGeorge, Harvey J. "Chemical Warfare: Seeking an East/West Balance." *Defense & Foreign Affairs* 15 (April 1987): 29–31.

McGeorge, Harvey J. "Feature Reports: Reversing the Trend on Terror." *Defense & Foreign Affairs* (April 1988): 16.

McGeorge, Harvey J. "Iraq's Secret Arsenal of Chemical and Biological Weapons." *Defense & Foreign Affairs*' Strategic Policy 19 no. 1-2 (January/February 1991): 6–9.

McGeorge, Harvey J. "Reversing the Trend on Terror." *Defense & Foreign Affairs* 16 (April 1988): 16–18.

McNaugher, Thomas L. "Ballistic Missiles and Chemical Weapons: The Legacy of the Iran-Iraq War." *International Security* 15 (Fall 1990): 5–34.

Merrifield, John T. "USAF Crews Increasing Proficiency in Chemical Defense Clothing." *Aviation Week & Space Technology* 125 (July 28, 1985): 73.

Meyer, Michael R. "A Self-inflicted Wound." *Newsweek* 113 (January 30, 1989): 42.

Miller, Charles. "Report on Mideast Threat." *Defense News* (April 1, 1996/April 7, 1996): 54.

Miller, Dennis. "Chemical Warfare—U.S. Policy and Capabilities." *NATO's Sixteen Nations* 30 (August 1985): 66–68.

Moffett, Donald L. "NBC and the Armor Crewman." *Armor* 90 (September/October 1981): 38–41.

Montgomery, Raymond H., II, and Stephen J. Demora, Jr. "Reinforcement in a Chemical Environment." *Army Logistician* (November/December 1989): 25–29.

Moon, John Ellis van Courtland. "Chemical Warfare: A Forgotten Lesson." *Bulletin of the Atomic Scientists* 45 (July/August 1989): 40–43.

Morris, Dee Dodson. "Chemical Protective Clothing." *United States Army Aviation Digest* 26 (October 1980): 34–35.

Morrison, David C. "Chemical Weapons Rerun." *National Journal* 18 (July 19, 1986): 1773–1777.

Mortley, James B. "Terrorism: The U.S. Targeted." *Defense & Foreign Affairs* 16 (July 1988): 34–38.

Mosier, Janice. "Middle of Nowhere." *Soldiers* 44 (May 1989): 37–41.

Nash, Colleen A. "Chemwar in the Third World." *Air Force Magazine* 73 (January 1990): 80–83.

Nason, Gardner M. "NBC Equipment—Good and Getting Better." *Soldiers* 35 (June 1980): 36–39.

Nordwall, Bruce D. "Navy Develops CB Protection for Pilots in Fast-paced Program." *Aviation Week & Space Technology* 134 (January 28, 1991): 59–60.

Norman, Colin. “Biowarfare Lab Faces Mounting Opposition.” *Science* 240 (April 8, 1988): 135.

Norman, Colin. “CIA Details Chemical Weapons Spread.” *Science* 243 (February 17, 1989): 888.

Nozaki, H., S. Hori, Y. Shinozawa, *et al.* Secondary Exposure of Medical Staff to Sarin Vapor in the Emergency Room. *Intensive Care Med.* 1995 Dec; 21(12): 1032–1035.

O’Keefe, Isabel. “The Silent Killer.” *New Statesman Society* 1 (October 28, 1988): 12–14.

Ooms, A. Jack. “Chemical Weapons: Is Revulsion a Safeguard?” *Atlantic Community Quarterly* 24 (Summer 1986): 157–166.

OSHA: Hospitals and Community Emergency Response–What You Need to Know. Emergency Response Safety Series. US Department of Labor, Occupational Safety and Health Administration. OSHA 3152: 1997.

OSHA: Regulations. Hazardous Waste Operations and Emergency Response. 29 CFR 1910.120. Washington DC: Government Printing Office; 1995.

Parker, Christopher J. “The Chemical Ingredients.” Field Artillery Journal 51 (May/ June 1983): 34–36.

Pasquill, Frank. *Atmospheric Diffusion: The Dispersion of Windborne Material from Industrial and Other Sources.* Ellis Horwood, Ltd.; 2nd edition. 1974.

Pendleton, Blaine D. “Just Another FM?” *United States Army Aviation Digest* 32 (April 1986): 22–24.

Peters, C.J., and K.M. Johnson. Arenaviridae: Lymphocytic Choriomeningitis Virus, Lassa Virus, and other Arenaviruses. In: Mandell, G.L.K., J.E. Bennett, and R. Dolin, eds. *Principles and Practice of Infectious Diseases*, 4th ed. New York: Churchill Livingston, Inc., 1995.

Petrovsky, Vladimir. “Mass Destruction Weapons: Who Opposes their Prohibition?” *New Times* no. 39 (September 1981): 5–7.

Pikalov, Vladimir. “Chemical Weapons.” *Soviet Military Review* No. 12 Supp. (December 1987): entire issue.

Pikalov, Vladimir. “Banning Chemical Weapons—A Proposal from the Ministry of Defense on a Critical Issue.” *Defense Science* 7 (April 1988): 6–7.

Pilat, Joseph F., and Walter L. Kirchner. “Counterproliferation and Beyond: The Technological Promise of Counterproliferation.” *Washington Quarterly* 18 (Winter 1995): 153.

Pillar, Charles. “DNA—Key to Biological Warfare?” *Nation* 237 (December 10, 1983): 585.

Pillar, Charles. “Test Site for Germ Warfare?” Nation 240 (March 9, 1985): 270–273.

Pluchinsky, Dennis A. “Academic Research on European Terrorist Developments: Pleas From a Government Terrorism Analyst.” *Studies in Conflict and Terrorism* 15 no. 1 (January/March 1992): 13–23.

Poyer, Joe. “Chemical Warfare.” *Combat Arms* 6 (November 1988): 70–80.

Prebel, Cecilia. “CWD Gear: Practice Pays Off.” *Flying Safety* 39 (July 1983): 20–23.

Pringle, Peter. “Yellow Rain: The Cost of Chemical Arms Control.” *SAIS Review* 5 (Winter/Spring 1985): 151–162.

Qian, Qichen. “China Urges Total Ban on Chemical Weapons.” *Beijing Review* 32 (January 23, 1989): 14–15.

Rathmell, Andrew. "Chemical Weapons in the Middle East: Syria, Iraq, Iran, and Libya." *Marine Corps Gazette* 74 (July 1990): 59–62.

Rawles, James W. "High-Technology Terrorism." *Defense Electronics* 22 no. 1 (January 1990): 74–78.

Resing, David C. "Averting Terrorist Forays Requires Prudent Planning." *National Defense* 80 (September 1995): 36–37.

Rhein, Reginald, and Mimi Bluestone. "Is the Pentagon Preparing for Biotech Warfare?" *Business Week* (August 10, 1987): 66–67.

Rhodes, Richard. *The Making of the Atomic Bomb*. Simon & Schuster, Inc.; New York, NY. 1986.

Rhodes, Richard. *Dark Sun*. Simon & Schuster, Inc.; New York, NY. 1995.

Robinson, J. P. Perry. "Disarmament and Other Options for Western Policy-Making on Chemical Warfare." *International Affairs* (London) 63 (Winter 1986-1987): 65–80.

Robinson, Julian, Jeanne Harley Guillemin, and Matthew Meselson. "Yellow Rain; The Story Collapses." *Foreign Policy* 68 (Fall 1987): 100–117.

Roos, John G. "Ultimate Nightmare (Nuclear, Biological or Chemical Terrorism)." *Armed Forces Journal International* 133 No. 3 (October 1995): 67–68.

Roosevelt, Edith Kermit. "Germ War—Terrorism and the 'New' Biology." *Combat Arms* 4 (July 1986): 38–42.

Rose, Stephen. "Coming Explosion of Silent Weapons." *Naval War College Review* 42 (Summer 1989): 6–29.

Rosser-Owen, David. "NBC Warfare and Anti-NBC Protection." *Armada International* 8 (January/February 1984): 78.

Rothwell, Nicholas. "Yellow Rain over Laos." *American Spectator* 15 (January 1982): 8–11.

Rutman, Robert J. "The Case against New Chemical Weapons." *Technology Review* 88 (November/December 1985): 18.

Salerno, Steve. "Chemical Warfare: America's Achilles' Heel." *American Legion Magazine* 122 (March 1987): 24–25.

Santoli, Al. "Little Girl in the Yellow Rain." *Reader's Digest* 124 (April 1984): 73–77.

Saunders, Barbara B., and Richard M. Price. "Tactical C^3I & the Chemical Warfare Environment." *Signal* 38 (November 1983): 67–71.

Savin, Alexander. "The Crime Goes On." *Soviet Military Review* 9 (September 1984): 42–44.

Schemmer, Benjamin F. "U.S. is in a 'Disaster Mode' vs U.S.S.R. Chemical Warfare Threat, DOD Warns." *Armed Forces Journal International* 119 (January 1982): 17.

Schwartzstein, Stuart J. D. "Chemical Warfare in Afghanistan: An Independent Assessment." *World Affairs* 145 (Winter 1982/1983): 267–272.

Seeley, Thomas D., Joan W. Nowicke, and Matthew Meselson. "Yellow Rain." *Scientific American* 253 (September 1985): 128–137.

Segal, David. "Soviet Union's Mighty Chemical Warfare Machine." *Army* 37 (August 1987): 26–29.

Seger, Karl A. "Iraqi Sponsored Terrorism: Target America?" *Military Technology* 17 no. 2 (April/June 1991): 24–28.

Shapiro, Jacob. *Radiation Protection: A Guide for Scientists and Physicians*. 2nd ed. Harvard University Press, Cambridge, MA. 1981.

Shestack, Jerome J., and Matthew Nimetz. "Reported Use of Lethal Chemical Weapons in Afghanistan and Indochina." *Department of State Bulletin* 80 (July 1980): 35–39.

Shulgin, Alexander. "Advancing through a Contaminated Area." *Soviet Military Review* 8 (August 1985): 26–27.

Shulman, Seth. "Biological Research and Military Funding." *Technology Review* 90 (April 1987): 13–14.

Shulman, Seth. "Bomb Burning in the Pacific." *Technology Review* 93 (October 1990): 18–20.

Shulman, Seth. "Poisons from the Pentagon." *Progressive* 51 (November 1987): 16–19.

Shultz, George E. "NBC Protection: A Personal Matter." *Army Logistician* 17 (May/June 1985): 22–24.

Siebert, George W., and Yearn H. Choi. Chemical Weapons: Dull Swords in the U.S. Armory." *Military Review* 65 (March 1985): 23–29.

Sims, Nicholas A. "Morality and Biological Warfare." *Arms Control* 8 (May 1987): 5–35.

Sinai, Joshua. "Next Stage in Middle East Terrorism." *Defense & Foreign Affairs*' Strategic Policy18 no. 11 (November 1990): 22–25.

Sloan, Stephen. "Understanding Terrorism Since the 60s—An Evaluation of Academic and Operational Perspectives: What Have We Learned and Where Are We Going?" *Special Warfare* 5 No. 1 (March 1992): 56–59.

Smith, R. Jeffrey. "Biowarfare Lab Approved Without Restrictions." *Science* 226 (December 21, 1984): 1405.

Smith, R. Jeffrey. "GAO Blasts BIGEYE Chemical Weapon." *Science* 232 (June 20, 1986): 1493–1494.

Smith, R. Jeffrey. "New *Army* Biowarfare Lab Raises Concerns." *Science* 226 (December 7, 1984): 1176–1178.

Smith, R. Jeffrey. "NRC Urges Destruction of Chemical Weapons." *Science* 226 (December 7, 1984): 1174–1175.

Smith, R. Jeffrey. "Soviet Biowarfare Effort Cited by Pentagon." *Science* 226 (May 17, 1985): 828.

Smith, William E. "Clouds of Desperation." *Time* 123 (March 19, 1984): 28–30.

Smolowe Jill. "Return of the Silent Killer." *Time* 132 (August 22, 1988): 46–49.

Snell, Albert E. and Edward J. Keusenkothen. Mass Destruction Weapons Enter Arsenal of Terrorists." *National Defense* 79 (January 1995): 20–21.

Starr, Barbara. "Chemical and Biological Terrorism." *Jane's Defence Weekly* 26 (August 14, 1996): 16.

Starr, Barbara. "Nightmare in the Making (Nuclear, Chemical and Biological Threats)." *Jane's Defence Weekly* 23 (June 3, 1995): 23–24.

Starr, Barbara. "The Jane's Interview." *Jane's Defence Weekly* 26 (August 14, 1996): 32.

Stelzmueller, H. "Difficult but Not Hopeless: Defence against NBC." *Asian Defence Journal* no. 8 (August 1983): 58–60.

Stelzmueller, H. "New Decontamination Systems for CBR-Defence." *Armada International* 5 (July/August 1981): 68.

Stringer, Hugh. "Deterring Chemical Warfare: U.S. Policy Options of the 1990s." *Atlantic Community Quarterly* 24 (Summer 1986): 167–168.

Teimourian, Hazhir. "Silence Deep as Death." *New Statesman Society* 1 (September 16, 1988): 20–22.

Ter Haar, Bas, and Piet De Klerk. "Verification of Non-Production: Chemical Weapons and Nuclear Weapons Compared." *Arms Control* 8 (December 1987): 197–212.

Terrill, W. Andrew, Jr. "Chemical Weapons in the Gulf War." *Strategic Review* 14 (Spring 1986): 51–58.

Tesko, Steven R. "Chemical Warfare Treaty—Chemical Warfare Threats—It's Not Just the Soviets Anymore." *National Defense* 74 (April 1989): 31–33.

Thanabalasingham, T., M.W. Beckett, and V. Murray. Hospital Response to a Chemical Incident: Report on Casualties of an Ethyldichlorosilane Spill. *BMJ* 1991 Jan 12; 302 (6768): 101 -102.

Thomas, F. R. "Employing Tanks with Collective NBC Protection." *Armor* 98 (July/August 1989): 11–17.

Thompson, Graham N., *et al.* "Fighting in a Toxic Environment: Chemical Defence Capability in the Soviet Ground Forces." *Armed Forces* 6 (September 1987): 400–404.

Tomilin, Y. "Washington Torpedoing a Chemical Weapon Ban." *International Affairs* (Moscow) (October 1983): 100–108.

Tucker, Jonathan B. "Gene Wars." *Foreign Policy* (Winter 1984/1985): 58–79.

Tucker, Jonathan B. Historical Trends Related to Bioterrorism: An Empirical Analysis. *Emerg. Infect. Dis.* 5 (1999): 498–504.

Turniville, Graham H., Jr. and Harold S. Orenstein. "Drugs and Terror: Eastern European and Transnational Security Threats." *Military Review* 71 no. 12 (December 1991): 57–68.

Turque, Bill. "The Specter of Iraq's Poison Gas." *Newsweek* 116 (August 20, 1990): 26.

Ulsamer, Edgar. "R&D at the Razor's Edge." *Air Force Magazine* 68 (May 1985): 50–56.

U.S. Army Field Manual 3-3. "Chemical and Biological Contamination Avoidance." September 29, 1994.

U.S. Army Field Manual 3-4, U.S. Marine Corp Fleet Marine Force Manual 11-09. "NBC Protection." February 21, 1996.

U.S. Army Field Manual 3-9, U.S. Navy Publication P-467, U.S. Air Force Manual 355-7. "Potential Military Chemical/Biological Agents and Compounds." December 12, 1990.

U.S. Army Field Manual 8-285, U.S. Navy Medical Publication P-5041, U.S. Air Force Joint Manual 44-149, U.S. Marine Corps Fleet Marine Force Manual 11-11. "Treatment Of Chemical Agent Casualties and Conventional Military Chemical Injuries." December 22, 1995.

U.S. Army Material Safety Data Sheet (MSDS) on HQ Mustard. Aberdeen Proving Ground, MD, June 30, 1995.

U.S. Army Medical Research Institute of Infectious Diseases (USAMRIID). Medical Management of Biological Casualties. 3rd ed. Fort Detrick, Frederick MD. U.S. *Army* Medical Research Institute of Infectious Diseases. 1998.

U.S. Army Office of the Surgeon General. Textbook of Military Medicine: Warfare, Weaponry, and the Casualty; Medical Aspects of Chemical and Biological Warfare. 1997.

U.S. Army Pamphlet 385-61. "Toxic Chemical Agent Safety Standards." Washington, DC. March 31, 1997.

U.S. Department of Defense (DoD) and U.S. Energy Research and Development Administration (ERDA). *Effects of Nuclear Weapons*. Compiled by Samuel Glasstone and P.J. Dolan, Washington, D.C., 1977.

USHSS: Managing Hazardous Materials Incidents. Vol. 2. Hospital Emergency Departments: A Planning Guide for the Management of Contaminated Patients. Atlanta: U.S. Department of Health and Human Services, Public Health Service, Agency for Toxic Substances and Disease Registration (ATSDR).

USHSS: Managing Hazardous Materials Incidents. Vol. 3. Hospital Emergency Departments: Management Guidelines for Acute Chemical Exposures. Atlanta: US Department of Health and Human Services, Public Health Service, Agency for Toxic Substances and Disease Registration (ATSDR).

Usvatov, Alexander. "Ban Chemical Weapons." *New Times* no. 26 (June 1983): 18–20.

Vachon, G. K. "Chemical Disarmament—A Regional Initiative?" *Millennium* 8 (Autumn 1979): 145–154.

Van Courtland, John Ellis. "Chemical Weapons and Deterrence: The World War II Experience." *International Security* 8 (Spring 1984): 3–35.

Vasilyev, S. "Links of the Same Chain." *New Times* no. 28 (July 1980): 29–30.

Verpoorten, Dennis M. "Chemical Reconnaissance." *Armor* 99 (November/December 1990): 32–35.

Vicary, A. G., and J. Wilson. "Nuclear Biological and Chemical Defense." *Royal United Services Institute for Defence Studies Jou*rnal 126 (December 1981): 7–12.

Wagner, Richard L., Jr., and Theodore S. Gold. "Banning Chemical Weapons." *Defense* (June 1984): 22–32.

Wagner, Richard L., Jr., and Theodore S. Gold. Why We Can't Avoid Developing Chemical Weapons." *Defense* (July 1982): 2–11.

Walker, Kevin M. "Enviro-Terrorism: SARA Title III and Its Impact on National Security." *Military Intelligence* (July/September 1992): 20–22.

Wall, Patrick. "Fallout from Chernobyl—A World Statesman Asks if Nuclear Disaster Could Lead to World Peace." *Sea Power* 29 (July 1986): 24–26.

Warner, Denis. "Mounting Dangers in Nuclear and Chemical Proliferation. *Asia-Pacific Defence Reporter* 16 (June 1990): 44–45.

Warner, Denis. "Weapons—The Gulf, Gas and The Gorbachev Initiative." *Asia-Pacific Defence Reporter* 14 (June 1988): 46–47.

Warner, John W. "The Case for Modernizing our Chemical Weapons." *Technology Review* 88 (November/December 1985): 19–20.

Warner, Margaret Garrard, and Theresa Waldrop. "Bonn Finally Comes Clean." *Newsweek* 113 (January 23, 1989): 32.

Waters, Lee. "Chemical Weapons in the Iran/Iraq War." *Military Review* 70 (October 1990): 56–63.

Watson, Gerald G., and Raymond L. Anderson. "An Urgent Need: Stockpiling Modern Chemical Munitions." *Military Review* 64 (January 1984): 58–67.

Watson, Russell. "Letting a Genie Out of a Bottle." *Newsweek* 112 (September 19, 1988): 30–31.

Webb, P.A., and J.I. Maiztegui. Argentine and Bolivian Hemorrhagic Fevers (South American Hemorrhagic Fevers). In: Gear, J.H.S., ed. *Handbook of viral and rickettsial Hemorrhagic Fevers*. Boca Raton, FL: CRC Press, Inc., 1988.

Webb, Philip H., Jr., and Timothy B. Savage. "Nuclear Biological Chemical Training and Development." *United States Army Aviation Digest* Part. 1, 27 (August 1981): 40–41; Part 2. "Decontamination Problems." 27 (October 1981): 34–36.

Weekly, Terry M. "Proliferation of Chemical Warfare: Challenge to Traditional Restraints." *Parameters* 19 (December 1989): 51–66.

Weeks, Albert A. "More than a Band-Aid Campaign Needed." *Defense Science* 4 (October/November 1985): 76–80.

Weickhardt, George G., and James M. Finberg. "New Push for Chemical Weapons." *Bulletin of the Atomic Scientists* 42 (November 1986): 28–33.

Weiss, Rick. "Neighbors Bugged by Germ Warfare Lab." *Science News* 133 (April 9, 1988): 229.

Welch, Thomas J. "Growing Global Menace of Chemical & Biological Warfare." *Defense* (July/August 1989): 19–27.

Whelan, Barbara. "Chemical Warfare: The USA and USSR Square Off." *Defense Science* 2 (December 1983): 54–57.

White, Terence, and Kathleen White. "Biological Weapons—How Big a Threat?" *International Defense Review* 23 (August 1990): 843–846.

Whiteside, Thomas. "The Yellow-Rain Complex (I)." *New Yorker* 66 (February 11, 1991): 38–42.

Whiteside, Thomas. "The Yellow-Rain Complex (II)." *New Yorker* 66 (February 18, 1991): 44–68.

Whitson, Thomas E. "The Chemical Environment and Army Aviation." *United States Army Aviation Digest* 29 (February 1983): 8–9.

Whyte, Stuart "Military Glasnost and Force Comparisons." *International Defense Review* 22 (May 1989): 559.

Wilkinson, Paul. "Terrorist Trends in the Middle East." *Jane's Intelligence Review* 5 no. 2 (February 1993): 73–75.

Willey, Fay. "War's Dirty Chemistry." *Newsweek* 103 (April 2, 1984): 55–56.

Wing, J.S., J.D. Brender, L.M. Sanderson, *et al.* Acute Health Effects in a Community after a Release of Hydrofluoric Acid. *Arch. Environ. Health* 1991 May-Jun; 46(3): 155–160.

Wohl, Richard. "Biological Warfare: Advances Breed New Dangers." *Defense Science* 3 (August 1984): 57–60.

Wohl, Richard. "Forgotten Threat: Biowar." *Defense Science* 8 (October 1989): 75–77.

Wright, Susan. "New Designs for Biological Weapons." *Bulletin of the Atomic Scientists* 43 (January/February 1987): 43–46.

Wright, Susan. "The Military and the New Biology." *Bulletin of the Atomic Scientists* 41 (May 1985): 10–16.

Yamamoto, Keith R. "Retargeting Research on Biological Weapons." *Technology Review* 92 (August/September 1989): 23–24.

Yegorov, Mikhail, and Grigory Khozin. "The Truth about Silent Death." *World Marxist Review* 29 (August 1986): 112–116.

Yi, Ping. "A Look at U.S.-Soviet Accord to Ban Chemical Weapons." *Beijing Review* 33 (October 8, 1990): 14–16.

Zajtchuk Russell, ed. "Decontamination." In: The Textbook of Military Medicine. Part I. Warfare, Weaponry and the Casualty. Vol. 3. 1997: 351–360.

Zanders, Jean Pascal. "Chemical Weapons: Beyond Emotional Concerns." *Bulletin of Peace Proposals* 21 (March 1990): 87–98.

Zhukov, Yuri. "Chemical Weapons." *New Times* no. 39 (September 1985): 26–27.

Ziemke, Earl F. "Superweapons." *Parameters* 12 (December 1982): 32–42.

Zilinskas, Raymond A. "Anthrax in Sverdlovsk?" *Bulletin of the Atomic Scientists* 39 (June/July 1983): 24–27.

GLOSSARY

Absorbed dose: the amount of radiation or substance taken into the body; for radiation the unit is the gray; for chemicals the most common unit is milligrams per kilograms of body weight (mg/kg).

Actinium series: a naturally occurring radioactive series of elements beginning with ^{235}U and ending with stable ^{207}Pb.

Aerosol: a solution of solid or liquid particles in a gas.

Aflatoxin: any of a number of toxic metabolites produced by certain fungi, mostly of the *Aspergillus* family.

Agent Blue: a military defoliant; cacodylic acid.

Agent BZ; a military psychotropic agent; ?-hydroxy-?-phenylbenzeneacetic acid, 1-azobicyclo[2,2,2]-oct-3-yl ester.

Agent CA: a military tear agent; bromobenzyl cyanide.

Agent CG: a military choking agent; carbonyl chloride.

Agent CK: a military blood agent; cyanogen chloride.

Agent CN: a military tear agent; α-chloroacetophenone.

Agent CNB: a military tear agent; chloroacetophenone in benzene and carbon tetrachloride.

Agent CNC: a military tear agent; chloroacetophenone in chloroform.

Agent CNS: a military tear agent; chloroacetophenone in chloroform.

Agent CS: a military tear agent; o-chlorobenzylmalononitrile.

Agent CX: a military blister agent; phosgene oxime dichloroformate.

Agent DF: a chemical which may be used as a military nerve agent, or, more commonly, as a component of a binary system for making Agent GB or Agent GD; methyl phosphonyl difluoride.

Agent DP: a military choking agent; diphosgene.

Agent ED: a military blister agent; ethyldichloroarsine.

Agent GA: a military nerve agent; ethyl dimethyl phosphoramidocyanide; also known as Tabun.

Agent GB: a military nerve agent; 1-methylethylmethylphosphonofluoride; also known as Sarin.

Agent GD: a military nerve agent; 1,2,2-trimethyl propylmethylphosphonofluoride; also known as Soman.

Agent GF: a military nerve agent; cyclohexyl methylphosphonofluoride; also known as Cyclosarin.

Agent Green: a military defoliant; a mixture of 2,4,5-trichlorophenoxyacetic acid (2,4,5-T) and its esters.

Agent HCN: a military blood agent; hydrogen cyanide.

Agent HD: a military blister agent; bis(chloroethyl) sulfide.

Agent HL: a military blister agent; a mixture of Lewisite and mustard (HD).

Agent HN-1: a military blister agent; bis(chloromethyl)ethylamine.

Agent HN-2: a military blister agent; bis(chloroethyl)methylamine.

Agent HN-3: a military blister agent; triethylamine.

Agent L: a military blister agent; bis(chloroethyl)chloroarsine; also known as Lewisite.

Agent MD: a military blister agent; methyldichloroarsine.

Agent OC: a military tear agent; capsicum oleoresin.

Agent OPA: a component of the binary system for making Agent GB; a mixture of isopropyl alcohol and isopropyl amine.

Agent Orange: a military defoliant; a mixture of 2,4,5-trichlorophenoxyacetic acid (2,4,5-T) and 2,4-dichlorophenoxyacetic acid (2,4-D), plus their esters.

Agent PD: a military blister agent; phenyldichloroarsine.

Agent PS: a military tear agent; chloropicrin.

Agent QL: a component of the binary system for making Agent VX; O-ethyl-O-2-diisopropylaminoethyl methylphosphonite.

Agent RP: a military battlefield smoke; red phosphorus (P_4).

Agent SA: a military blood agent; arsenic trihydride.

Agent V: a military nerve agent; S-(2-(diethylamino)ethyl)-O-2-methylphosphonothioate.

Agent VX: a military nerve agent; S-(2-(bis(1-methylethyl)amino)ethyl)-O-ethyl-methylphosphonothioate.

Agent WP: a military battlefield smoke; white phosphorus (P).

Alpha particle: a free helium nucleus, composed of two protons and two neutrons.

Anatomy: the study of the structure of plants and animals.

Anthrax: a disease caused by the bacterium *Bacillus anthracis*, which may present cold or flu-like symptoms or skin lesions; the inhalational form is often fatal.

Antidote: a chemical, drug, or pharmaceutical that counteracts the adverse effects of a chemical agent.

Atom: the simplest unit of an element, and composed of a nucleus of protons, neutrons, and other subatomic particles encircled by orbital electrons.

Atomic bomb: a common name for a nuclear fission device.

Atomic mass: the relative mass (or weight) of an atom or subatomic particle expressed in atomic mass units (amu) equal to 1.66×10^{-24} grams.

Atomic number: a unique number assigned to an element equal to the number of protons in each nucleus of its atoms.

Barn: a unit of neutron-absorbing cross sectional area equal to 1.0×10^{-24} square centimeters.

Beta particle: a free electron (not bound to an atom).

Binary agent: a chemical agent formed during the flight to the target by mixing two (or more) less toxic or less hazardous chemicals.

Binding energy: the energy holding the atomic nucleus together and allowing the positively charged protons to exist in proximity to each other.

Biological agent: any bacterium, virus, other organism, or toxin used deliberately to cause injury or disease; an equivalent term is "bioweapon."

Bioweapon: any bacterium, virus, other organism, or toxin used deliberately to cause injury or disease; an equivalent term is "biological agent."

Blast: the combined effects of high velocity winds and rapidly expanding gases resulting from an explosion.

Blister agent: any of a number of chemicals capable of causing skin burns, lesions, and destruction; also known as vesicants.

Blood agent: any of a number of chemicals capable of damage to, or interference with, the blood's ability to carry oxygen to the cells or remove gaseous waste products from the cells.

Bomb: a military weapon containing an explosive charge most commonly dropped from an aerial vehicle (aircraft); some modern bombs have guidance mechanisms to improve accuracy.

Botulism: a disease caused by a metabolic toxin of the bacterium *Clostridium botulinum* that presents nerve impairment, paralysis, and often death.

Bremsstrahlung: secondary radiation resulting from the slowing down of charged particles.

British anti-Lewisite (BAL): a military decontamination agent originally for mustard gas, but today for a number of blister agents; 2,3-dimercaptopropanol.

Broken Arrow: the code term for an accident or missing nuclear weapon.

Brucellosis: a disease caused by several varieties of *Brucella* bacteria; it is seldom fatal in humans, and usually presents fever, headache, and general malaise.

Bubo: a swollen lymph node that may be caused by the plague virus.

Burster: a small explosive device generally used to break a barrier between materials.

CRUD: originally an acronym for Chalk River Unidentified Deposits; now a common term for anything dirty or messy.

Calutron: isotope separation device using electromagnetic fields; the name is coined from the University of California, where it was designed.

Cancer: the uncontrolled growth of cells.

Carcinogen: a chemical or radiation capable of inducing or causing uncontrolled growth of cells.

Casualty: anyone injured or killed.

Cell: the basic building unit of organisms, consisting of (in most cases) a nucleus, protoplasm, and cellular membrane (cell wall in plants).

Central nervous system: system composed of the brain, spinal cord, and related neural networks.

Centrifuge: a device utilizing centrifugal force to separate materials of different masses or densities; in nuclear separation, a device using gaseous uranium hexafluoride to separate isotopes based upon differing masses.

Chemical agent: any substance used deliberately to cause injury.

Chikungunya: a disease caused by an arbovirus carried by several mosquito species, and exhibiting severe joint pain, fever, and rash; it is sometimes fatal.

Choking agent: any of a number of chemicals capable of producing a choking sensation, resulting in the inability to breathe properly.

Cholera: a disease cause by the bacterium *Vibrio cholerae* presenting gastroenteritis and watery diarrhea; the untreated fatality rate can be as high as 50%.

Cholinesterase: an enzyme produced at nerve synapses which acts to shut down the flow of electrical impulses following discharge; a number of chemicals and toxins can prevent the enzyme from functioning, leading to, essentially, a short circuit across the nerve synapse.

Chromatography: a number of laboratory methods for separating materials for analysis.

Chromosome: the carrier of genetic information, according to the types of DNA composing the structure.

Circulatory system: system consisting of the heart, lungs, and connecting blood vessels.

Coccidioidomycosis: a disease caused by fungi of the genus *Coccidioides* and presenting flu-like symptoms of chills and fever; the untreated fatality rate can be as high as 50%.

Compound: two or more elements chemically combined by sharing of electrons or electrostatic attraction by transfer of electrons; the simplest form is the molecule.

Concentration: mass per unit volume; for airborne materials, often expressed as milligrams per cubic meter (mg/m^3).

Corneum: the tough outer layer of the skin, composed of a mixture of flattened dead skin cells, oil, free fatty acids, and dirt; also known as the stratum corneum.

Cosmic radiation: highly energetic particles and radiation produced by stars.

Covalent bonding: the process by which two or more atoms are chemically combined as a result of the sharing of electrons.

Critical mass: the minimum amount of a fissile material necessary to continue a self-sustaining fission reaction.

Cross section: the mathematical probability of an atom absorbing a neutron in order to undergo fission.

Curie (Ci): an older unit of the amount of radioactivity equal to 3.7×10^{10} nuclear transformations per second; roughly equal to the activity of one gram of pure radium.

Dalton: a unit of atomic mass equal to one atomic mass unit, and approximately equal to the mass of a proton.

Decontamination: the process of removing harmful materials from surfaces, including the human body.

Decontamination Solution 2 (DS-2): a commonly used liquid material for removing chemical agents from hard or semi-porous surfaces (not human skin).

Delta radiation: secondary radiation resulting from a charged particle passing nearby an atom.

Dengue: a disease caused by an arbovirus carried by *Aedes* mosquitoes and presenting fever, severe headaches, and joint pain; it is rarely fatal.

Deoxyribonucleic acid (DNA): a polynucleotide chain formed by nucleotides bonded together, each nucleotide containing three parts: a phosphate, a sugar, and a nitrogenous base. DNA is the chemical making up chromosomes, and has a double helical structure.

Dermal absorption: the uptake of a chemical or biological agent through the skin.

Dermis: the layer of the skin consisting of connective tissue, hair roots, sweat glands, oil glands, and fat deposits.

Detonator: a device used to initiate an explosion.

Detoxification: the process by which a toxic material is rendered less hazardous.

Deuterium (H-2 or ^{2}H or D): an isotopic form of hydrogen whose atoms contain one proton and one neutron.

Deuteron: a hydrogen atom containing one proton and one neutron in the nucleus.

Diffusion: the movement of atoms by which they spread out and separate based upon their masses; the term is also used in optics for spreading of spectral bands.

Digestive system: system consisting of the mouth, stomach, anus, rectum, and connecting passages such as the small intestines and colon.

Dimethyl sulfate: a military blister agent; also known as DMSO4.

Dirty bomb: a common name for a bomb containing radioactive materials dispersed by conventional explosives.

Dispersion: the spreading or movement of materials in a gas or liquid.

Dose: the amount of material or radiation actually taken into the body as opposed to the exposure, which is the amount of material or radiation surrounding the body.

Dose rate: the time function of the amount of a material or radiation taken into the body; often expressed as milligrams per hour (mg/hr).

Dose-response curve: a graphical representation of the effect of increasing dose.

Dosimeter: a device for the detection and measurement of radiation (or other) doses.

Double strand break: damage or breaking of both strands of the DNA helix.

Dull Sword: the code term for minor damage to a nuclear weapon.

Ebola: a disease caused by a virus of the *Filoviridae* family and presenting high fever, headache, and abdominal pain followed by diarrhea and bleeding externally and internally; the fatality rate is quite high (maybe 70%).

Electromagnetic pulse (EMP): an intense discharge of electromagnetic energy resulting from a nuclear detonation, capable of disrupting or destroying electronic equipment and communications over a wide area.

Electrometer: a device for the detection and measurement of electrostatic fields.

Electron: a negatively-charged subatomic particle of very small mass.

Electron volt: an amount of energy equal to 1.6×10^{-19} joules often used to express the energies of nuclear processes.

Element: one of the hundred plus or so unique substances; the simplest form is an atom.

Encephalitis: any of a number of related diseases caused by a number of viruses, often an arthropod carried from mammal hosts; symptoms may vary but often there is a neurological aspect; human fatality rates are low from most of these diseases.

Endocrine system: system consisting of the ductless glands of the body that release their hormones into the blood.

Entry: going into; the act of entering.

Epidemic typhus: a disease caused by the rickettsial organism *Rickettsia prowazekii* and presenting fever, weakness, and joint pain; the untreated fatality rate is as high as 30%.

Epidermis: the outer layer of the skin, composed of two parts: the outermost corneum, a tough mixture of flattened dead skin cells, oil, free fatty acids, and dirt, and the germinativum just below and made of living skin cells.

Exposure: the amount of material or radiation surrounding a body; the amount of a chemical, biological, or radioactive substance, or particulate or electromagnetic radiation, available for ingestion, injection, or inhalation.

Fallout: the term for the mixture of fission fragments, induced radioactive particles, and bomb debris that fall to earth following a nuclear detonation.

Fast neutron: a neutron having an energy greater than 0.1 MeV.

Fat Man: the code name for the first implosion design atomic bomb.

Fireball: the term used for the extremely hot (180,000,000 °F) gases and bomb debris resulting from a nuclear detonation.

Fissile material: material composed of atoms capable of absorbing neutrons and, subsequently, undergoing fission.

Fission fragments: those atoms (largely radioactive) resulting from the splitting of atoms.

Fission initiator: a device used to begin the fission process.

Flash blindness: the loss of sight due to the highly intense visible and UV radiation from a nuclear detonation; the loss may be temporary or permanent.

Fog oil: a military battlefield smoke and obscurant; a mixture of short-chain hydrocarbons.

Free radical: two or more atoms chemically combined having neutral electrical charge.

Gadget: the code name for the first test device using the implosion design in Project Trinity.

Gamma radiation: high-energy electromagnetic radiation resulting from nuclear transformations.

Gas: the state of matter able to fill any container into which it is put; a gas has neither fixed shape nor volume.

Gas chromatograph: an instrument that may be used to identify chemical compounds based upon differential travel times of molecules adsorbed onto and from substrate packing.

Germinativum: the layer of the outer skin made of living skin cells located below the stratum corneum.

Graphite fibers: carbon fibers sometimes used in mixtures of battlefield smokes and obscurants to block transmission of infrared and visible light.

Gray (Gy): the modern unit of radiation dose equal to 1 joule absorbed per kilogram of matter; it replaced the rad.

Gun tube design: the earliest design for an atomic bomb by which two separated masses are forced together down a linear tube to form a critical mass.

HTH: a common name for calcium hypochlorite.

Half life: the average length of time it takes for one-half of a given amount of a radioactive substance to decay.

Hantaviral diseases: a number of related diseases caused by several hantaviruses carried by rodents, and presenting fever, headache, and cough; the untreated fatality rate can be as high as 65%.

Heavy water: hydrogen oxide composed of molecules containing deuterons and oxygen atoms.

Hematopoetic effects: generally adverse effects on the blood-forming organs and processes in the body.

Hemorrhagic fever: any of a number of viral diseases showing fever, renal shutdown, and blood loss.

Herbicide: any of a number of chemicals capable of altering growth, affecting metabolism, or killing plants.

Hiroshima: the Japanese city on southwest Honshu which was bombed with "Little Boy" on August 6, 1945.

Hydrogen: the element possessing one proton in each atom's nucleus.

Hydrogen bomb: a common name for a thermonuclear device.

Implosion design: a design for an atomic bomb by which separated pieces of fissile material are forced into a critical mass in a spherical form.

Incapacitating agent: any of a number of non-lethal chemicals capable of altering perception or ability to perform tasks.

Incendiary: any material capable of causing heat or producing fire.

Induced radioactivity: the release of energy from atoms as a result of instability generated within the nucleus by external causes.

Infective dose - 50% (ID_{50}): the dose at which 50% of a population will be infected.

Influenza: a number of related diseases caused by a number of viruses, and presenting fever, malaise, respiratory distress, and headache; untreated fatality rates of current strains are low, but more dangerous strains are possible.

Infrared radiation (IR): electromagnetic radiation between 0.8 and 1,000 micrometers wavelength.

Infrared spectrometer: an instrument that may be used to identify chemical compounds based upon absorbance of infrared radiation by chemical bonds between atoms.

Ion mobility spectrometer: an instrument that may be used to detect chemical agents based upon differential travel times for ions of different masses.

Ionic bonding: the process by which two or more atoms are chemically combined as a result of transfer of electrons from a metal to a non-metal.

Ionization: the formation of one or more charged particles due to the loss or gain of electrons from the constituent atoms.

Ionization chamber: a device for the detection and measurement of radiation utilizing the process of ionization of gas molecules.

Isopleth: a graphical representation of lines of equal quantities on a map.

Isotope: atoms of same element differing in the number of neutrons, resulting in different atomic masses.

Kilodalton (kd): a unit of atomic mass equal to 1,000 amu.

Kinetic energy released in matter (KERMA): the amount of energy of motion released upon passing through matter.

LSD: a psychotropic agent; lysergic acid diethylamide.

Lacrimatory agent: any of a number of chemicals capable of causing crying and uncontrolled tearing of the eyes; also known as tear agents or tear gases.

Lethal does - 50% (LD_{50}): the dose at which 50% of a population will be killed.

Lethality: the potential for causing death.

Linear energy transfer: the amount of particulate or electromagnetic energy deposited upon passing through matter.

Liquid: the state of matter having fixed volume but not shape.

Liquid chromatograph: an instrument which may be used to identify chemical compounds based upon differential travel times of volatile and semi-volatile molecules absorbed onto and from substrate packing, commonly with heating and/or high pressure.

Little Boy: the code name for the first gun tube atomic bomb.

Liver: the organ that detoxifies chemicals in the body and produces bile to aid in digestion of fats.

Logarithm: a mathematical expression of the exponent that indicates the power to which a number must be raised to produce a given number; for example, the base ten logarithm of 100 is 2 since $10^2 = 100$.

Lymphatic system: system consisting of the lymph nodes and connecting vessels that carry the lymph.

M-8 paper: a chemically-impregnated test strip for detection of blister and nerve agents.

M-9 paper: a chemically-impregnated test strip with an adhesive back for detection of blister and nerve agents.

MIRV: an acronym for Multiple Impact Re-entry Vehicle.

MOPP: an acronym for Mission-Oriented Protective Posture (U.S. military term).

Mass casualties: the presence of large numbers of victims of an event.

Mass spectrometer: an instrument that may be used to identify chemical compounds based upon differential travel times or radii of curvature of ions of different masses.

Megaton (MT): a unit of mass equal to one million tons; for bomb damage, the equivalent effects of one million tons of trinitrotoluene (TNT) explosive power.

Metabolism: the total of all the chemical reactions taking place in the body.

Mine: a buried military weapon commonly used against small moving targets (personnel and vehicles).

Mission-Oriented Protective Posture (MOPP): a U.S. military term for any of four categories of protective gear wear, based upon likely threat scenarios.

Mixture: a physical combination of two or more materials; the components can be physically separated.

Molecule: the simplest form of a compound consisting of two or more atoms chemically combined.

Multiple Impact Re-entry Vehicle (MIRV): a missile warhead capable of targeting a number of targets simultaneously.

Murine typhus: a disease caused by the rickettsial organism *Rickettsia typhi* carried by fleas and lice from rodents, and presenting chills, fever, cough, and generalized pain; the untreated fatality rate is very low.

Mustard: a common name for any of several blister agents; originally because of the faint odor of Agent HD.

Mutagen: a chemical or radiation capable of causing changes in succeeding generations (second generation and beyond).

Mutation: offspring changed due to alterations in the DNA.

Nagasaki: the Japanese city which was bombed with "Fat Man" on August 8, 1945.

Natural radioactivity: the release of energy from atoms as a result of instability within the nucleus of certain naturally-occurring materials.

Nerve agent: any of a number of chemicals capable of damage to, or interference with, normal nerve transmission; many act as cholinesterase inhibitors.

Nervous system: system consisting of the brain, spinal cord, ganglia, and connecting nerves as well as all the nerves and synapses of the body.

Neutron: an uncharged subatomic particle approximately the same mass as a proton.

Nitrogen mustard: a common name for any of several blister agents (HN-1, HN-2, and HN-3).

Nuclear fission: the splitting of certain unstable atoms usually as the result of bombardment with neutrons or other particles.

Nuclear fusion: the formation of atoms from simpler atoms being brought together under extreme heat and pressure.

Nuclear Non-proliferation Treaty: the common name for the Treaty on the Non-Proliferation of Nuclear Weapons which entered into force in 1970 and limited the expansion of nations possessing nuclear weapons.

Nuclear reactor: a device utilizing controlled fission for the purpose of producing heat to generate power, or to produce radioactive isotopes for research or medicine.

Nucleic acids: polynucleotide chains formed by many nucleotides bonded together, with each nucleotide containing three parts: a phosphate, a sugar, and a nitrogenous base.

Nucleus (atomic): the central core of an atom, consisting of protons, neutrons, and other particles.

Nucleus (biological): the central part of a cell, containing the cellular DNA (or RNA) and mitochondria.

Obscurant: any aerosol used to block the observation of troops or an event; these may reduce transmission in the visible, radiofrequency, or infrared bands.

Ocular absorption: the uptake of a chemical or biological agent through the eye.

Ocular system: system consisting of the eye, optic nerve, and related connecting tissue.

O'Nyong-nyong: a disease caused by an arbovirus carried by *Anopheles* mosquitoes, and presenting severe back pain, chills, headache, joint pain, and rash; fatality rates are essentially zero.

Overpressure: any pressure greater than normal atmospheric pressure (approximately 15 pounds per square inch or 760 millimeters of mercury).

Oxidation state: the ability of an atom to take in or release electrons (undergo oxidation-reduction).

Pasquill category: a set of weather conditions defining the dispersion of materials.

Periodic chart of the elements: a chart having the elements arranged according to their atomic number; repeating chemical characteristics aligning the elements into certain patterns.

Peripheral nervous system: those parts of the nervous system not composing the central nervous system; that is, those nerves and related tissue not a part of the brain, spinal cord, or ganglia.

Physiology: the study of how plants and animals function.

Pile: a term coined by Enrico Fermi for his graphite-moderated nuclear reactor.

Plague: a disease caused by the bacillus *Yersinia pestis* from fleas carried by rodents, and presenting headache, weakness, and coughing which progresses to buboes in some, but not all, cases; the untreated fatality rate can be as high as 95%.

Plutonium: an artificially produced element having several isotopes capable of absorbing neutrons and, subsequently, undergoing fission.

Polymerase chain replication (PCR): the process of duplicating DNA or RNA in order to increase the concentration for analysis; also called polymerase chain reaction..

Population response curve: a graphical representation of the number of people affected by exposure to a toxic material or radiation.

Primer: a material used to initiate an explosion within a detonator.

Prion: any of a number of protein organisms without a nucleic acid, suspected of being the causative agent in "mad cow" disease and a number of other diseases leading to the destruction of brain tissue and nerve cells.

Probit: a statistical measurement of probability based upon deviations from the mean of a normal frequency distribution.

Prompt radiation: the short-lived initial radiation resulting from a nuclear detonation, composed primarily of neutrons but also possessing alpha, beta, gamma, and soft x-ray radiation.

Proton: a positively-charged subatomic particle approximately the same mass as a neutron.

Psychogenic agent: any of a number of chemicals capable of inducing confusion or altering perception of reality.

Q fever: a disease caused by the rickettsial organism *Coxiella burnetti*, and presenting symptoms such as high fever, chest pain, cough, and nausea in only about 50% of the cases; the untreated fatality rate can be as high as 65%.

RADIAC: an acronym for Radiation Detection, Identification, and Computation.

Rad: an older unit of absorbed radiation equal to 100 ergs of energy per gram of matter; it has been replaced by the gray (Gy).

Radical: two or more atoms usually functioning as a single unit.

Radioactivity: the release of energy from atoms as a result of nuclear transformations.

Radioprotectant: a chemical that might be of value in reducing the effects of radiation.

Reactor waste: highly radioactive isotopes resulting from nuclear processes in a reactor having no practical uses at this time.

Rem (Roentgen Equivalent Man): an older unit of biological effectiveness of radiation dose equal to the dose in rads modified by the type of radiation and its energy; it has been replaced by the sievert (Sv).

Reproductive system: system consisting of the gonads, uterus in females, and connecting vessels allowing for the male sperm to reach the female egg.

Respiratory system: system consisting of the nares, lungs, and connecting passages.

Response: any effect being looked at; this can range from death to irritation.

Ribonucleic acid (RNA): a polynucleotide chain formed by nucleotides bonded together, each nucleotide contains three parts: a phosphate, a sugar, and a nitrogenous base. RNA is involved in telling DNA when and how to replicate, and may supply material to the DNA.

Ricin: a globular protein found in the castor bean; it is highly toxic when ingested or injected. Death is common from shock.

Rift Valley fever: a disease caused by an arbovirus carried by a number of species of mosquitoes; while the symptoms may be severe, including fever, malaise, vomiting, and muscle pain, the disease is rarely fatal.

Riot control agent: a chemical agent which produces tears, eye pain, or other symptoms in order to render large numbers of people unable to coordinate actions.

Rocket: a self-propelled projectile commonly containing an explosive warhead; usually either solid-fueled or liquid-fueled.

Rocky Mountain spotted fever: a disease caused by the rickettsial organism *Rickettsia rickettsii* carried by ticks from a number of mammals, and presenting fever, severe headache, fatigue, swollen eyes, and muscle pain; the untreated fatality rate is around 20%.

Roentgen (R): an older unit of radiation exposure to gamma and x-rays equal to 2.58×10^{-4} coulombs per kilogram of dry air.

Saxitoxin: a toxin produced by several types of bacteria, and having cholinesterase inhibition characteristics, leading to numbness, collapse, and often death.

Scintillation: the release of energy in the form of light.

Semiconductor: a material capable of transmission of electricity intermediate between a conductor and a non-conductor; it commonly consists of silicon-based chips having a small amount of another material (doping).

Shielding: the blocking of the path of radiation.

Sievert (Sv): the modern unit of biological effectiveness of radiation dose equal to the dose in grays modified by the type and energy of the radiation.

Single strand break: damage or breaking of only one of the two strands of the DNA helix.

Skin: the outer covering of the human body.

Slow neutron: neutron having an energy between that of thermal neutrons and about 0.1 MeV.

Smallpox: a disease caused by the bacterium *Variola major*, and presenting skin pustules, fever, swollen lymph nodes, rigors, vomiting, and backache; the untreated fatality rate can be as high as 50%.

Solid: the state of matter having fixed shape and volume.

Source term: the initial composition and amount of material released from an event such as a nuclear detonation or incinerator stack.

Staphylococcal enterotoxin B (SEB): an exotoxin produced by the bacterium *Staphylococcus aureus*, and presenting high fever, chills, headache, chest pain, pulmonary edema, vomiting, diarrhea, and possibly shock; the untreated fatality rate is low.

Stopping power: the average amount of radiation energy absorbed by a material per unit distance, often expressed as MeV per centimeter (MeV/cm).

Strategic Arms Reduction Talks (SALT): a long series of negotiations which culminated in 2002 (SALT II) leading to drastic reductions in the number of nuclear weapons between the United States and the Russian Federation.

Subatomic particle: any of a number of particles which make up atoms; examples are protons, neutrons, electrons, mesons, muons, etc.

Sulfur mustard: a common name for any of several blister agents (HD, HL, and DMSO4).

Supertropical Bleach (STB): a commonly used solid material for removing chemical agents from hard surfaces (not human skin).

Tear agent: any of a number of chemicals capable of causing crying and uncontrolled tearing of the eyes; also known as lacrimatory agents or tear gases.

Teratogen: a chemical or radiation capable of producing changes in first-generation offspring.

Thermal: having to do with heat.

Thermal neutron: the slowest moving neutron.

Thermoluminescent dosimeter: a device for the detection and measurement of radiation utilizing the release of stored energy in the form of light.

Thermonuclear device: commonly called a "hydrogen bomb"; a fusion weapon which produces enormous amounts of energy from the formation of elements from very light elements under conditions initiated by a fission bomb.

Thin-layer chromatograph: an instrument that may be used to identify chemical compounds based upon differential travel times of molecules on thin films.

Thorium series: a naturally occurring radioactive series of elements beginning with ^{232}U and ending with stable ^{208}Pb.

Threshold level: the minimum amount of a chemical, biological agent, or radiation necessary to have an effect.

Toxic: being able to have an adverse effect; the effect can range from irritation to death.

Transient psychological incapacitation: an adverse mental condition involving the higher brain functions following exposure to high dose rates of ionizing radiation.

Trauma: injury or wound to a living body from the application of external forces or violence.

Trichothecene: any of a number of mycotoxins from fungi of the genus *Fusarium*, and presenting vomiting and bleeding, loss of consciousness, and even death.

Tritium (H-3 or ^{3}H or T): an isotopic form of hydrogen whose atoms contain one proton and two neutrons.

Triton: a hydrogen atom containing one proton and two neutrons in the nucleus.

Tularemia: a disease caused by the bacterium *Francisella tularensis*, and presenting skin ulcers, fever, chills, muscle aches, joint pain, dry cough, and bloody sputum; the untreated fatality rate is low, usually less than 5%.

Typhoid fever: a disease caused by the bacterium *Salmonella typhi* usually from contaminated soil or water containing human feces, and presenting sustained fever, constipation, and fatigue; the untreated fatality rate is only about 10%.

Ultraviolet (UV) radiation: electromagnetic radiation of energies greater than visible light.

Ultraviolet spectrometer: an instrument that may be used to identify chemical compounds based upon absorbance of ultraviolet radiation by molecules.

Uranium: a naturally occurring element having several isotopes capable of absorbing neutrons and, subsequently, undergoing fission.

Uranium series: a naturally occurring radioactive series of elements beginning with ^{238}U and ending with stable ^{206}Pb.

Urinary system: the kidneys, bladder, and connecting vessels that carry away liquid wastes.

Vapor: the gaseous form of a material, normally a liquid, at the given temperature and pressure.

Vesicant: any of a number of chemicals capable of causing skin burns, lesions, and destruction; also known as blister agents.

Viability: the potential for being able to thrive, often expressed as the ability to live or reproduce.

Visible light: electromagnetic radiation capable of being seen by the human eye, between 0.38 and 0.76 micrometers wavelength.

Vomiting agent: any of a number of chemicals capable of inducing involuntary emesis or vomiting.

Weapon of mass destruction: any material or device capable of causing extensive casualties.

X-ray: high-energy electromagnetic radiation resulting from electron orbital transitions.

Yellow rain: a term used to describe a trichothecene mycotoxin used in southeast Asia during and following the Viet Nam War.

INDEX